W9-AXK-013

RAND McNALLY

GOODE'S

ATLAS OF Latin America

distributed by
John Wiley & Sons, Inc.

Howard Veregin, Ph.D., Editor

Editorial Advisory Board

Byron Augustin, D.A., Texas State University-San Marcos

Joshua Comenetz, Ph.D., University of Florida

Francis Galgano, Ph.D., United States Military Academy

Sallie A. Marston, Ph.D., University of Arizona

Virginia Thompson, Ph.D., Towson University

Abridgement of
21ST Edition

John Wiley & Sons, Inc. and RAND McNALLY

Working together to bring you the best in geography education

Few publishers can claim as rich a history as John Wiley & Sons, Inc. (publishers since 1807) and Rand McNally & Company (publishers since 1856). Even fewer can claim as long-standing a commitment to geographic education.

Wiley's partnership with the geographic community began at the very beginning of the 20th century with the publication of textbooks on surveying. Rand McNally's partnership began even earlier, with the publication of the first Rand McNally maps in 1872. Since then, both companies have worked in parallel to help students visualize spatial relationships and appreciate the earth's dynamic landscapes and diverse cultures.

Now these two publishers have combined their efforts to bring you this new atlas, which represents the very best in educational resources for geography.

Based on the 21st edition of the *Goode's World Atlas*, the *Goode's Atlas of Latin America* features:

- An emphasis on map accuracy and legibility, and the mixture of maps of different types and scales to facilitate interpretation of geographic phenomena.

- World, continental, and regional population density maps, which have been created using LandScan, a digital population database developed using satellite and computer-mapping technology.

- Graphs accompanying many of the maps, to show important statistical information, trends over time, and relationships between variables.

- Maps and graphs that have been updated, based on the most current available data in accordance with the high standards and quality that have always been a defining feature of the *Goode's World Atlas*.

Wiley and Rand McNally are currently offering seven new course-specific atlases, which can be packaged with any of Wiley's best-selling textbooks, or sold separately as stand-alones. These atlases include:

Rand McNally Goode's Atlas of Political Geography	0-471-70694-9
Rand McNally Goode's Atlas of Latin America	0-471-70697-3
Rand McNally Goode's Atlas of North America	0-471-70696-5
Rand McNally Goode's Atlas of Asia	0-471-70699-X
Rand McNally Goode's Atlas of Urban Geography	0-471-70695-7
Rand McNally Goode's Atlas of Physical Geography	0-471-70693-0
Rand McNally Goode's Atlas of Human Geography	0-471-70692-2

This book was set by GGS Book Services and printed and bound by Walsworth Press. The cover was printed by Phoenix Color.

To order books or for customer service please, call 1-800-CALL WILEY (225-5945).

ISBN 0471-70697-3

Printed in the United States

10 9 8 7 6 5 4 3 2 1

Table of Contents

Introduction

Basic Earth Properties

The subject matter of **geography** includes people, landforms, climate, and all the other physical and human phenomena that make up the earth's environments and give unique character to different places. Geographers construct maps to visualize the **spatial distributions** of these phenomena: that is, how the phenomena vary over geographic space. Maps help geographers understand and explain phenomena and their interactions.

To better understand how maps portray geographic distributions, it is helpful to have an understanding of the basic properties of the earth.

The earth is essentially **spherical** in shape. Two basic reference points — the **North and South Poles** — mark the locations of the earth's axis of rotation. Equidistant between the two poles and encircling the earth is the **equator**. The equator divides the earth into two halves, called the **northern and southern hemispheres**. (See the figures to the right.)

Latitude and longitude are used to identify the locations of features on the earth's surface. They are measured in degrees, minutes and seconds. There are 60 minutes in a degree and 60 seconds in a minute. Latitude is the angle north or south of the equator. The symbols °, ', and " represent degrees, minutes and seconds, respectively. The N means north of the equator. For latitudes south of the equator, S is used. For example, the Rand McNally head office in Skokie, Illinois, is located at 42°1'51" N. The minimum latitude of 0° occurs at the equator. The maximum latitudes of 90° N and 90° S occur at the North and South Poles.

A **line of latitude** is a line connecting all points on the earth having the same latitude. Lines of latitude are also called **parallels**, as they run parallel to each other. Two parallels of special importance are the **Tropic of Cancer** and the **Tropic of Capricorn**, at approximately 23°30' N and S respectively. This angle coincides with the inclination of the earth's axis relative to its orbital plane around the sun. These tropics are the lines of latitude where the noon sun is directly overhead on the solstices. (See figure on page 66.) Two other important parallels are the **Arctic Circle** and the **Antarctic Circle**, at approximately 66°30' N and S respectively. These lines mark the most northerly and southerly points at which the sun can be seen on the solstices.

While latitude measures locations in a north-south direction, longitude measures them east-west. Longitude is the angle east or west of the **Prime Meridian**. A **meridian** is a line of longitude, a straight line extending from the North Pole to the South Pole. The Prime Meridian is the meridian passing through the Royal Observatory in Greenwich, England. For this reason the Prime Meridian is sometimes referred to as the **Greenwich Meridian**. This location for the Prime Meridian was adopted at the International Meridian Conference in Washington, D.C., in 1884.

Like latitude, longitude is measured in degrees, minutes, and seconds. For example, the Rand McNally head office is located at 87°43'6" W. The qualifiers E and W indicate whether a location is east or west of the Greenwich Meridian. Longitude ranges from 0° at Greenwich to 180° E or W. The meridian at 180° E is the same as the meridian at 180° W. This meridian, together with the Greenwich Meridian, divides the earth into **eastern and western hemispheres**.

Any circle that divides the earth into equal hemispheres is called a **great circle**. The equator is an example. The shortest distance between any two points on the earth is along a great circle. Other circles, including all other lines of latitude, are called **small circles**. Small circles divide the earth into two unequal pieces.

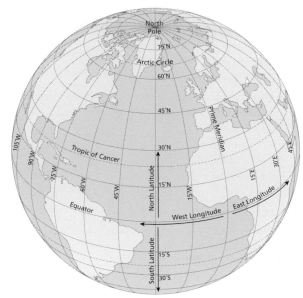

View of earth centered on 30° N, 30° W

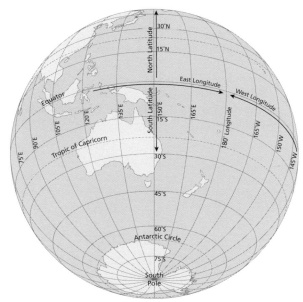

View of earth centered on 30° S, 150° E

The Geographic Grid

The grid of lines of latitude and longitude is known as the **geographic grid**. The following are some important characteristics of the grid.

All lines of longitude are equal in length and meet at the North and South Poles. These lines are called meridians.

All lines of latitude are parallel and equally spaced along meridians. These lines are called parallels.

The length of parallels increases with distance from the poles. For example, the length of the parallel at 60° latitude is one-half the length of the equator.

Meridians get closer together with increasing distance from the equator, and finally converge at the poles.

Parallels and meridians meet at right angles.

Map Scale

To use maps effectively it is important to have a basic understanding of map scale.

Map scale is defined as the ratio of distance on the map to distance on the earth's surface. For example, if a map shows two towns as separated by a distance of 1 inch, and these towns are actually 1 mile apart, then the scale of the map is 1 inch to 1 mile.

The statement "1 inch to 1 mile" is called a **verbal scale**. Verbal scales are simple and intuitive, but a drawback is that they are tied to the specific set of map and real-world units in the numerator and denominator of the ratio. This makes it difficult to compare the scales of different maps.

A more flexible way of expressing scale is as a **representative fraction**. In this case, both the numerator and denominator are converted to the same unit of measurement. For example, since there are 63,360 inches in a mile, the verbal scale "1 inch to 1 mile" can be expressed as the representative fraction 1:63,360. This means that 1 inch on the map represents 63,360 inches on the earth's surface. The advantage of the representative fraction is that it applies to any linear unit of measurement, including inches, feet, miles, meters, and kilometers.

Map scale can also be represented in graphical form. Many maps contain a **graphic scale** (or **bar scale**) showing real-world units such as miles or kilometers. The bar scale is usually subdivided to allow easy calculation of distance on the map.

Map scale has a significant effect on the amount of detail that can be portrayed on a map. This concept is illustrated here using a series of maps of the Washington, D.C., area. (See the figures to the right.) The scales of these maps range from 1:40,000,000 (top map) to 1:4,000,000 (center map) to 1:25,000 (bottom map). The top map has the **smallest scale** of the three maps, and the bottom map has the **largest scale**.

Note that as scale increases, the area of the earth's surface covered by the map decreases. The smallest-scale map covers thousands of square miles, while the largest-scale map covers only a few square miles within the city of Washington. This means that a given feature on the earth's surface will appear larger as map scale increases. On the smallest-scale map, Washington is represented by a small dot. As scale increases the dot becomes an orange shape representing the built-up area of Washington. At the largest scale Washington is so large that only a portion of it fits on the map.

Because small-scale maps cover such a large area, only the largest and most important features can be shown, such as large cities, major rivers and lakes, and international boundaries. In contrast, large-scale maps contain relatively small features, such as city streets, buildings, parks, and monuments.

Small-scale maps depict features in a more simplified manner than large-scale maps. As map scale decreases, the shapes of rivers and other features must be simplified to allow them to be depicted at a highly reduced size. This simplification process is known as **map generalization**.

Maps in *Goode's Atlas of Latin America* have a wide range of scales. The smallest scales are used for the world thematic map series, where scales range from approximately 1:200,000,000 to 1:75,000,000. Reference map scales range from a minimum of 1:100,000,000 for world maps to a maximum of 1:1,000,000 for city maps. Most reference maps are regional views with a scale of 1:4,000,000.

1:40,000,000 scale

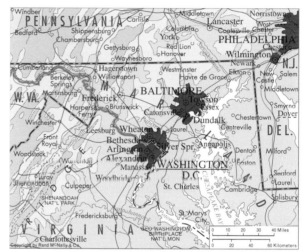

1:4,000,000 scale

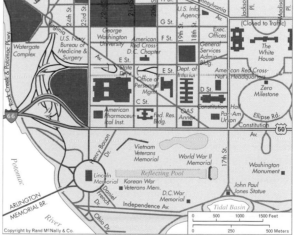

1:25,000 scale

Map Projections

Map projections influence the appearance of features on the map and the ability to interpret geographic phenomena.

A **map projection** is a geometric representation of the earth's surface on a flat or plane surface. Since the earth's surface is curved, a map projection is needed to produce any flat map, whether a page in this atlas or a computer-generated map of driving directions on www.randmcnally.com. Hundreds of projections have been developed since the dawn of mapmaking. A limitation of all projections is that they distort some geometric properties of the earth, such as shape, area, distance, or direction. However, certain properties are preserved on some projections.

If shape is preserved, the projection is called **conformal**. On conformal projections the shapes of features agree with the shapes these features have on the earth. A limitation of conformal projections is that they necessarily distort area, sometimes severely.

Equal-area projections preserve area. On equal area projections the areas of features correspond to their areas on the earth. To achieve this effect, equal-area projections distort shape.

Some projections preserve neither shape nor area, but instead balance shape and area distortion to create an aesthetically-pleasing result. These are often referred to as **compromise** projections.

Distance is preserved on **equidistant** projections, but this can only be achieved selectively, such as along specific meridians or parallels. No projection correctly preserves distance in all directions at all locations. As a result, the stated scale of a map may be accurate for only a limited set of locations. This problem is especially acute for small-scale maps covering large areas.

The projection selected for a particular map depends on the relative importance of different types of distortion, which often depends on the purpose of the map. For example, world maps showing phenomena that vary with area, such as population density or the distribution of agricultural crops, often use an equal-area projection to give an accurate depiction of the importance of each region.

Map projections are created using mathematical procedures. To illustrate the general principles of projections without using mathematics, we can view a projection as the geometric transfer of information from a globe to a flat projection surface, such as a sheet of paper. If we allow the paper to be rolled in different ways, we can derive three basic types of map projections: **cylindrical, conic,** and **azimuthal**. (See the figures to the right.)

For cylindrical projections, the sheet of paper is rolled into a tube and wrapped around the globe so that it is **tangent** (touching) along the equator. Information from the globe is transferred to the tube, and the tube is then unrolled to produce the final flat map.

Conic projections use a cone rather than a cylinder. The figure shows the cone tangent to the earth along a line of latitude with the apex of the cone over the pole. The line of tangency is called the **standard parallel** of the projection.

Azimuthal projections use a flat projection surface that is tangent to the globe at a single point, such as one of the poles.

The figures show the **normal orientation** of each type of surface relative to the globe. The **transverse orientation** is produced when the surface is rotated 90 degrees from normal. For azimuthal projections this orientation is usually called **equatorial** rather than transverse. An **oblique orientation** is created if the projection surface is oriented at an angle between normal and transverse. In general, map distortion increases with distance away from the point or line of tangency. This is why the normal orientations of the cylindrical, conic, and azimuthal projections are often used for mapping equatorial, mid-latitude, and polar regions, respectively.

The projection surface model is a visual tool useful for illustrating how information from the globe can be projected to the map. However, each of the three projection surfaces actually represents scores of individual projections. There are, for example, many projections with the term "cylindrical" in the name, each of which has the same basic rectangular shape, but different spacings of parallels and meridians. The projection surface model does not account for the numerous mathematical details that differentiate one cylindrical, conic, or azimuthal projection from another.

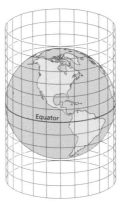
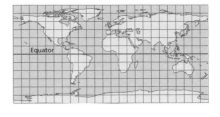

Cylindrical Projection

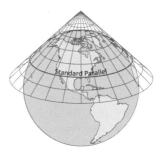

Conic Projection

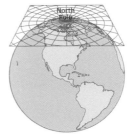

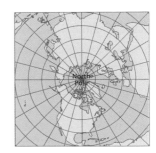

Azimuthal Projection

Map Projections Used in *Goode's Atlas of Latin America*

Of the hundreds of projections that have been developed, only a fraction are in everyday use. The main projections used in *Goode's Atlas of Latin America* are described below.

Simple Conic

Type: Conic   Conformal: No   Equal-area: No

Notes: Shape and area distortion on the Simple Conic projection are relatively low, even though the projection is neither conformal nor equal-area. The origins of the Simple Conic can be traced back nearly two thousand years, with the modern form of the projection dating to the 18th century.

Uses in *Goode's Atlas of Latin America*: Larger-scale reference maps of North America, Europe, Asia, and other regions.

Lambert Conformal Conic

Type: Conic   Conformal: Yes   Equal-area: No

Notes: On the Lambert Conformal Conic projection, spacing between parallels increases with distance away from the standard parallel, which allows the property of shape to be preserved. The projection is named after Johann Lambert, an 18th century mathematician who developed some of the most important projections in use today. It became widely used in the United States in the 20th century following its adoption for many statewide mapping programs.

Uses in *Goode's Atlas of Latin America*: Thematic maps of the United States and Canada, and reference maps of parts of Asia.

Albers Equal-Area Conic

Type: Conic   Conformal: No   Equal-area: Yes

Notes: On the Albers Equal-Area Conic projection, spacing between parallels decreases with distance away from the standard parallel, which allows the property of area to be preserved. The projection is named after Heinrich Albers, who developed it in 1805. It became widely used in the 20th century, when the United States Coast and Geodetic Survey made it a standard for equal area maps of the United States.

Uses in *Goode's Atlas of Latin America*: Thematic maps of North America and Asia.

Polyconic

Type: Conic   Conformal: No   Equal-area: No

Notes: The term polyconic — literally "many-cones" — refers to the fact that this projection is an assemblage of different cones, each tangent at a different line of latitude. In contrast to many other conic projections, parallels are not concentric, and meridians are curved rather than straight. The Polyconic was first proposed by Ferdinand Hassler, who became Head of the United States Survey of the Coast (later renamed the Coast and Geodetic Survey) in 1807. The United States Geological Survey used this projection exclusively for large-scale topographic maps until the mid-20th century.

Uses in *Goode's Atlas of Latin America*: Reference maps of North America and Asia.

Lambert Azimuthal Equal-Area

Type: Azimuthal   Conformal: No   Equal-area: Yes

Notes: This projection (another named after Johann Lambert) is useful for mapping large regions, as area is correctly preserved while shape distortion is relatively low. All orientations — polar, equatorial, and oblique — are common.

Uses in *Goode's Atlas of Latin America*: Thematic and reference maps of North and South America, Asia, Africa, Australia, and polar regions.

Simple Conic Projection

Lambert Conformal Conic Projection

Albers Equal-Area Conic Projection

Polyconic Projection

Lambert Azimuthal Equal-Area Projection

Miller Cylindrical

Type: Cylindrical   **Conformal:** No   **Equal-area:** No

Notes: This projection is useful for showing the entire earth in a simple rectangular form. However, polar areas exhibit significant exaggeration of area, a problem common to many cylindrical projections. The projection is named after Osborn Miller, Director of the American Geographical Society, who developed it in 1942 as a compromise projection that is neither conformal nor equal-area.

Uses in *Goode's Atlas of Latin America*: World climate and time zone maps.

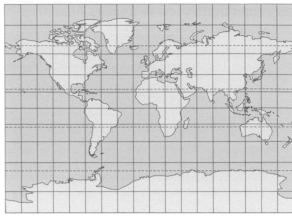

Miller Cylindrical Projection

Sinusoidal

Type: Pseudocylindrical   **Conformal:** No   **Equal-area:** Yes

Notes: The straight, evenly spaced parallels on this projection resemble the parallels on cylindrical projections. Unlike cylindrical projections, however, meridians are curved and converge at the poles. This causes significant shape distortion in polar regions. The Sinusoidal is the oldest-known pseudocylindrical projection, dating to the 16th century.

Uses in *Goode's Atlas of Latin America*: Reference maps of equatorial regions.

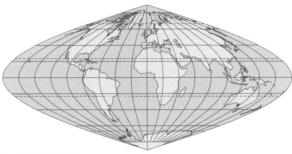

Sinusoidal Projection

Mollweide

Type: Pseudocylindrical   **Conformal:** No   **Equal-area:** Yes

Notes: The Mollweide (or Homolographic) projection resembles the Sinusoidal but has less shape distortion in polar areas due to its elliptical (or oval) form. One of several pseudocylindrical projections developed in the 19th century, it is named after Karl Mollweide, an astronomer and mathematician.

Uses in *Goode's Atlas of Latin America*: Oceanic reference maps.

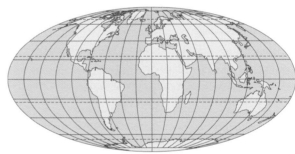

Mollweide Projection

Goode's Interrupted Homolosine

Type: Pseudocylindrical   **Conformal:** No   **Equal-area:** Yes

Notes: This projection is a fusion of the Sinusoidal between 40º44'N and S, and the Mollweide between these parallels and the poles. The unique appearance of the projection is due to the introduction of discontinuities in oceanic regions, the goal of which is to reduce distortion for continental landmasses. A condensed version of the projection also exists in which the Atlantic Ocean is compressed in an east-west direction. This modification helps maximize the scale of the map on the page. The Interrupted Homolosine projection is named after J. Paul Goode of the University of Chicago, who developed it in 1923. Goode was an advocate of interrupted projections and, as editor of *Goode's School Atlas*, promoted their use in education.

Uses in *Goode's Atlas of Latin America*: Small-scale world thematic and reference maps. Both condensed and non-condensed forms are used. An uninterrupted example is used for the Pacific Ocean map.

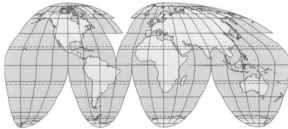

Goode's Interrupted Homolosine Projection

Robinson

Type: Pseudocylindrical   **Conformal:** No   **Equal-area:** No

Notes: This projection resembles the Mollweide except that polar regions are flattened and stretched out. While it is neither conformal nor equal-area, both shape and area distortion are relatively low. The projection was developed in 1963 by Arthur Robinson of the University of Wisconsin, at the request of Rand McNally.

Uses in *Goode's Atlas of Latin America*: World maps where the interrupted nature of Goode's Homolosine would be inappropriate, such as the World Oceanic Environments map.

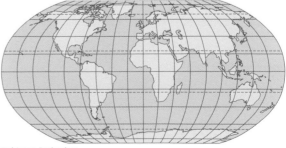

Robinson Projection

Thematic Maps in *Goode's Atlas of Latin America*

Thematic maps depict a single "theme" such as population density, agricultural productivity, or annual precipitation. The selected theme is presented on a base of locational information, such as coastlines, country boundaries, and major drainage features. The primary purpose of a thematic map is to convey an impression of the overall geographic distribution of the theme. It is usually not the intent of the map to provide exact numerical values. To obtain such information, the graphs and tables accompanying the map should be used.

Goode's Atlas of Latin America contains many different types of thematic maps. The characteristics of each are summarized below.

Point Symbol Maps

Point symbol maps are perhaps the simplest type of thematic map. They show features that occur at discrete locations. Examples include earthquakes, nuclear power plants, and minerals-producing areas. The Precious Metals map is an example of a point symbol map showing the locations of areas producing gold, silver, and platinum. A different color is used for each type of metal, while symbol size indicates relative importance.

Area Symbol Maps

Area symbol maps are useful for delineating regions of interest on the earth's surface. For example, the Tobacco and Fisheries map shows major tobacco-producing regions in one color and important fishing areas in another. On some area symbol maps, different shadings or colors are used to differentiate between major and minor areas.

Dot Maps

Dot maps show a distribution using a pattern of dots, where each dot represents a certain quantity or amount. For example, on the Sugar map, each dot represents 20,000 metric tons of sugar produced. Different dot colors are used to distinguish cane sugar from beet sugar. Dot maps are an effective way of representing the variable density of geographic phenomena over the earth's surface. This type of map is used extensively in *Goode's Atlas of Latin America* to show the distribution of agricultural commodities.

Area Class Maps

On area class maps, the earth's surface is divided into areas based on different classes or categories of a particular geographic phenomenon. For example, the Ecoregions map differentiates natural landscape categories, such as Tundra, Savanna, and Prairie. Other examples of area class maps in *Goode's Atlas of Latin America* include Landforms, Climatic Regions, Natural Vegetation, Soils, Agricultural Areas, Languages and Religions.

Isoline Maps

Isoline maps are used to portray quantities that vary smoothly over the surface of the earth. These maps are frequently used for climatic variables such as precipitation and temperature, but a variety of other quantities — from crop yield to population density — can also be treated in this way.

An isoline is a line on the map that joins locations with the same value. For example, the Summer (May to October) Precipitation map contains isolines at 5, 10, 20, and 40 inches. On this map, any 10-inch isoline separates areas that have less than 10 inches of precipitation from areas that have more than 10 inches. Note that the areas between isolines are given different colors to assist in map interpretation.

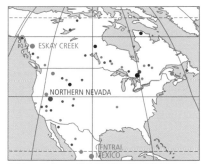
Point symbol map: Detail of Precious Metals

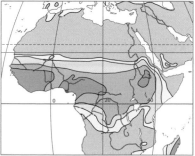
Area symbol map: Detail of Tobacco and Fisheries

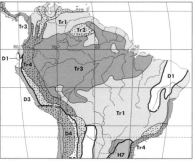
Dot map: Detail of Sugar

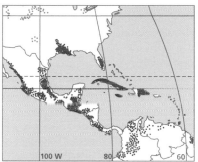
Area class map: Detail of Ecoregions

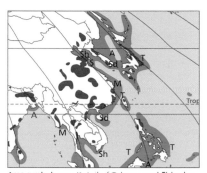
Isoline map: Detail of Precipitation

Proportional Symbol Maps

Proportional symbol maps portray numerical quantities, such as the total population of each state, the total value of agricultural goods produced in different regions, or the amount of hydroelectricity generated in different countries. The symbols on these maps — usually circles —- are drawn such that the size of each is proportional to the value at that location. For example the Exports map shows the value of goods exported by each country in the world, in millions of U.S. dollars.

Proportional symbols are frequently subdivided based on the percentage of individual components making up the total. The Exports map uses wedges of different color to show the percentages of various types of exports, such as manufactured articles and raw materials.

Flow Line Maps

Flow line maps show flows between locations. Usually, the thickness of the flow lines is proportional to flow volume. Flows may be physical commodities like petroleum, or less tangible quantities like information. The flow lines on the Mineral Fuels map represent movement of petroleum measured in billions of U.S. dollars. Note that the locations of flow lines may not represent actual physical routes.

Choropleth Maps

Choropleth maps apply distinctive colors to predefined areas, such as counties or states, to represent different quantities in each area. The quantities shown are usually rates, percentages, or densities. For example, the Birth Rate map shows the annual number of births per one thousand people for each country.

Digital Images

Some maps are actually digital images, analogous to the pictures captured by digital cameras. These maps are created from a very fine grid of cells called **pixels**, each of which is assigned a color that corresponds to a specific value or range of values. The population density maps in this atlas are examples of this type. The effect is much like an isoline map, but the isolines themselves are not shown and the resulting geographic patterns are more subtle and variable. This approach is increasingly being used to map environmental phenomena observable from remote sensing systems.

Cartograms

Cartograms deliberately distort map shapes to achieve specific effects. On **area cartograms**, the size of each area, such as a country, is made proportional to its population. Countries with large populations are therefore drawn larger than countries with smaller populations, regardless of the actual size of these countries on the earth.

The world cartogram series in this atlas depicts each country as a rectangle. This is a departure from cartograms in earlier editions of the atlas, which attempted to preserve some of the salient shape characteristics for each country. The advantage of the rectangle method is that it is easier to compare the area of countries when their shapes are consistent.

The cartogram series incorporates choropleth shading on top of the rectangular cartogram base. In this way map readers can make inferences about the relationship between population and another thematic variable, such as HIV-infection rates.

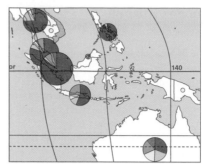

Proportional symbol map: Detail of Exports

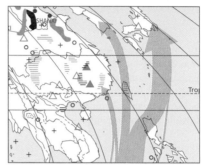

Flow line map: Detail of Mineral Fuels

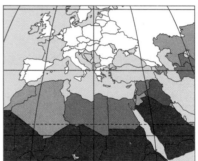

Choropleth map: Detail of Birth Rate

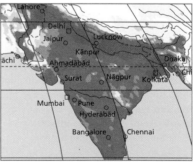

Digital image map: Detail of Population Density

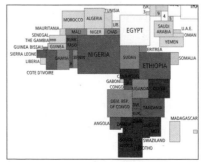

Cartogram: Detail of HIV Infection

GOODE'S

ATLAS OF Latin America

Map Legend

Political Boundaries

Political maps	Physical maps	
-----	=====	International (Demarcated, Undemarcated, and Administrative)
-·-·-	-··-··-	Disputed de facto
▪▪▪	▪▪▪	Indefinite or Undefined
------	------	Secondary, State, Provincial, etc.
▯		Parks, Indian Reservations
		City Limits
		Urbanized Areas

Transportation

Political maps	Physical maps	
———	———	Railroads
--------	--------	Railroad Ferries
	———	Major Roads
	———	Minor Roads
	··········	Caravan Routes
	✈	Airports

Cultural Features

�	Dams
··········	Pipelines
▲	Points of Interest
∴	Ruins

Populated Places

◉	1,000,000 and over
◎	250,000 to 1,000,000
☉	100,000 to 250,000
•	25,000 to 100,000
○	Under 25,000
▫	Neighborhoods, Sections of Cities
TŌKYŌ	National Capitals
Boise	Secondary Capitals

Note: On maps at 1:20,000,000 and smaller, symbols do not follow the population classification shown above. Some other maps use a slightly different classification, which is shown in a separate legend in the map margin. On all maps, type size indicates the relative importance of the city.

Land Features

△	Peaks, Spot Heights
⤬	Passes
	Sand
	Contours

Elevation

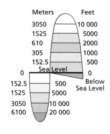

Meters		Feet
3050		10 000
1525		5000
610		2000
305		1000
152.5		500
0	Sea Level	0
152.5	500	Below Sea Level
1525	5000	
3050	10 000	
6100	20 000	

Lakes and Reservoirs

	Fresh Water
	Fresh Water: Intermittent
	Salt Water
	Salt Water: Intermittent

Other Water Features

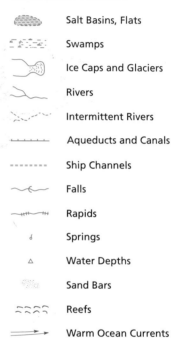

	Salt Basins, Flats
	Swamps
	Ice Caps and Glaciers
	Rivers
	Intermittent Rivers
	Aqueducts and Canals
--------	Ship Channels
	Falls
	Rapids
	Springs
△	Water Depths
	Sand Bars
	Reefs
⟶	Warm Ocean Currents
⟶	Cold Ocean Currents

The legend above shows the symbols used for the political and physical reference maps in *Goode's Atlas of Latin America*.

To portray relative areas correctly, uniform map scales have been used wherever possible:

    Continents – 1:40,000,000
    Countries and regions – between 1:4,000,000 and 1:20,000,000
    World, polar areas and oceans – between 1:50,000,000
        and 1:100,000,000
    Urbanized areas – 1:1,000,000

Elevations on the maps are shown using a combination of shaded relief and hypsometric tints. Shaded relief (or hill-shading) gives a three-dimensional impression of the landscape, while hypsometric tints show elevation ranges in different colors.

The choice of names for mapped features is complicated by the fact that a variety of languages and alphabets are used throughout the world. A local-names policy is used in *Goode's Atlas of Latin America* for populated places and local physical features. For some major features, an English form of the name is used with the local name given below in parentheses. Examples include Moscow (Moskva), Vienna (Wien) and Naples (Napoli). In countries where more than one official language is used, names are given in the dominant local language. For large physical features spanning international borders, the conventional English form of the name is used. In cases where a non-Roman alphabet is used, names have been transliterated according to accepted practice.

Selected features are also listed in the Index (pp. 39-55), which includes a pronunciation guide. A list of foreign geographic terms is provided in the Glossary (p. 36).

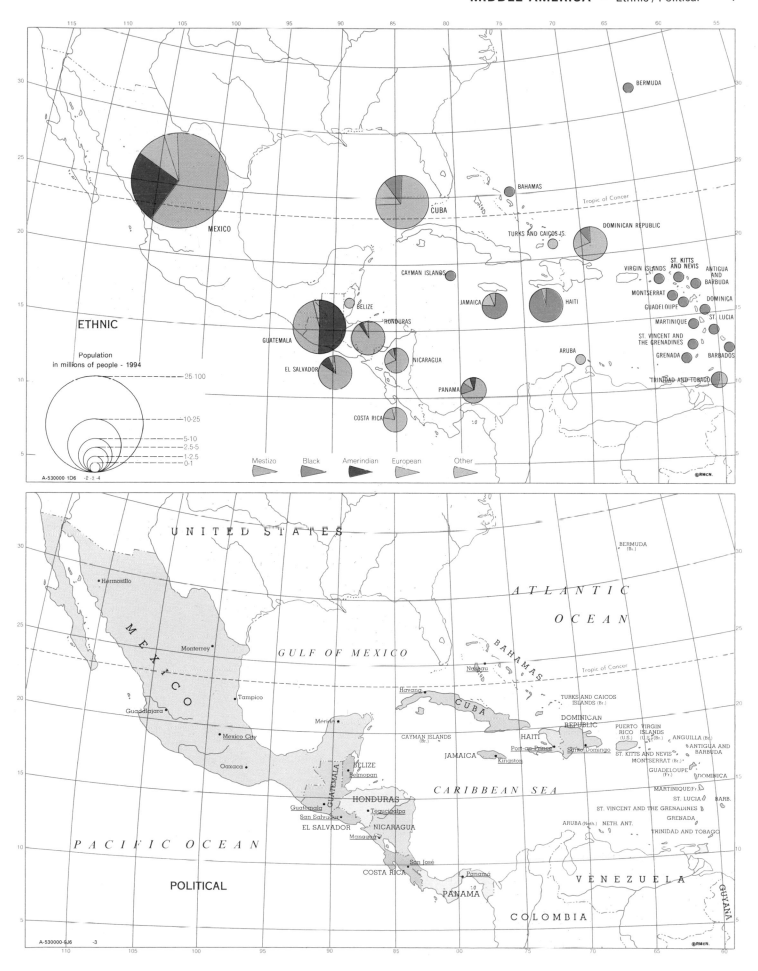

ETHNIC

Population
in millions of people - 1994

25-100
10-25
5-10
2.5-5
1-2.5
0-1

A-530000 1D6   -2 -2 -4

Mestizo   Black   Amerindian   European   Other

MEXICO

BAHAMAS

CUBA

Tropic of Cancer

TURKS AND CAICOS IS.

DOMINICAN REPUBLIC

CAYMAN ISLANDS

BELIZE

HONDURAS

GUATEMALA

EL SALVADOR

NICARAGUA

COSTA RICA

PANAMA

JAMAICA

HAITI

VIRGIN ISLANDS

ST. KITTS AND NEVIS

ANTIGUA AND BARBUDA

MONTSERRAT

GUADELOUPE

DOMINICA

MARTINIQUE

ST. LUCIA

ST. VINCENT AND THE GRENADINES

GRENADA

BARBADOS

ARUBA

TRINIDAD AND TOBAGO

BERMUDA

©RMcN.

POLITICAL

UNITED STATES

MEXICO

• Hermosillo

Monterrey •

• Tampico

Guadalajara •

• Mexico City

Oaxaca •

Merida •

BELIZE
Belmopan

GUATEMALA

HONDURAS

Guatemala •   • Tegucigalpa

San Salvador •

EL SALVADOR

NICARAGUA

Managua •

COSTA RICA

San José •

PANAMA

Panama •

GULF OF MEXICO

BAHAMAS

Nassau •

Tropic of Cancer

CUBA

Havana •

CAYMAN ISLANDS
(Br.)

JAMAICA

Kingston •

HAITI

Port-au-Prince •

DOMINICAN REPUBLIC

Santo Domingo •

TURKS AND CAICOS
ISLANDS (Br.)

PUERTO VIRGIN
RICO ISLANDS
(U.S.) (U.S.) (Br.)

ANGUILLA (Br.)

ANTIGUA AND
BARBUDA

ST. KITTS AND NEVIS

MONTSERRAT (Br.)

GUADELOUPE (Fr.)

DOMINICA

MARTINIQUE (Fr.)

ST. LUCIA

ST. VINCENT AND THE GRENADINES

BARB.

GRENADA

ARUBA (Neth.)   NETH. ANT.

TRINIDAD AND TOBAGO

CARIBBEAN SEA

ATLANTIC OCEAN

BERMUDA
(Br.)

PACIFIC OCEAN

VENEZUELA

COLOMBIA

GUYANA

A-530000-9J6   -3

©RMcN.

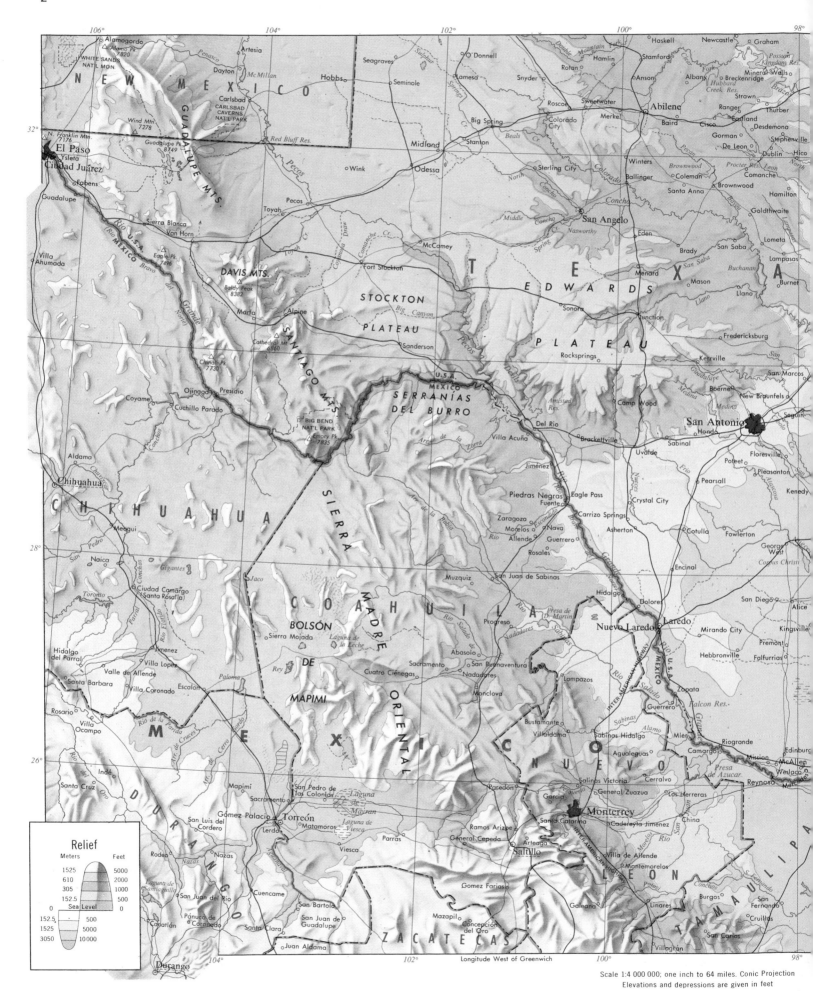

Scale 1:4 000 000; one inch to 64 miles. Conic Projection
Elevations and depressions are given in feet

Longitude West of Greenwich

Relief

Meters		Feet
1525		5000
610		2000
305		1000
152.5		500
0	Sea Level	0
152.5		500
1525		5000
3050		10000

Cities
and
Towns

0 to 50,000

50,000 to 500,000

500,000 to 1,000,000

1,000,000 and over

Scale 1:1 000 000

A-511007-76   5⌐ 5-8

COPYRIGHT BY
RAND McNALLY & COMPANY
MADE IN U.S.A.

4

Scale 1:16 000 000; one inch to 250 miles. Polyconic Projection
Elevations and depressions are given in feet

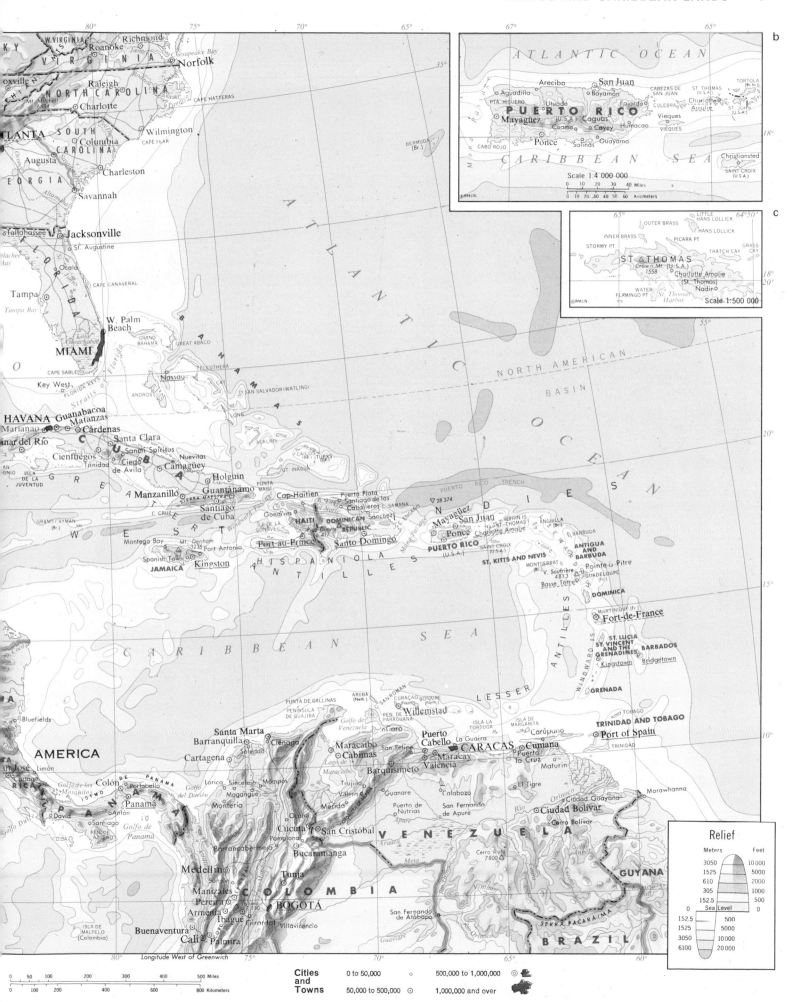

b

ATLANTIC OCEAN

Arecibo   San Juan
o Aguadilla   Bayamón   CABEZAS DE   ST. THOMAS   TORTOLA (Br.)
PTA. HIGUERO   Utuado   SAN JUAN   (U.S.A.)   Charlotte   ST. JOHN
PUERTO RICO   CULEBRA   Amalie   (U.S.A.)
Mayagüez   (U.S.A.)   Caguas   Vieques
Coamo   Cayey Humacao   VIEQUES
CABO ROJO   Ponce   Guayama
Salinas
CARIBBEAN   SEA   Christiansted
SAINT CROIX   18°
(U.S.A.)

Scale 1:4 000 000
0   10   20   30   40 Miles
0 10 20 .30 40 50 60 Kilometers

c

LITTLE   64°50'
OUTER BRASS   HANS LOLLICK   65°
INNER BRASS   HANS LOLLICK
STORMY PT.   PICARA PT   THATCH CAY   GRABS
        CAY
ST THOMAS
Crown Mt. (U.S.A.)   18°
1558   Charlotte Amalie   20'
(St. Thomas)
WATER   Nadir
FLAMINGO PT   St. Thomas
Harbor
Scale 1:500 000

Relief

Meters		Feet
3050		10 000
1525		5000
610		2000
305		1000
152.5		500
Sea Level		0
152.5		500
1525		5000
3050		10 000
6100		20 000

0   50   100   200   300   400   500 Miles
0   100   200   400   600   800 Kilometers

Longitude West of Greenwich

Cities and Towns
0 to 50,000   o   500,000 to 1,000,000
50,000 to 500,000   ⊙   1,000,000 and over

6

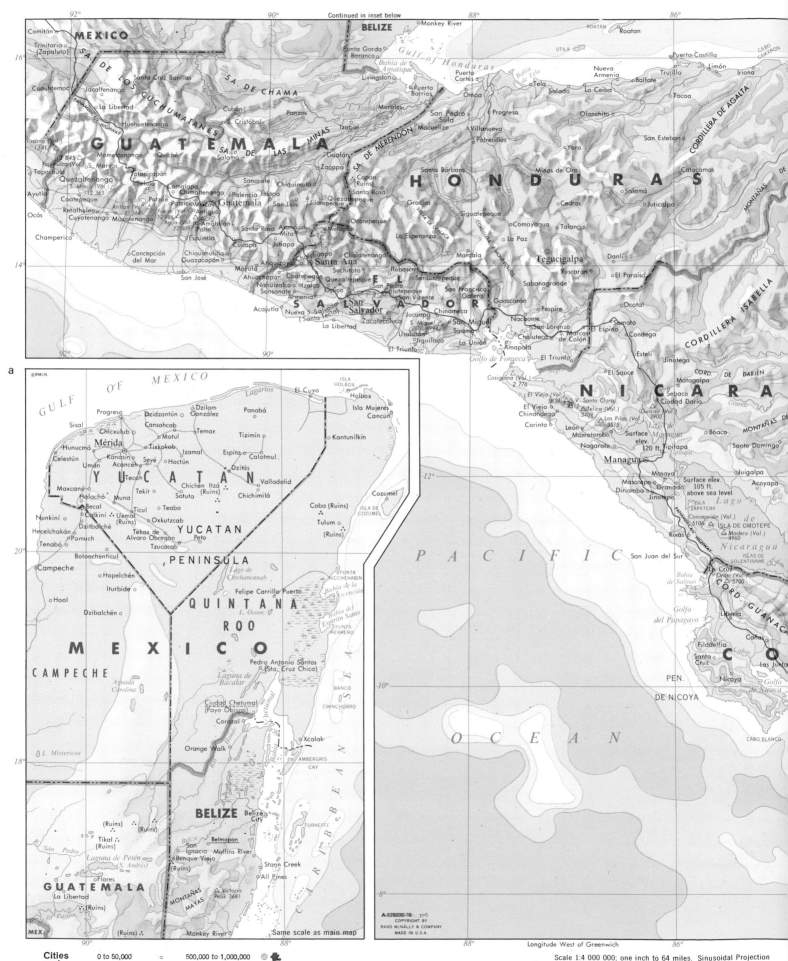

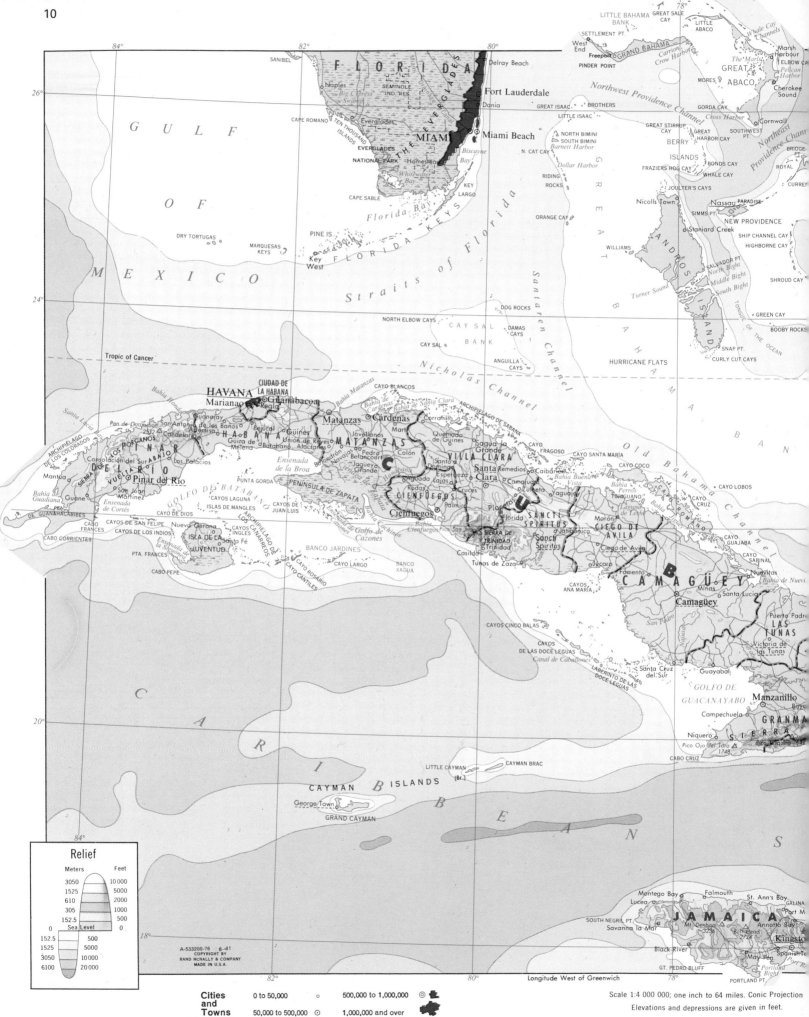

FLORIDA

Naples
Big Cypress
Swamp
SEMINOLE
IND. RES.
CAPE ROMANO
TEN THOUSAND
ISLANDS
Everglades
NATIONAL PARK
Homestead
CAPE SABLE
Whitewater
Bay

Delray Beach
Fort Lauderdale
Dania
MIAMI
Miami Beach
Biscayne
Bay
KEY
LARGO

G U L F

O F

M E X I C O

SANIBEL

THE EVERGLADES

Florida Bay

DRY TORTUGAS
MARQUESAS
KEYS
PINE IS.
Key West
FLORIDA KEYS

Straits of Florida

LITTLE BAHAMA BANK
SETTLEMENT PT.
West End
Freeport
GRAND BAHAMA
PINDER POINT
GREAT ISAAC
LITTLE ISAAC
BROTHERS
NORTH BIMINI
SOUTH BIMINI
Barnett Harbor
N. CAT CAY
Dollar Harbor
RIDING ROCKS
ORANGE CAY

78°
GREAT SALE CAY
LITTLE ABACO
Whale Cay Channels
The Marls
Marsh Harbor
ELBOW CAY
GREAT
ABACO
MORES
Cherokee Sound
GORDA CAY
Cross Harbor
GREAT STIRRUP CAY
GREAT HARBOR CAY
BERRY
ISLANDS
BONDS CAY
FRAZIERS HOG CAY
WHALE CAY
JOULTER'S CAYS
Nicolls Town
SIMMS PT.
Staniard Creek

Northwest Providence Channel

Pelican Harbor
SOUTHWEST
BRIDGE
ROYAL
CURREN
Nassau
PARADISE
NEW PROVIDENCE
SHIP CHANNEL CAY
HIGHBORNE CAY

Tropic of Cancer

NORTH ELBOW CAYS
DOG ROCKS
DAMAS CAYS
CAY SAL
BANK
ANGUILLA CAYS

CAY SAL

Santaren Channel

Nicholas Channel

HURRICANE FLATS

Old Bahama Channel

WILLIAMS
ANDROS ISLAND
Turner Sound
North Bight
Middle Bight
South Bight
SALVADOR PT.
SNAP PT.
GREEN CAY
BOOBY CAY
CURLY CUT CAYS
TONGUE OF THE OCEAN

HAVANA
CIUDAD DE
LA HABANA
Marianao
Guanabacoa
Regla
HABANA
Guanajay
San Antonio de los Baños
Bejucal
Güira de Melena
Candelaria
Los Palacios
PINAR
DEL RIO
Pan de Guajaibon
2532
SIERRA
ABAJO
VUELTA
Pinar del Rio
Consolación del Sur
San Juan y Martinez
Guane
Mantua
ARCHIPIÉLAGO DE LOS COLORADOS
Bahia Honda
Santa Lucia
Bahia de Guadiana
PEN. DE GUANAHACABIBES
CABO CORRIENTES
CABO FRANCES
PTA. FRANCES
CABO PEPE

Matanzas
MATANZAS
Cárdenas
Corralillo
Martí
Quemado de Güines
Bahía Matanzas
Bahía de Cárdenas
Bahía de Santa Clara
Cayo Blancos
ARCHIPIÉLAGO DE SABANA
Jovellanos
Unión de Reyes
Colón
Jagüey Grande
Pedro Betancourt
Bolondrón
Navajas
Santo Domingo
Lajas
Esperanza
Sagua la Grande
VILLA CLARA
Santa Clara
Cruces
Palmira
CIENFUEGOS
Cienfuegos
Rodas
Aguada
CAYO FRAGOSO
Remedios
Caibarién
Camajuaní
Zulueta
Yaguajay
CAYO SANTA MARÍA
Bahía Buena Vista
CAYO COCO
CAYO GUILLERMO
CAYO LOBOS

Güines
Güira de Melena

PENÍNSULA DE ZAPATA
Ensenada de la Broa
PUNTA GORDA
Guayabal
Ensenada de Cochinos
Bahía Cochinos
Bahía de Cienfuegos
Pico San Juan
SIERRA DE TRINIDAD
Casilda
Trinidad
Tunas de Zaza

SANCTI
SPIRITUS
Sancti
Spiritus
Jatibonico
Florida
CIEGO DE
AVILA
Ciego de Avila
Júcaro
Fomento
Placetas

CAMAGÜEY
Camagüey
Minas
Santa Lucía
San Pedro
CAYOS ANA MARÍA
CAYOS CINCO BALAS
CAYOS DE LAS DOCE LEGUAS
LABERINTO DE LAS DOCE LEGUAS
Santa Cruz del Sur
Nuevitas
Bahía de Nuevi
CAYO GUAJABA
CAYO SABINAL
CAYO CRUZ
CAYO ROMANO
Morón
Laguna de Leche
CAYO PAREDÓN GRANDE
TRIGUANO
Laguna
PERROS

LAS TUNAS
Victoria de las Tunas

GOLFO DE
GUACANAYABO
Guayabal
Manzanillo
Campechuela
GRANMA
SIERRA
Niquero
Pico Ojo del Toro
1748
CABO CRUZ

ISLA DE LA
JUVENTUD
Nueva Gerona
Santa Fé
ARCHIPIÉLAGO
DE LOS CANARREOS
CAYOS DE SAN FELIPE
CAYOS DE LOS INDIOS
Ensenada de la Siguanea
GOLFO DE BATABANÓ
CAYOS LAGUNA
ISLAS DE MANGLES
CAYO DE DIOS
CAYO DE JUAN LUIS
Batabanó
Melena
Golfo de Cazones
BANCO JARDINES
CAYO ROSARIO
CAYO CANTILES
CAYO LARGO
BANCO XAGUA

LITTLE CAYMAN
CAYMAN BRAC
(Br.)
CAYMAN ISLANDS
George Town
GRAND CAYMAN

C A R I B B E A N   S E A

MONTEGO BAY
Montego Bay
Falmouth
St. Ann's Bay
GALINA
Lucea
Port M
SOUTH NEGRIL PT.
Savanna la Mar
JAMAICA
Mt. Denham
2256
Annotto Bay
May Pen
Spanish T
Kingsto
Black River
Savanna la Mar
GT. PEDRO BLUFF
PORTLAND PT.
Portland Bight

82°
80°
78°
Longitude West of Greenwich

Relief

Meters		Feet
3050		10 000
1525		5000
610		2000
305		1000
152.5		500
0	Sea Level	0
152.5		500
1525		5000
3050		10 000
6100		20 000

Cities and Towns

0 to 50,000 ○
50,000 to 500,000 ⊙
500,000 to 1,000,000 ◎
1,000,000 and over

Scale 1:4 000 000; one inch to 64 miles. Conic Projection
Elevations and depressions are given in feet.

HAVANA (La Habana)

Scale 1:1 000 000

GULF OF MEXICO

Playa de Guanab·
Cojimar
Guanabacoa
Regla
Campo Florido
Baracoa
San Francisco de Paula
Marianao
Cotorro
Arroyo Arena
Calabazar
Rancho Boyeros
Cuatro Caminos
Bauta
Santiago de las Vegas
Managua
Caimito del Guayabal
Bejucal
San José de las Lajas
La Sabina
L. de Ariguanabo
San Antonio de los Baños
Buenaventura
Ceiba del Agua
San Antonio de las Vegas
∧ 950

ATLANTIC OCEAN

JAMES PT.
Governor's Harbour
PALMETTO PT.
ELEUTHERA
arpum Bay
Rock Sound
WELL PT.
ELEUTHERA PT.
Arthur's Town
NORTHEAST PT.
LITTLE SAN SALVADOR
CAT
Old Bight
SAN SALVADOR (WATLING)
(Columbus Oct. 12, 1492)
HAWKS NEST PT.
COLUMBUS PT.
SOUTHWEST PT.
GREAT GUANA CAY
CONCEPTIÓN
DARBY
LEE STOCKING
CAPE STA. MARIA
Rolleville
RUM CAY
Tropic of Cancer
GREAT EXUMA
George Town
LITTLE EXUMA
HOG CAY
LONG
JUMENTO CAYS
Clarence Town
SAMANA OR ATWOOD CAY
WATER CAY
FLAMINGO CAY
CAP VERDE
BIRD ROCK
CROOKED
JAMAICA CAY
NORTHEAST PT.
PLANA OR FLAT CAYS
SEAL CAYS
FORTUNE
The Bight of Acklins
Man of War Channel
NURSE CAY
DIANA BANK
FISH CAY
ACKLINS
RACCOON CAY
SALINA PT.
CASTLE
MAYAGUANA
Abraham's Bay
GREAT RAGGED
COLUMBUS BANK
MIRA POR VOS ISLETS
CAY VERDE
COCHINOS BANKS
CAY STA. DOMINGO
HOGSTY REEF
Caicos Passage
PROVIDENCIALES
NORTH CAICOS
GRAND CAICOS
CAPE COMETE
EAST CAICOS
BROWN BANK
WEST CAICOS
CAICOS IS. (Br.)
CAICOS BANK
SOUTH CAICOS
GRAND TURK
LITTLE INAGUA
Grand Turk
TURKS IS. (Br.)
AMBERGRIS CAYS
GALT CAY
PALMETTO PT.
NORTHEAST PT.
WEST SAND SPIT
SEAL CAYS
MOUCHOIR BANK
Gibara
CABO LUCRECIA
Ocean Bight
The Lake
Mouchoir Passage
Banes
GREAT INAGUA
Antilla
Matthew Town
Man of War Bay
SILVER BANK
Holguin
Bahia de Nipe
Mayari
South Bay
HOLGUÍN
Sagua de Tánamo
CUCHILLAS DE TOA
NAVIDAD BANK
SANTIAGO DE CUBA
Baracoa
SA. DEL CRISTAL
GUANTANAMO
SA. DE PURIAL
PUNTA MAISI
Alto Songo
SA. MAESTRA
San Luis
Bahia de Ovando
ILE DE LA TORTUE
CABO ISABELA
Santiago de Cuba
Coney
Caimanera
Canal de la Tortue
Monte Cristi
Puerto Plata
CABO FRANCES VIEJO
Guantánamo
Port de Paix
CORDILLERA SEPTENTRIONAL
Yateras
Le Borgne
Cap-Haitien
Guayubin
Gaspar Hernández
Naval Station (U.S.A.)
CAP ST. NICOLAS
Le Môle
Limbé
Fort Liberté
Dajabón
Santiago Rodriguez
Bahia de Guantánamo
PTE. PLATEFORME
Grande Rivière du Nord
Ouanaminthe
Santiago de los Caballeros
Bahia Escocesa
Gonaïves
St. Michel-de-l'Atalaye
Vallière
Moca
Nagua
CABO SAMANÁ
Windward Passage
GOLFE DES GONAÏVES
St. Marc
Hinche
DOMINICAN
La Vega
Riva
Bahia de Samaná
CABO SAN RAFAEL
Sánchez
Samaná
Pic Bonhomme
Miches
CORDILLERA CENTRAL
Sabana de la Mar
HAITI
Pico Duarte
CORDILLERA ORIENTAL
POINT OUEST
ILE DE LA GONAVE
SIERRA DE NEIBA
Jarabacoa
Hato Mayor
Jérémie
Mirebalais
San Juan
Yamasá
Seibo
ILE GRANDE CAYEMITE
Canal du Sud
Baie des Baradères
Port-au-Prince
Petionville
Bánica
Neiba
Azua
San Cristóbal
Los Llanos
Higüey
CAP DAME MARIE
Léogâne
La Romana
Anse d'Hainault
Miragoâne
CORDILLERA
Baní
CAP DES IROIS
Petit Goâve
MASSIF DE LA HOTTE
MASSIF DE LA SELLE
SIERRA DE BAHORUCO
Santo Domingo
CATALINA
Tiburon
Duverge
Barahona
PTA. PALENQUE
Cateaux
Aquin
Enriquillo
SAONA
Les Cayes
Jacmel
Belle-Anse
BAHORUCO
S. Pedro de Macoris
FORMIGAS BANK
Roche à Bateau
ILE À VACHE
HISPANIOLA
NAVASSA (U.S.A.)
POINTE À GRAVOIS
Oviedo
Trujin
MORANT PT.
CABO FALSO
Port Antonio
BEATA
CABO BEATA
ALTO VELO

0 10 20 30 40 50 60 70 80 90 100 110 120 Miles
0 20 40 60 80 100 120 140 160 180 200 Kilometers

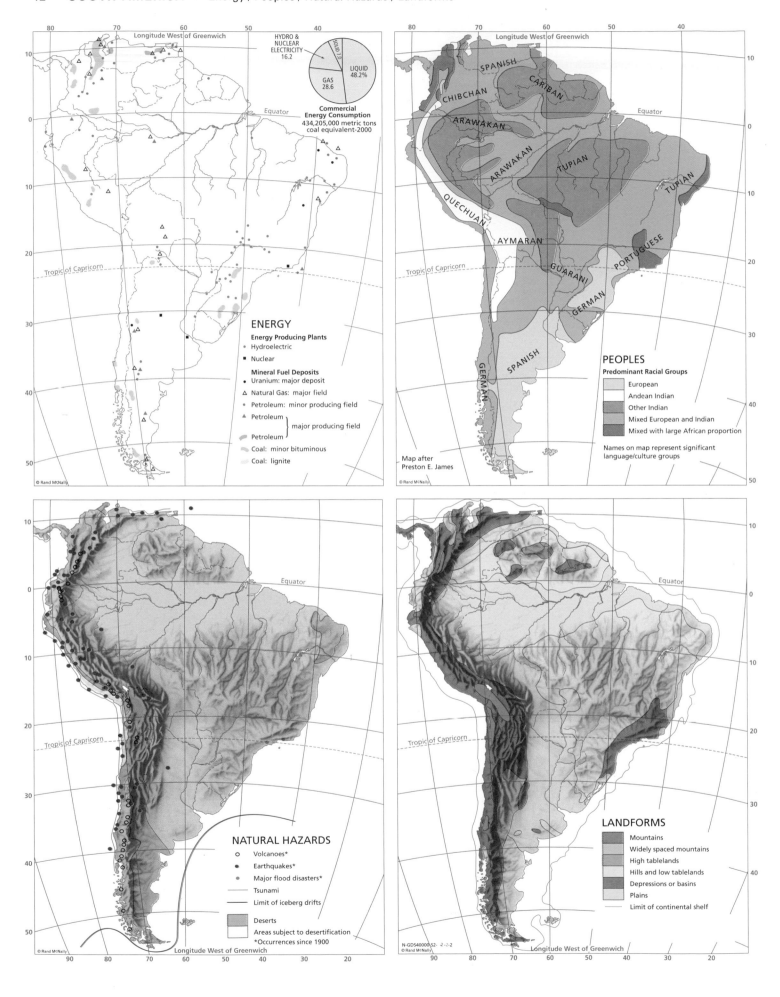

ENERGY

Energy Producing Plants
- • Hydroelectric
- ■ Nuclear

Mineral Fuel Deposits
- • Uranium: major deposit
- △ Natural Gas: major field
- • Petroleum: minor producing field
- ▲ Petroleum ⎫
    } major producing field
-    Petroleum ⎭
-    Coal: minor bituminous
-    Coal: lignite

HYDRO & NUCLEAR ELECTRICITY 16.2
SOLID 7.0
LIQUID 48.2%
GAS 28.6

Commercial Energy Consumption
434,205,000 metric tons coal equivalent-2000

© Rand McNally

PEOPLES

Predominant Racial Groups
-   European
-   Andean Indian
-   Other Indian
-   Mixed European and Indian
-   Mixed with large African proportion

Names on map represent significant language/culture groups

Map after Preston E. James

© Rand McNally

NATURAL HAZARDS

- ○ Volcanoes*
- ● Earthquakes*
- ● Major flood disasters*
- ⎯ Tsunami
- ⎯ Limit of iceberg drifts
-   Deserts
-   Areas subject to desertification

*Occurrences since 1900

© Rand McNally

LANDFORMS

-   Mountains
-   Widely spaced mountains
-   High tablelands
-   Hills and low tablelands
-   Depressions or basins
-   Plains
- ⎯ Limit of continental shelf

N-GDS40000 S2- -2-2-2
© Rand McNally

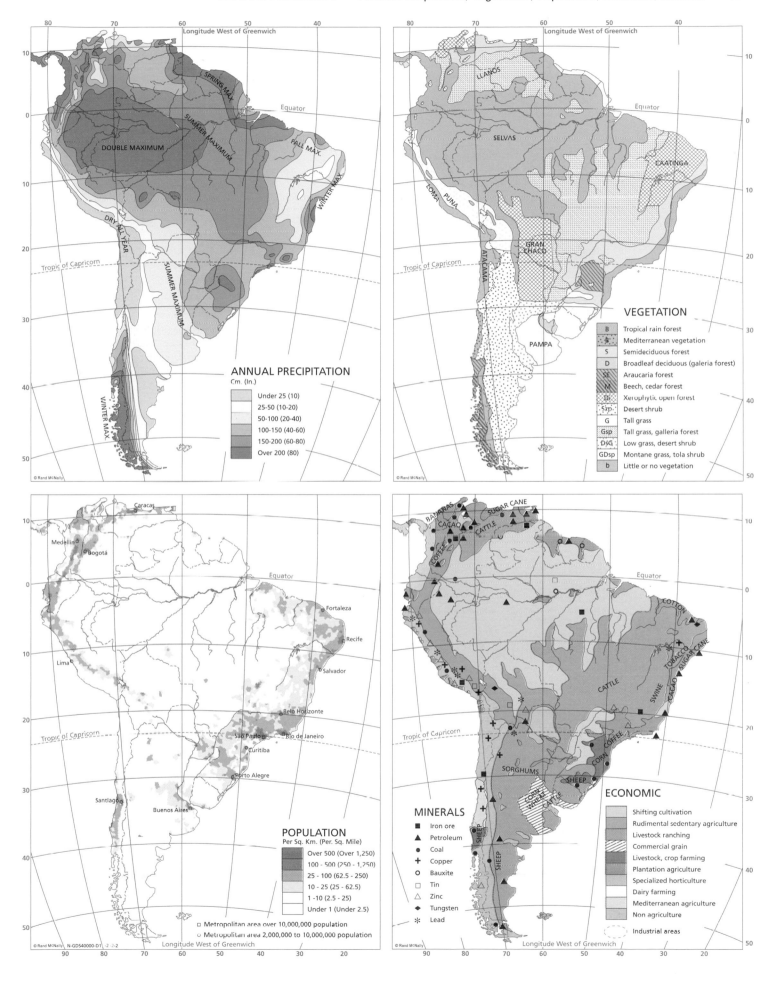

ANNUAL PRECIPITATION

Cm. (In.)

- Under 25 (10)
- 25-50 (10-20)
- 50-100 (20-40)
- 100-150 (40-60)
- 150-200 (60-80)
- Over 200 (80)

SPRING MAX.
SUMMER MAXIMUM
FALL MAX.
DOUBLE MAXIMUM
WINTER MAX.
DRY ALL YEAR
SUMMER MAXIMUM
WINTER MAX.

© Rand McNally

VEGETATION

B	Tropical rain forest
B	Mediterranean vegetation
S	Semideciduous forest
D	Broadleaf deciduous (galeria forest)
SE	Araucaria forest
M	Beech, cedar forest
Di	Xerophytic open forest
Szp	Desert shrub
G	Tall grass
Gsp	Tall grass, galleria forest
DsG	Low grass, desert shrub
GDsp	Montane grass, tola shrub
b	Little or no vegetation

LLANOS
SELVAS
CAATINGA
LOMA
PUNA
ATACAMA
GRAN CHACO
PAMPA

© Rand McNally

POPULATION

Per Sq. Km. (Per. Sq. Mile)

- Over 500 (Over 1,250)
- 100 - 500 (250 - 1,250)
- 25 - 100 (62.5 - 250)
- 10 - 25 (25 - 62.5)
- 1 - 10 (2.5 - 25)
- Under 1 (Under 2.5)

□ Metropolitan area over 10,000,000 population
○ Metropolitan area 2,000,000 to 10,000,000 population

Caracas
Medellín
Bogotá
Lima
Fortaleza
Recife
Salvador
Belo Horizonte
São Paulo
Rio de Janeiro
Curitiba
Porto Alegre
Santiago
Buenos Aires

© Rand McNally   N-GDS40000-D1--2-2-2

MINERALS

- ■ Iron ore
- ▲ Petroleum
- ● Coal
- + Copper
- ○ Bauxite
- □ Tin
- △ Zinc
- ◆ Tungsten
- ✳ Lead

ECONOMIC

- Shifting cultivation
- Rudimental sedentary agriculture
- Livestock ranching
- Commercial grain
- Livestock, crop farming
- Plantation agriculture
- Specialized horticulture
- Dairy farming
- Mediterranean agriculture
- Non agriculture

- ⬭ Industrial areas

BANANAS
CACAO
SUGAR CANE
CATTLE
COFFEE
COTTON
TOBACCO
SUGAR CANE
CACAO
SWINE
CATTLE
COFFEE
CORN
SORGHUMS
CORN WHEAT
CATTLE
SHEEP
SHEEP
SHEEP

© Rand McNally

Longitude West of Greenwich
Equator
Tropic of Capricorn

Scale 1:40,000,000; one inch to 630 miles. Lambert's Azimuthal. Equal Area Projection
Elevations and depressions are given in feet

40,000 SQ MI
AREA

0   300   600
Miles

0   200   400   600   800   1000 Miles
0   400   800   1200   1600 Kilometers

A-540000-26 -47-16
COPYRIGHT BY
RAND McNALLY & COMPANY
MADE IN U.S.A.

Longitude West of Greenwich

ATLANTIC OCEAN

HAVANA
CUBA
Bahía de Campeche
PEN. DE YUCATÁN
Gulf of Honduras
JAMAICA
HISPANIOLA
San Juan
PUERTO RICO (U.S.A.)
GUADELOUPE (Fr.)
MARTINIQUE (Fr.)
CARIBBEAN SEA
WEST INDIES
BARBADOS
CENTRAL
Lago de Nicaragua
AMERICA
Panamá
Golfo de Panamá
PUNTA DE GALLINAS
Barranquilla
Cartagena
TRINIDAD AND TOBAGO
Port of Spain
Maracaibo
La Guaira
Valencia CARACAS
Mérida
Ciudad Bolívar
Orinoco
Georgetown
Paramaribo
Cayenne
LLANOS
VENEZUELA
GUYANA
SURINAME FR. GUIANA
Medellín
BOGOTÁ
Boa Vista do Rio Branco
GUIANA HIGHLANDS
ISLA DEL COCO (Costa Rica)
ISLA DE MALPELO (Colombia)
Nevado del 17 110
COLOMBIA
ILHA DE MARAJÓ
Quito
Equator
ROCEDOS SÃO PEDRO E SÃO PAULO (Brazil)
ARCHIPIÉLAGO DE COLÓN (GALÁPAGOS ISLANDS) (Ec.)
ECUADOR
Cotopaxi 19 347
Chimborazo 20 702
Guayaquil
Golfo de Guayaquil
Iquitos
Leticia
Manaus (Manáos)
Belém (Pará)
São Luís (Maranhão)
Fortaleza (Ceará)
ARQUIPÉLAGO DE FERNANDO DE NORONHA (Brazil)
Chiclayo
Trujillo
Nev. Huascarán 22 139
Río Branco
Porto Velho
Teresina
Natal
João Pessoa (Paraíba)
RECIFE (Pernambuco)
Maceió
LIMA
Callao
Cusco
Volcán Misti 19 101
Arequipa
Mollendo
La Paz
Nev. Illimani 21 274
CHAPADA DE MATO GROSSO
Cuiabá
Brasília
Diamantina
Salvador (Bahia)
BOLIVIA
Sucre
Potosí
Belo Horizonte
Iquique
Antofagasta
Cerro 19 947
Salta
Copiapó
Tucumán
Asunción
SÃO PAULO
Pico da Bandeira 9482
Vitória
CABO FRIO
Santos
RIO DE JANEIRO
ISLA DE SAN FÉLIX (Chile)
ISLA DE SAN AMBROSIO (Chile)
PARAGUAY
Corrientes
Florianópolis
GRAN CHACO
Coquimbo
Cerro Aconcagua 22 835
Córdoba
Santa Fe
Salto
Porto Alegre
Valparaíso
SANTIAGO
Mendoza
Rosario
URUGUAY
Río Grande
ISLAS DE JUAN FERNÁNDEZ (Chile)
BUENOS AIRES
La Plata
MONTEVIDEO
Río de la Plata
PAMPAS
Concepción
Colorado
ATLANTIC OCEAN
Valdivia
Bahía Blanca
Puerto Montt
Viedma
Golfo San Matías
ISLA DE CHILOÉ
ARCHIPIÉLAGO DE LOS CHONOS
Mte. 314 San Valentín
Comodoro Rivadavia
Golfo San Jorge
FALKLAND IS. (ISLAS MALVINAS) (Br.)
WELLINGTON I.
HANOVER I.
Río Gallegos
Stanley
Punta Arenas
DESOLACIÓN I.
Estrecho de Magallanes
TIERRA DEL FUEGO
MT. 8100
CABO DE HORNOS (CAPE HORN)
Drake Passage
SOUTH GEORGIA (Br.)
SOUTH SANDWICH ISLANDS (Br.)
SOUTH SHETLAND ISLANDS (Br.)
SOUTH ORKNEY IS (Br.)
ANTARCTIC PENINSULA
JAMES ROSS I.
JOINVILLE I.
PACIFIC OCEAN
Tropic of Capricorn
PERU
ANDES
CHILE
ARGENTINA
PATAGONIA
BRAZIL
BRAZILIAN HIGHLANDS
Amazon (Amazonas)
Negro
Japurá
Putumayo
Juruá
Purús
Madeira
Tapajós
Xingú
Tocantins
São Francisco
SERRA DO MAR
SERRA DO ESPINHAÇO
SERRA DO PARNAÍBA
Paraná
Paraguay
Pilcomayo

Antarctic Circle

NORTH AMERICAN BASIN
PUERTO RICO TRENCH
PERU-CHILE TRENCH

Relief		
Meters		Feet
3050		10 000
1525		5000
610		2000
305		1000
	Sea Level	0
152.5		500
1525		5000
3050		10 000
6100		20 000

A-540000-76
COPYRIGHT BY
RAND MCNALLY & COMPANY
MADE IN U.S.A.

| 0 | 200 | 400 | 600 | 800 | 1000 Miles |

| 0 | 400 | 800 | 1200 | 1600 Kilometers |

Scale 1:40 000 000; one inch to 630 miles. Lambert's Azimuthal. Equal Area Projection
Elevations and depressions are given in feet

CUBA
JAMAICA
Kingston
San Juan
PUERTO
RICO
HISPANIOLA

Caribbean Sea

ATLANTIC

OCEAN

Barranquilla
Maracaibo
CARACAS
Port of Spain
TRINIDAD

Panamá

LLANOS
Orinoco
Georgetown

BOGOTÁ

Quito
Negro
Equator
Belém

Iquitos
Amazon
Manaus
Fortaleza

SELVAS

Rio Branco
São Francisco
Recife

LIMA
Salvador

La Paz
Cuiaba
MATO
GROSSO
Brasília

Iquique
Paraná
Belo Horizonte

Tropic of Capricorn
GRAN
CHACO
Asunción
**SÃO
PAULO**
RIO DE JANEIRO

ANDES

San Miguel
de Tucumán

Porto Alegre

Córdoba

SANTIAGO
BUENOS AIRES
Montevideo

PAMPA

Bahía Blanca

PACIFIC

OCEAN

ATLANTIC

OCEAN

Puerto Montt

PATAGONIA

FALKLAND
ISLANDS

Punta Arenas
TIERRA
DEL FUEGO

SOUTH
GEORGIA

Drake Passage

Legend:
- Urban
- Cropland
- Cropland & Woodland
- Cropland & Grazing Land
- Grassland, Grazing Land
- Forest, Woodland
- Swamp, Marshland
- Shrub, Sparse Grass, Wasteland
- Barren Land

A-540000-36
COPYRIGHT BY
RAND McNALLY & COMPANY
MADE IN U.S.A.

Scale 1:36,000,000; one inch to 570 miles Lambert Azimuthal Equal-Area Projection

0   100   200   400   600   800 Miles
0   150   300   600   900   1200 Kilometers

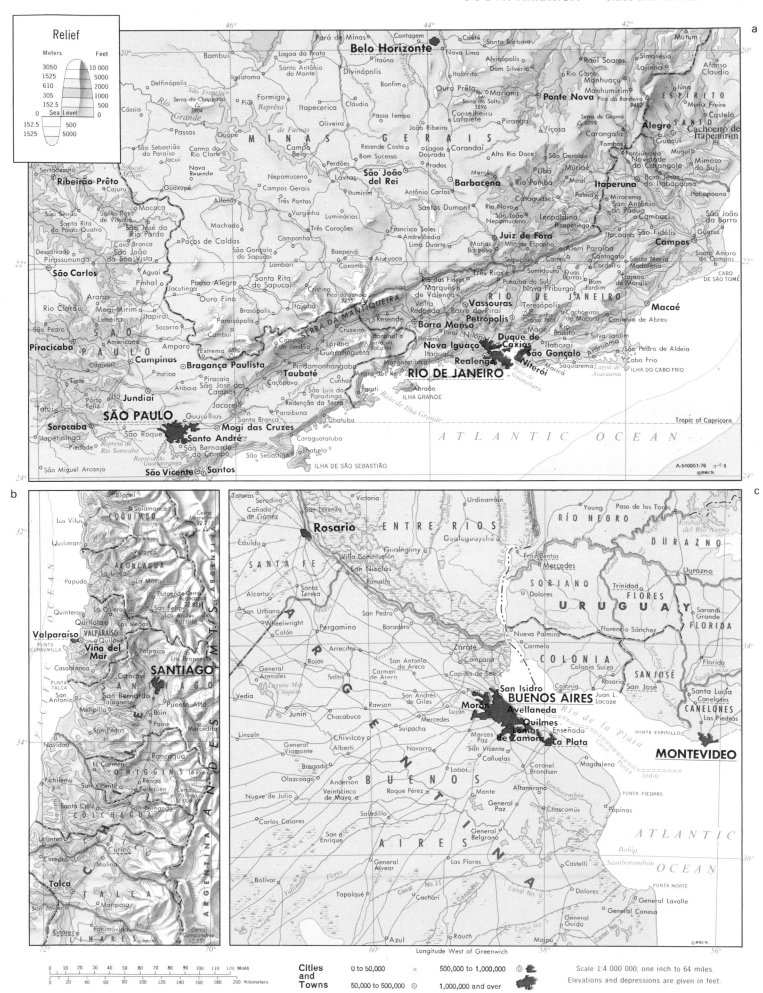

Relief

Meters	Feet
3050	10 000
1525	5000
610	2000
305	1000
152.5	500
0 Sea Level	0
152.5	500
1525	5000

MINAS GERAIS

Belo Horizonte

Ribeirão Prêto

São Carlos

Piracicaba Campinas

Jundiaí

Sorocaba

SÃO PAULO

Mogi das Cruzes

Santo André

São Vicente Santos

SÃO PAULO

RIO DE JANEIRO

Nova Iguaçu Duque de Caxias

Realenga São Gonçalo

Niterói

RIO DE JANEIRO

Barra Mansa Petrópolis

Vassouras

Juiz de Fora

Barbacena

São João del Rei

Ponte Nova

Alegre

Macaé

Campos

Tropic of Capricorn

ATLANTIC OCEAN

ILHA DE SÃO SEBASTIÃO

A-540051-76
©RMCN.

b

PACIFIC OCEAN

COQUIMBO

ACONCAGUA

Valparaíso VALPARAÍSO

Viña del Mar

SANTIAGO

O'HIGGINS

COLCHAGUA

Talca

TALCA

ARGENTINA ANDES MTS. S.

CHILE

c

Rosario

ENTRE RIOS

RÍO NEGRO

DURAZNO

SORIANO

URUGUAY

FLORIDA

COLONIA

SAN JOSÉ

CANELONES

Florida

Santa Lucía

San Isidro BUENOS AIRES

Morón Avellaneda

Quilmes

Lomas de Zamora La Plata

MONTEVIDEO

Río de la Plata

A R G E N T I N A

B U E N O S

A I R E S

ATLANTIC

OCEAN

Longitude West of Greenwich

©RMCN.

	Miles											
0	10	20	30	40	50	60	70	80	90	100	110	120
0	20	40	60	80	100	120	140	160	180	200	Kilometers	

Cities and Towns

0 to 50,000	○	500,000 to 1,000,000	◎
50,000 to 500,000	⊙	1,000,000 and over	▓

Scale 1:4 000 000; one inch to 64 miles.
Elevations and depressions are given in feet.

Continued on page 20

b

Inset map (Caracas region):

CARIBBEAN SEA

ISLA DE MARGARITA
Boca del Pozo △ 2303
PUNTA ARENAS
Punta de Piedras
NUEVA ESPARTA
ISLA CUBAGUA

Tocuyo de la Costa
Chichiriviche
CAYO SOMBRERO
Tucacas
Golfo Triste
Puerto Cabello
Morón
El Cambur
Montalbán
Guacara
San Joaquín
Miranda
Valencia
Güigüe
CARABOBO
Tinaquillo
COJEDES

Maiquetía La Guaira Naiguatá
Carayaca DISTRITO FEDERAL La Sabana
CARACAS Guatire
Petare Santa Lucía
Los Teques MIRANDA Santa Teresa
Pico Naiguatá 9072
La Victoria
Cúa Ocumare del Tuy San Francisco de Macaira
San Sebastián San Casimiro Altagracia de Orituco
San Juan de los Morros
Camatagua
Dos Caminos
Barbacoas

Higuerote
Río Chico
Boca de Uchire
Puerto Píritu Barcelona
El Hatillo Guanta
Soublette
San José de Guaribe
San Miguel
San Antonio de Tamanaco Guanape San Pablo
Onoto
Aragua de Barcelona
Libertad de Orituco

ISLA LA TORTUGA
CABO CODERA
ISLA LA BORRACHA

PUNTA DE ARAYA
Manicuare
Cumaná SUCRE
El Pilar
Bergantín △ 8000
Santa Inés
ANZOÁTEGUI
Anaco Santa Rosa

GUÁRICO

Scale 1:4 000 000
0 10 20 30 40 Miles
0 10 20 30 40 50 60 Kilometers
© R.M.C.N.

Main map labels (selection):

Port of Spain
TRINIDAD AND TOBAGO
TRINIDAD

Boca Grande
Morawhanna
Georgetown
Bartica New Amsterdam
Paramaribo
SURINAME FRENCH GUIANA
Cayenne
Saint-Georges

ACARAI MTS.
TUMUC-HUMAC MTS.
AMAPÁ
Amapá
Macapá
Mazagão
ILHA CAVIANA
ILHA DE MARAJO

ATLANTIC OCEAN

Equator

Manaus (Manáos)
Óbidos
Santarém
Amazon
Belém (Pará)
Abaetetuba
São Luís (Maranhão)
Parnaíba
FORTALEZA (Ceará)

PARÁ
MARANHÃO
Teresina
PIAUÍ
CEARÁ
RIO GRANDE DO NORTE
Natal
PARAÍBA
João Pessoa (Paraíba)
PERNAMBUCO
RECIFE (Pernambuco)
Olinda
Caruaru
ALAGOAS
Maceió
SERGIPE
Aracaju

BRAZIL

TOCANTINS
Palmas
BAHIA
Feira de Santana
SALVADOR (Bahia)

CHAPADA DE MATO GROSSO
MATO GROSSO
Cuiabá
GOIÁS
Brasília
Goiânia
MINAS GERAIS
BELO HORIZONTE
Vitória
ESPÍRITO SANTO

MATO GROSSO DO SUL
Campo Grande
SÃO PAULO
São José do Rio Preto
Ribeirão Preto
Uberlândia
Uberaba
Araçatuba
Marília
Bauru
Piracicaba
Campinas
Jundiaí
Sorocaba
SÃO PAULO
Santos
RIO DE JANEIRO
Niterói
Petrópolis
Nova Friburgo
Campos

PARANÁ
Londrina
Ponta Grossa
Curitiba

PARAGUAY
GRAN CHACO

Tropic of Capricorn

Relief		
Meters		**Feet**
3050		10 000
1525		5000
610		2000
305		1000
152.5		500
0	Sea Level	0
152.5		500
1525		5000
3050		10 000
6100		20 000

0 50 100 200 300 400 500 Miles
0 100 200 400 600 800 Kilometers

Continued on pages 18-19

Relief

Meters	Feet
3050	10 000
1525	5000
610	2000
305	1000
152.5	500
Sea Level	0
0	Level
152.5	500
1525	5000
3050	10 000
6100	20 000

Below Sea Level

a

BUENOS AIRES

Scale 1:1 000 000

0   4   8   12   16 Kilometers

b

Scale 1:1 000 000

0   4   8   12   16 Kilometers

A-549200-76   -11-8-14

COPYRIGHT BY
RAND McNALLY & COMPANY
MADE IN U.S.A.

Longitude West of Greenwich

Scale 1:16 000 000; one inch to 250 miles. Sinusoidal Projection
Elevations and depressions are given in feet

0   50   100   200   300   400   500 Miles

0   100   200   400   600   800 Kilometers

WORLD POLITICAL INFORMATION TABLE

This table gives the area, population, population density, political status, capital, and predominant languages for every country in the world. The political units listed are categorized by political status in the form of government column of the table, as follows: A—independent countries; B—internally independent political entities which are under the protection of another country in matters of defense and foreign affairs; C—colonies and other dependent political units; and D—the major administrative subdivisions of Australia, Canada, China, the United Kingdom, and the United States. For comparison, the table also includes the continents and the world. All footnotes appear at the end of the table.

The populations are estimates for January 1, 2004, made by Rand McNally on the basis of official data, United States Census Bureau estimates, and other available information. Area figures include inland water.

REGION OR POLITICAL DIVISION	Area Sq. Mi.	Est. Pop. 1/1/04	Pop. Per Sq. Mi.	Form of Government and Ruling Power	Capital	Predominant Languages	International Organizations
Afars and Issas see Djibouti							
Afghanistan	251,773	29,205,000	116	Transitional A	Kābul	Dari, Pashto, Uzbek, Turkmen	UN
Africa	11,700,000	866,305,000	74	D			
Alabama	52,419	4,515,000	86	State (U.S.) D	Montgomery	English	
Alaska	663,267	650,000	1.0	State (U.S.) D	Juneau	English, indigenous	
Albania	11,100	3,535,000	318	Republic A	Tiranë	Albanian, Greek	NATO/PP, UN
Alberta	255,541	3,215,000	13	Province (Canada) D	Edmonton	English	
Algeria	919,595	33,090,000	36	Republic A	Algiers (El Djazaïr)	Arabic, Berber dialects, French	AL, AU, OPEC, UN
American Samoa	77	58,000	753	Unincorporated territory (U.S.) C	Pago Pago	Samoan, English	
Andorra	181	70,000	387	Parliamentary co-principality (Spanish and French) B	Andorra	Catalan, Spanish (Castilian), French, Portuguese	UN
Angola	481,354	10,875,000	23	Republic A	Luanda	Portuguese, indigenous	AU, COMESA, UN
Anguilla	37	13,000	351	Overseas territory (U.K.) C	The Valley	English	
Anhui	53,668	61,215,000	1,141	Province (China) D	Hefei	Chinese (Mandarin)	
Antarctica	5,400,000	(¹)					
Antigua and Barbuda	171	68,000	398	Parliamentary state A	St. John's	English, local dialects	OAS, UN
Aomen (Macau)	6.9	445,000	64,493	Special administrative region (China) D	Macau (Aomen)	Chinese (Cantonese), Portuguese	
Argentina	1,073,519	38,945,000	36	Republic A	Buenos Aires	Spanish, English, Italian, German, French	MERCOSUR, OAS, UN
Arizona	113,998	5,600,000	49	State (U.S.) D	Phoenix	English	
Arkansas	53,179	2,735,000	51	State (U.S.) D	Little Rock	English	
Armenia	11,506	3,325,000	289	Republic A	Yerevan	Armenian, Russian	CIS, NATO/PP, UN
Aruba	75	71,000	947	Self-governing territory (Netherlands protection) B	Oranjestad	Dutch, Papiamento, English, Spanish	
Ascension	34	1,000	29	Dependency (St. Helena) C	Georgetown	English	
Asia	17,300,000	3,839,320,000	222				
Australia	2,969,910	19,825,000	6.7	Federal parliamentary state A	Canberra	English, indigenous	ANZUS, UN
Australian Capital Territory	911	325,000	357	Territory (Australia) D	Canberra	English	
Austria	32,378	8,170,000	252	Federal republic A	Vienna (Wien)	German	EU, NATO/PP, UN
Azerbaijan	33,437	7,850,000	235	Republic A	Baku (Bakı)	Azeri, Russian, Armenian	CIS, NATO/PP, UN
Bahamas	5,382	300,000	56	Parliamentary state A	Nassau	English, Creole	OAS, UN
Bahrain	267	675,000	2,528	Monarchy A	Al Manāmah	Arabic, English, Persian, Urdu	AL, UN
Bangladesh	55,598	139,875,000	2,516	Republic A	Dhaka (Dacca)	Bangla, English	UN
Barbados	166	280,000	1,687	Parliamentary state A	Bridgetown	English	OAS, UN
Beijing (Peking)	6,487	14,135,000	2,179	Autonomous city (China) D	Beijing (Peking)	Chinese (Mandarin)	
Belarus	80,155	10,315,000	129	Republic A	Minsk	Belarussian, Russian	CIS, NATO/PP, UN
Belau see Palau							
Belgium	11,787	10,340,000	877	Constitutional monarchy A	Brussels (Bruxelles)	Dutch (Flemish), French, German	EU, NATO, UN
Belize	8,867	270,000	30	Parliamentary state A	Belmopan	English, Spanish, Mayan, Garifuna, Creole	OAS, UN
Benin	43,484	7,145,000	164	Republic A	Porto-Novo and Cotonou	French, Fon, Yoruba, indigenous	AU, UN
Bermuda	21	65,000	3,095	Overseas territory (U.K. protection) B	Hamilton	English, Portuguese	
Bhutan	17,954	2,160,000	120	Monarchy (Indian protection) B	Thimphu	Dzongkha, Tibetan and Nepalese dialects	UN
Bolivia	424,165	8,655,000	20	Republic A	La Paz and Sucre	Aymara, Quechua, Spanish	OAS, UN
Bosnia and Herzegovina	19,767	4,000,000	202	Republic A	Sarajevo	Bosnian, Serbian, Croatian	UN
Botswana	224,607	1,570,000	7.0	Republic A	Gaborone	English, Tswana	AU, UN
Brazil	3,300,172	183,080,000	55	Federal republic A	Brasília	Portuguese, Spanish, English, French	MERCOSUR, OAS, UN
British Columbia	364,764	4,245,000	12	Province (Canada) D	Victoria	English	
British Indian Ocean Territory	23	(¹)		Overseas territory (U.K.) C		English	
British Virgin Islands	58	22,000	379	Overseas territory (U.K.) C	Road Town	English	
Brunei	2,226	360,000	162	Monarchy A	Bandar Seri Begawan	Malay, English, Chinese	ASEAN, UN
Bulgaria	42,855	7,550,000	176	Republic A	Sofia (Sofiya)	Bulgarian, Turkish	NATO, UN
Burkina Faso	105,869	13,400,000	127	Republic A	Ouagadougou	French, indigenous	AU, UN
Burma see Myanmar							
Burundi	10,745	6,165,000	574	Republic A	Bujumbura	French, Kirundi, Swahili	AU, COMESA, UN
California	163,696	35,590,000	217	State (U.S.) D	Sacramento	English	
Cambodia	69,898	13,245,000	189	Constitutional monarchy A	Phnom Penh (Phnum Pénh)	Khmer, French, English	ASEAN, UN
Cameroon	183,568	15,905,000	87	Republic A	Yaoundé	English, French, indigenous	AU, UN
Canada	3,855,103	32,360,000	8.4	Federal parliamentary state A	Ottawa	English, French, other	NAFTA, NATO, OAS, UN
Cape Verde	1,557	415,000	267	Republic A	Praia	Portuguese, Crioulo	AU, UN
Cayman Islands	102	43,000	422	Overseas territory (U.K.) C	George Town	English	
Central African Republic	240,536	3,715,000	15	Republic A	Bangui	French, Sango, indigenous	AU, UN
Ceylon see Sri Lanka							
Chad	495,755	9,395,000	19	Republic A	N'Djamena	Arabic, French, indigenous	AU, UN
Channel Islands	75	155,000	2,067	Two crown dependencies (U.K. protection)		English, French	
Chile	291,930	15,745,000	54	Republic A	Santiago	Spanish	OAS, UN
China (excl. Taiwan)	3,690,045	1,298,720,000	352	Socialist republic A	Beijing (Peking)	Chinese dialects	UN
Chongqing	31,815	31,600,000	993	Autonomous city (China) D	Chongqing (Chungking)	Chinese (Mandarin)	
Christmas Island	52	400	7.7	External territory (Australia) C	Settlement	English, Chinese, Malay	
Cocos (Keeling) Islands	5.4	600	111	External territory (Australia) C	West Island	English, Cocos-Malay	
Colombia	439,737	41,985,000	95	Republic A	Bogotá	Spanish	OAS, UN
Colorado	104,094	4,565,000	44	State (U.S.) D	Denver	English	
Comoros (excl. Mayotte)	863	640,000	742	Republic A	Moroni	Arabic, French, Shikomoro	AL, AU, COMESA, UN
Congo	132,047	2,975,000	23	Republic A	Brazzaville	French, Lingala, Monokutuba, indigenous	AU, UN
Congo, Democratic Republic of the (Zaire)	905,446	57,445,000	63	Republic A	Kinshasa	French, Lingala, indigenous	AU, COMESA, UN
Connecticut	5,543	3,495,000	631	State (U.S.) D	Hartford	English	

REGION OR POLITICAL DIVISION	Area Sq. Mi.	Est. Pop. 1/1/04	Pop. Per Sq. Mi.	Form of Government and Ruling Power	Capital	Predominant Languages	International Organizations
Cook Islands	91	21,000	231	Self-governing territory (New Zealand protection) ... B	Avarua	English, Maori	......
Costa Rica	19,730	3,925,000	199	Republic ... A	San José	Spanish, English	OAS, UN
Cote d'Ivoire (Ivory Coast)	124,504	17,145,000	138	Republic ... A	Abidjan and Yamoussoukro	French, Dioula and other indigenous	AU, UN
Croatia	21,829	4,430,000	203	Republic ... A	Zagreb	Croatian	NATO/PP, UN
Cuba	42,804	11,290,000	264	Socialist republic ... A	Havana (La Habana)	Spanish	OAS, UN
Cyprus	3,572	775,000	217	Republic ... A	Nicosia	Greek, Turkish, English	EU, UN
Czech Republic	30,450	10,250,000	337	Republic ... A	Prague (Praha)	Czech	EU, NATO, UN
Delaware	2,489	820,000	329	State (U.S.) ... D	Dover	English	......
Denmark	16,640	5,405,000	325	Constitutional monarchy ... A	Copenhagen (København)	Danish	EU, NATO, UN
District of Columbia	68	565,000	8,309	Federal district (U.S.) ... D	Washington	English	......
Djibouti	8,958	460,000	51	Republic ... A	Djibouti	French, Arabic, Somali, Afar	AL, AU, COMESA, UN
Dominica	290	69,000	238	Republic ... A	Roseau	English, French	OAS, UN
Dominican Republic	18,730	8,775,000	468	Republic ... A	Santo Domingo	Spanish	OAS, UN
East Timor	5,743	1,010,000	176	Republic ... A	Dili	Portuguese, Tetum, Bahasa Indonesia (Malay), English	UN
Ecuador	109,484	13,840,000	126	Republic ... A	Quito	Spanish, Quechua, indigenous	OAS, UN
Egypt	386,662	75,420,000	195	Republic ... A	Cairo (Al Qāhirah)	Arabic	AL, AU, CAEU, COMESA, UN
Ellice Islands see Tuvalu	......	......	......	......	......	......	......
El Salvador	8,124	6,530,000	804	Republic ... A	San Salvador	Spanish, Nahua	OAS, UN
England	50,356	50,360,000	1,000	Administrative division (U.K.) ... D	London	English	......
Equatorial Guinea	10,831	515,000	48	Republic ... A	Malabo	French, Spanish, indigenous, English	AU, UN
Eritrea	45,406	4,390,000	97	Republic ... A	Asmera	Afar, Arabic, Tigre, Kunama, Tigrinya, other	AU, COMESA, UN
Estonia	17,462	1,405,000	80	Republic ... A	Tallinn	Estonian, Russian, Ukrainian, Finnish, other	EU, NATO, UN
Ethiopia	426,373	67,210,000	158	Federal republic ... A	Addis Ababa (Adis Abeba)	Amharic, Tigrinya, Orominga, Guaraginga, Somali, Arabic	AU, COMESA, UN
Europe	3,800,000	729,330,000	192	......	......	......	......
Falkland Islands (²)	4,700	3,000	0.6	Overseas territory (U.K.) ... C	Stanley	English	......
Faroe Islands	540	47,000	87	Self-governing territory (Danish protection) ... B	Tórshavn	Danish, Faroese	......
Fiji	7,056	875,000	124	Republic ... A	Suva	English, Fijian, Hindustani	UN
Finland	130,559	5,210,000	40	Republic ... A	Helsinki (Helsingfors)	Finnish, Swedish, Sami, Russian	EU, NATO/PP, UN
Florida	65,755	17,070,000	260	State (U.S.) ... D	Tallahassee	English	......
France (excl. Overseas Departments)	208,482	60,305,000	289	Republic ... A	Paris	French	EU, NATO, UN
French Guiana	32,253	190,000	5.9	Overseas department (France) ... C	Cayenne	French	......
French Polynesia	1,544	265,000	172	Overseas territory (France) ... C	Papeete	French, Tahitian	......
Fujian	46,332	35,495,000	766	Province (China) ... D	Fuzhou	Chinese dialects	......
Gabon	103,347	1,340,000	13	Republic ... A	Libreville	French, Fang, indigenous	AU, UN
Gambia, The	4,127	1,525,000	370	Republic ... A	Banjul	English, Malinke, Wolof, Fula, indigenous	AU, UN
Gansu	173,746	26,200,000	151	Province (China) ... D	Lanzhou	Chinese (Mandarin), Mongolian, Tibetan dialects	......
Gaza Strip	139	1,300,000	9,353	Israeli territory with limited self-government	......	Arabic, Hebrew	(²)
Georgia	59,425	8,710,000	147	State (U.S.) ... D	Atlanta	English	......
Georgia	26,911	4,920,000	183	Republic ... A	Tbilisi	Georgian, Russian, Armenian, Azeri, other	NATO/PP, UN
Germany	137,847	82,415,000	598	Federal republic ... A	Berlin	German	EU, NATO, UN
Ghana	92,098	20,615,000	224	Republic ... A	Accra	English, Akan and other indigenous	AU, UN
Gibraltar (²)	2.3	28,000	12,174	Overseas territory (U.K.) ... C	Gibraltar	English, Spanish, Italian, Portuguese	......
Gilbert Islands see Kiribati	......	......	......	......	......	......	......
Golan Heights	454	37,000	81	Occupied by Israel	......	Arabic, Hebrew	......
Great Britain see United Kingdom	......	......	......	......	......	......	......
Greece	50,949	10,635,000	209	Republic ... A	Athens (Athína)	Greek, English, French	EU, NATO, UN
Greenland	836,331	56,000	0.07	Self-governing territory (Danish protection) ... B	Godthåb (Nuuk)	Danish, Greenlandic, English	......
Grenada	133	89,000	669	Parliamentary state	St. George's	English, French	OAS, UN
Guadeloupe (incl. Dependencies)	687	440,000	640	Overseas department (France) ... C	Basse-Terre	French, Creole	......
Guam	212	165,000	778	Unincorporated territory (U.S.) ... C	Hagåtña (Agana)	English, Chamorro, Japanese	......
Guangdong	68,649	88,375,000	1,287	Province (China) ... D	Guangzhou (Canton)	Chinese dialects, Miao-Yao	......
Guangxi Zhuangzu	91,236	45,905,000	503	Autonomous region (China) ... D	Nanning	Chinese dialects, Thai, Miao-Yao	......
Guatemala	42,042	14,095,000	335	Republic ... A	Guatemala	Spanish, indigenous	OAS, UN
Guernsey (incl. Dependencies)	30	65,000	2,167	Crown dependency (U.K. protection) ... B	St. Peter Port	English, French	......
Guinea	94,926	9,135,000	96	Republic ... A	Conakry	French, indigenous	AU, UN
Guinea-Bissau	13,948	1,375,000	99	Republic ... A	Bissau	Portuguese, Crioulo, indigenous	AU, UN
Guizhou	65,637	36,045,000	549	Province (China) ... D	Guiyang	Chinese (Mandarin), Thai, Miao-Yao	......
Guyana	83,000	705,000	8.5	Republic ... A	Georgetown	English, indigenous, Creole, Hindi, Urdu	OAS, UN
Hainan	13,205	8,050,000	610	Province (China) ... D	Haikou	Chinese, Min, Tai	......
Haiti	10,714	7,590,000	708	Republic ... A	Port-au-Prince	Creole, French	OAS, UN
Hawaii	10,931	1,260,000	115	State (U.S.) ... D	Honolulu	English, Hawaiian, Japanese	......
Hebei	73,359	68,965,000	940	Province (China) ... D	Shijiazhuang	Chinese (Mandarin)	......
Heilongjiang	181,082	37,725,000	208	Province (China) ... D	Harbin	Chinese dialects, Mongolian, Tungus	......
Henan	64,479	94,655,000	1,468	Province (China) ... D	Zhengzhou	Chinese (Mandarin)	......
Holland see Netherlands	......	......	......	......	......	......	......
Honduras	43,277	6,745,000	156	Republic ... A	Tegucigalpa	Spanish, indigenous	OAS, UN
Hubei	72,356	61,645,000	852	Province (China) ... D	Wuhan	Chinese dialects	......
Hunan	81,082	65,855,000	812	Province (China) ... D	Changsha	Chinese dialects, Miao-Yao	......
Hungary	35,919	10,045,000	280	Republic ... A	Budapest	Hungarian	EU, NATO, UN
Iceland	39,769	280,000	7.0	Republic ... A	Reykjavík	Icelandic, English, other	EFTA, NATO, UN
Idaho	83,570	1,370,000	16	State (U.S.) ... D	Boise	English	......
Illinois	57,914	12,690,000	219	State (U.S.) ... D	Springfield	English	......
India (incl. part of Jammu and Kashmir)	1,222,510	1,057,415,000	865	Federal republic ... A	New Delhi	English, Hindi, Telugu, Bengali, indigenous	UN
Indiana	36,418	6,215,000	171	State (U.S.) ... D	Indianapolis	English	......
Indonesia	735,310	236,680,000	322	Republic ... A	Jakarta	Bahasa Indonesia (Malay), English, Dutch, indigenous	ASEAN, OPEC, UN
Iowa	56,272	2,955,000	53	State (U.S.) ... D	Des Moines	English	......
Iran	636,372	68,650,000	108	Islamic republic ... A	Tehrān	Persian, Turkish dialects, Kurdish, other	OPEC, UN
Iraq	169,235	25,025,000	148	Republic ... A	Baghdād	Arabic, Kurdish, Assyrian, Armenian	AL, CAEU, OPEC, UN
Ireland	27,133	3,945,000	145	Republic ... A	Dublin (Baile Átha Cliath)	English, Irish Gaelic	EU, NATO/PP, UN
Isle of Man	221	74,000	335	Crown dependency (U.K. protection) ... B	Douglas	English, Manx Gaelic	......

REGION OR POLITICAL DIVISION	Area Sq. Mi.	Est. Pop. 1/1/04	Pop. Per Sq. Mi.	Form of Government and Ruling Power	Capital	Predominant Languages	International Organizations
Israel (excl. Occupied Areas)	8,019	6,160,000	768	Republic ... A	Jerusalem (Yerushalayim)	Hebrew, Arabic	UN
Italy	116,342	58,030,000	499	Republic ... A	Rome (Roma)	Italian, German, French, Slovene	EU, NATO, UN
Ivory Coast see Cote d'Ivoire							
Jamaica	4,244	2,705,000	637	Parliamentary state ... A	Kingston	English, Creole	OAS, UN
Japan	145,850	127,205,000	873	Constitutional monarchy ... A	Tōkyō	Japanese	UN
Jersey	45	90,000	2,000	Crown dependency (U.K. protection) ... B	St. Helier	English, French	
Jiangsu	39,614	76,065,000	1,920	Province (China) ... D	Nanjing (Nanking)	Chinese dialects	
Jiangxi	64,325	42,335,000	658	Province (China) ... D	Nanchang	Chinese dialects	
Jilin	72,201	27,895,000	386	Province (China) ... D	Changchun	Chinese (Mandarin), Mongolian, Korean	
Jordan	34,495	5,535,000	160	Constitutional monarchy ... A	'Ammān	Arabic	AL, CAEU, UN
Kansas	82,277	2,730,000	33	State (U.S.) ... D	Topeka	English	
Kazakhstan	1,049,156	16,780,000	16	Republic ... A	Astana (Aqmola)	Kazakh, Russian	CIS, NATO/PP, UN
Kentucky	40,409	4,130,000	102	State (U.S.) ... D	Frankfort	English	
Kenya	224,961	31,840,000	142	Republic ... A	Nairobi	English, Swahili, indigenous	AU, COMESA, UN
Kiribati	313	100,000	319	Republic ... A	Bairiki	English, I-Kiribati	UN
Korea, North	46,540	22,585,000	485	Socialist republic ... A	P'yŏngyang	Korean	UN
Korea, South	38,328	48,450,000	1,264	Republic ... A	Seoul (Sŏul)	Korean	UN
Kuwait	6,880	2,220,000	323	Constitutional monarchy ... A	Kuwait (Al Kuwayt)	Arabic, English	AL, CAEU, OPEC, UN
Kyrgyzstan	77,182	4,930,000	64	Republic ... A	Bishkek	Kirghiz, Russian	CIS, NATO/PP, UN
Laos	91,429	5,995,000	66	Socialist republic ... A	Viangchan (Vientiane)	Lao, French, English	ASEAN, UN
Latvia	24,942	2,340,000	94	Republic ... A	Riga	Latvian, Lithuanian, Russian, other	EU, NATO, UN
Lebanon	4,016	3,755,000	935	Republic ... A	Beirut (Bayrūt)	Arabic, French, Armenian, English	AL, UN
Lesotho	11,720	1,865,000	159	Constitutional monarchy ... A	Maseru	English, Sesotho, Zulu, Xhosa	AU, UN
Liaoning	56,255	43,340,000	770	Province (China) ... D	Shenyang (Mukden)	Chinese (Mandarin), Mongolian	
Liberia	43,000	3,345,000	78	Republic ... A	Monrovia	English, indigenous	AU, UN
Libya	679,362	5,565,000	8.2	Socialist republic ... A	Tripoli (Ṭarābulus)	Arabic	AL, AU, CAEU, OPEC, UN
Liechtenstein	62	33,000	532	Constitutional monarchy ... A	Vaduz	German	EFTA, UN
Lithuania	25,213	3,590,000	142	Republic ... A	Vilnius	Lithuanian, Polish, Russian	EU, NATO, UN
Louisiana	51,840	4,510,000	87	State (U.S.) ... D	Baton Rouge	English	
Luxembourg	999	460,000	460	Constitutional monarchy ... A	Luxembourg	French, Luxembourgish, German	EU, NATO, UN
Macedonia	9,928	2,065,000	208	Republic ... A	Skopje	Macedonian, Albanian, other	NATO/PP, UN
Madagascar	226,658	17,235,000	76	Republic ... A	Antananarivo	French, Malagasy	AU, COMESA, UN
Maine	35,385	1,310,000	37	State (U.S.) ... D	Augusta	English	
Malawi	45,747	11,780,000	258	Republic ... A	Lilongwe	Chichewa, English, indigenous	AU, COMESA, UN
Malaysia	127,320	23,310,000	183	Federal constitutional monarchy ... A	Kuala Lumpur and Putrajaya (')	Bahasa Melayu, Chinese dialects, English, other	ASEAN, UN
Maldives	115	335,000	2,913	Republic ... A	Male'	Dhivehi	UN
Mali	478,841	11,790,000	25	Republic ... A	Bamako	French, Bambara, indigenous	AU, UN
Malta	122	400,000	3,279	Republic ... A	Valletta	English, Maltese	EU, UN
Manitoba	250,116	1,190,000	4.8	Province (Canada) ... D	Winnipeg	English	
Marshall Islands	70	57,000	814	Republic (U.S. protection) ... B	Majuro (island)	English, indigenous, Japanese	UN
Martinique	425	430,000	1,012	Overseas department (France) ... C	Fort-de-France	French, Creole	
Maryland	12,407	5,525,000	445	State (U.S.) ... D	Annapolis	English	
Massachusetts	10,555	6,455,000	612	State (U.S.) ... D	Boston	English	
Mauritania	397,956	2,955,000	7.4	Republic ... A	Nouakchott	Arabic, Wolof, Pular, Soninke, French	AL, AU, CAEU, UN
Mauritius (incl. Dependencies)	788	1,215,000	1,542	Republic ... A	Port Louis	English, French, Creole, other	AU, COMESA, UN
Mayotte (')	144	180,000	1,250	Departmental collectivity (France) ... C	Mamoudzou	French, Swahili (Mahorian)	
Mexico	758,452	104,340,000	138	Federal republic ... A	Mexico City (Ciudad de México)	Spanish, indigenous	NAFTA, OAS, UN
Michigan	96,716	10,110,000	105	State (U.S.) ... D	Lansing	English	
Micronesia, Federated States of	271	110,000	406	Republic (U.S. protection) ... B	Palikir	English, indigenous	UN
Midway Islands	2.0	(')		Unincorporated territory (U.S.) ... C		English	
Minnesota	86,939	5,075,000	58	State (U.S.) ... D	St. Paul	English	
Mississippi	48,430	2,890,000	60	State (U.S.) ... D	Jackson	English	
Missouri	69,704	5,720,000	82	State (U.S.) ... D	Jefferson City	English	
Moldova	13,070	4,440,000	340	Republic ... A	Chişinău (Kishinev)	Romanian (Moldovan), Russian, Gagauz	CIS, NATO/PP, UN
Monaco	0.8	32,000	40,000	Constitutional monarchy ... A	Monaco	French, English, Italian, Monegasque	UN
Mongolia	604,829	2,730,000	4.5	Republic ... A	Ulan Bator (Ulaanbaatar)	Khalkha Mongol, Turkish dialects, Russian	UN
Montana	4,095	920,000	225	State (U.S.) ... D	Helena	English	
Montserrat	39	9,000	231	Overseas territory (U.K.) ... C	Plymouth	English	
Morocco (excl. Western Sahara)	172,414	31,950,000	185	Constitutional monarchy ... A	Rabat	Arabic, Berber dialects, French	AL, UN
Mozambique	309,496	18,695,000	60	Republic ... A	Maputo	Portuguese, indigenous	AU, UN
Myanmar (Burma)	261,228	42,620,000	163	Provisional military government ... A	Rangoon (Yangon)	Burmese, indigenous	ASEAN, UN
Namibia	317,818	1,940,000	6.1	Republic ... A	Windhoek	English, Afrikaans, German, indigenous	AU, COMESA, UN
Nauru	8.1	13,000	1,605	Republic ... A	Yaren District	Nauruan, English	UN
Nebraska	77,354	1,745,000	23	State (U.S.) ... D	Lincoln	English	
Nei Mongol (Inner Mongolia)	456,759	24,295,000	53	Autonomous region (China) ... D	Hohhot	Mongolian	
Nepal	56,827	26,770,000	471	Constitutional monarchy ... A	Kathmandu	Nepali, indigenous	UN
Netherlands	16,164	16,270,000	1,007	Constitutional monarchy ... A	Amsterdam and The Hague ('s-Gravenhage)	Dutch, Frisian	EU, NATO, UN
Netherlands Antilles	309	215,000	696	Self-governing territory (Netherlands protection) ... B	Willemstad	Dutch, Papiamento, English, Spanish	
Nevada	110,561	2,250,000	20	State (U.S.) ... D	Carson City	English	
New Brunswick	28,150	770,000	27	Province (Canada) ... D	Fredericton	English, French	
New Caledonia	7,172	210,000	29	Territorial collectivity (France) ... C	Nouméa	French, indigenous	
Newfoundland and Labrador	156,453	535,000	3.4	Province (Canada) ... D	St. John's	English	
New Hampshire	9,350	1,290,000	138	State (U.S.) ... D	Concord	English	
New Hebrides see Vanuatu							
New Jersey	8,721	8,665,000	994	State (U.S.) ... D	Trenton	English	
New Mexico	121,590	1,880,000	15	State (U.S.) ... D	Santa Fe	English, Spanish	
New South Wales	309,129	6,665,000	22	State (Australia) ... D	Sydney	English	
New York	54,556	19,245,000	353	State (U.S.) ... D	Albany	English	
New Zealand	104,454	3,975,000	38	Parliamentary state ... A	Wellington	English, Maori	ANZUS, UN
Nicaragua	50,054	5,180,000	103	Republic ... A	Managua	Spanish, English, indigenous	OAS, UN
Niger	489,192	11,210,000	23	Republic ... A	Niamey	French, Hausa, Djerma, indigenous	AU, UN
Nigeria	356,669	135,570,000	380	Transitional military government ... A	Abuja	English, Hausa, Fulani, Yoruba, Ibu, indigenous	AU, OPEC, UN
Ningxia Huizu	25,637	5,745,000	224	Autonomous region (China) ... D	Yinchuan	Chinese (Mandarin)	
Niue	100	2,000	20	Self-governing territory (New Zealand protection) ... B	Alofi	Niuean, English	
Norfolk Island	14	2,000	143	External territory (Australia) ... C	Kingston	English, Norfolk	

REGION OR POLITICAL DIVISION	Area Sq. Mi.	Est. Pop. 1/1/04	Pop. Per Sq. Mi.	Form of Government and Ruling Power	Capital	Predominant Languages	International Organizations
North America	9,500,000	505,780,000	53				
North Carolina	53,819	8,430,000	157	State (U.S.) D	Raleigh	English	
North Dakota	70,700	635,000	9.0	State (U.S.) D	Bismarck	English	
Northern Ireland	5,242	1,725,000	329	Administrative division (U.K.) D	Belfast	English	
Northern Mariana Islands	179	77,000	430	Commonwealth (U.S. protection) B	Saipan (island)	English, Chamorro, Carolinian	
Northern Territory	520,902	200,000	0.4	Territory (Australia) D	Darwin	English, indigenous	
Northwest Territories	519,735	43,000	0.08	Territory (Canada) D	Yellowknife	English, indigenous	
Norway (incl. Svalbard and Jan Mayen)	125,050	4,565,000	37	Constitutional monarchy A	Oslo	Norwegian, Sami, Finnish	EFTA, NATO, UN
Nova Scotia	21,345	965,000	45	Province (Canada) D	Halifax	English	
Nunavut	808,185	30,000	0.04	Territory (Canada) D	Iqaluit	English, indigenous	
Oceania (incl. Australia)	3,300,000	32,170,000	9.7				
Ohio	44,825	11,470,000	256	State (U.S.) D	Columbus	English	
Oklahoma	69,898	3,520,000	50	State (U.S.) D	Oklahoma City	English	
Oman	119,499	2,855,000	24	Monarchy A	Muscat (Masqaṭ)	Arabic, English, Baluchi, Urdu, Indian dialects	AL, UN
Ontario	415,599	12,495,000	30	Province (Canada) D	Toronto	English	
Oregon	98,381	3,570,000	36	State (U.S.) D	Salem	English	
Pakistan (incl. part of Jammu and Kashmir)	339,732	152,210,000	448	Federal Islamic republic A	Islāmābād	English, Urdu, Punjabi, Sindhi, Pashto, other	UN
Palau (Belau)	188	20,000	106	Republic (U.S. protection) B	Koror and Melekeok (5)	Angaur, English, Japanese, Palauan, Sonsorolese, Tobi	UN
Panama	29,157	2,980,000	102	Republic A	Panamá	Spanish, English	OAS, UN
Papua New Guinea	178,704	5,360,000	30	Parliamentary state A	Port Moresby	English, Motu, Pidgin, indigenous	UN
Paraguay	157,048	6,115,000	39	Republic A	Asunción	Guarani, Spanish	MERCOSUR, OAS, UN
Pennsylvania	46,055	12,400,000	269	State (U.S.) D	Harrisburg	English	
Peru	496,225	28,640,000	58	Republic A	Lima	Quechua, Spanish, Aymara	OAS, UN
Philippines	115,831	85,430,000	738	Republic A	Manila	English, Filipino, indigenous	ASEAN, UN
Pitcairn Islands (incl. Dependencies)	19	100	5.3	Overseas territory (U.K.) C	Adamstown	English, Pitcairnese	
Poland	120,728	38,625,000	320	Republic A	Warsaw (Warszawa)	Polish	EU, NATO, UN
Portugal	35,516	10,110,000	285	Republic A	Lisbon (Lisboa)	Portuguese, Mirandese	EU, NATO, UN
Prince Edward Island	2,185	140,000	64	Province (Canada) D	Charlottetown	English	
Puerto Rico	3,515	3,890,000	1,107	Commonwealth (U.S. protection) B	San Juan	Spanish, English	
Qatar	4,412	830,000	188	Monarchy A	Ad Dawḥah (Doha)	Arabic	AL, OPEC, UN
Qinghai	277,994	5,295,000	19	Province (China) D	Xining	Tibetan dialects, Mongolian, Turkish dialects, Chinese (Mandarin)	
Quebec	595,391	7,675,000	13	Province (Canada) D	Québec	French, English	
Queensland	668,208	3,785,000	5.7	State (Australia) D	Brisbane	English	
Reunion	969	760,000	784	Overseas department (France) C	Saint-Denis	French, Creole	
Rhode Island	1,545	1,080,000	699	State (U.S.) D	Providence	English	
Rhodesia see Zimbabwe							
Romania	91,699	22,370,000	244	Republic A	Bucharest (Bucureşti)	Romanian, Hungarian, German	NATO, UN
Russia	6,592,849	144,310,000	22	Federal republic A	Moscow (Moskva)	Russian, other	CIS, NATO/PP, UN
Rwanda	10,169	7,880,000	775	Republic A	Kigali	English, French, Kinyarwanda, Kiswahili	AU, COMESA, UN
St. Helena (incl. Dependencies)	121	7,500	62	Overseas territory (U.K.) C	Jamestown	English	
St. Kitts and Nevis	101	39,000	386	Parliamentary state A	Basseterre	English	OAS, UN
St. Lucia	238	165,000	693	Parliamentary state A	Castries	English, French	OAS, UN
St. Pierre and Miquelon	93	7,000	75	Territorial collectivity (France) C	Saint-Pierre	French	
St. Vincent and the Grenadines	150	115,000	767	Parliamentary state A	Kingstown	English, French	OAS, UN
Samoa	1,093	180,000	165	Constitutional monarchy A	Apia	English, Samoan	UN
San Marino	24	28,000	1,167	Republic A	San Marino	Italian	UN
Sao Tome and Principe	372	180,000	484	Republic A	São Tomé	Portuguese	AU, UN
Saskatchewan	251,366	1,025,000	4.1	Province (Canada) D	Regina	English	
Saudi Arabia	830,000	24,690,000	30	Monarchy A	Riyadh (Ar Riyāḍ)	Arabic	AL, OPEC, UN
Scotland	30,167	5,135,000	170	Administrative division (U.K.) D	Edinburgh	English, Scots Gaelic	
Senegal	75,951	10,715,000	141	Republic A	Dakar	French, Wolof and other indigenous	AU, UN
Serbia and Montenegro (Yugoslavia)	39,449	10,660,000	270	Republic A	Belgrade (Beograd)	Serbian, Albanian	UN
Seychelles	176	81,000	460	Republic A	Victoria	English, French, Creole	AU, COMESA, UN
Shaanxi	79,151	36,865,000	466	Province (China) D	Xi'an (Sian)	Chinese (Mandarin)	
Shandong	59,074	92,845,000	1,572	Province (China) D	Jinan	Chinese (Mandarin)	
Shanghai	2,394	17,120,000	7,151	Autonomous city (China) D	Shanghai	Chinese (Wu)	
Shanxi	60,232	33,715,000	560	Province (China) D	Taiyuan	Chinese (Mandarin)	
Sichuan	188,263	85,175,000	452	Province (China) D	Chengdu	Chinese (Mandarin), Tibetan dialects, Miao-Yao	
Sierra Leone	27,699	5,815,000	210	Republic A	Freetown	English, Krio, Mende, Temne, indigenous	AU, UN
Singapore	264	4,685,000	17,746	Republic A	Singapore	Chinese (Mandarin), English, Malay, Tamil	ASEAN, UN
Slovakia	18,924	5,420,000	286	Republic A	Bratislava	Slovak, Hungarian	EU, NATO, UN
Slovenia	7,821	1,935,000	247	Republic A	Ljubljana	Slovenian, Croatian, Serbian	EU, NATO, UN
Solomon Islands	10,954	515,000	47	Parliamentary state A	Honiara	English, indigenous	UN
Somalia	246,201	8,165,000	33	Transitional A	Mogadishu (Muqdisho)	Arabic, Somali, English, Italian	AL, AU, CAEU, UN
South Africa	470,693	42,770,000	91	Republic A	Pretoria, Cape Town, and Bloemfontein	Afrikaans, English, Xhosa, Zulu, other indigenous	AU, UN
South America	6,900,000	366,600,000	53				
South Australia	379,724	1,525,000	4.0	State (Australia) D	Adelaide	English	
South Carolina	32,020	4,160,000	130	State (U.S.) D	Columbia	English	
South Dakota	77,117	765,000	9.9	State (U.S.) D	Pierre	English	
South Georgia and the South Sandwich Islands (2)	1,450	(1)		Overseas territory (U.K.) C		English	
South West Africa see Namibia							
Spain	194,885	40,250,000	207	Constitutional monarchy A	Madrid	Spanish (Castilian), Catalan, Galician, Basque	EU, NATO, UN
Spanish North Africa (7)	12	140,000	11,667	Five possessions (Spain) C		Spanish, Arabic, Berber dialects	
Spanish Sahara see Western Sahara							
Sri Lanka	25,332	19,825,000	783	Socialist republic A	Colombo and Sri Jayewardenepura Kotte	English, Sinhala, Tamil	UN
Sudan	967,500	38,630,000	40	Provisional military government A	Khartoum (Al Kharṭūm)	Arabic, Nubian, and other indigenous, English	AL, AU, CAEU, COMESA, UN
Suriname	63,037	435,000	6.9	Republic A	Paramaribo	Dutch, Sranan Tongo, English, Hindustani, Javanese	OAS, UN

REGION OR POLITICAL DIVISION	Area Sq. Mi.	Est. Pop. 1/1/04	Pop. Per Sq. Mi.	Form of Government and Ruling Power	Capital	Predominant Languages	International Organizations
Swaziland	6,704	1,165,000	174	Monarchy ... A	Mbabane and Lobamba	English, siSwati	AU, COMESA, UN
Sweden	173,732	8,980,000	52	Constitutional monarchy ... A	Stockholm	Swedish, Sami, Finnish	EU, NATO/PP, UN
Switzerland	15,943	7,430,000	466	Federal republic ... A	Bern (Berne)	German, French, Italian, Romansch	EFTA, NATO/PP, UN
Syria	71,498	17,800,000	249	Republic ... A	Damascus (Dimashq)	Arabic, Kurdish, Armenian, Aramaic, Circassian	AL, CAEU, UN
Taiwan	13,901	22,675,000	1,631	Republic ... A	T'aipei	Chinese (Mandarin), Taiwanese (Min), Hakka	
Tajikistan	55,251	6,935,000	126	Republic ... A	Dushanbe	Tajik, Russian	CIS, NATO/PP, UN
Tanzania	364,900	36,230,000	99	Republic ... A	Dar es Salaam and Dodoma	English, Swahili, indigenous	AU, UN
Tasmania	26,409	475,000	18	State (Australia) ... D	Hobart	English	
Tennessee	42,143	5,860,000	139	State (U.S.) ... D	Nashville	English	
Texas	268,581	22,185,000	83	State (U.S.) ... D	Austin	English, Spanish	
Thailand	198,115	64,570,000	326	Constitutional monarchy ... A	Bangkok (Krung Thep)	Thai, indigenous	ASEAN, UN
Tianjin (Tientsin)	4,363	10,235,000	2,346	Autonomous city (China) ... D	Tianjin (Tientsin)	Chinese (Mandarin)	
Togo	21,925	5,495,000	251	Republic ... A	Lomé	French, Ewe, Mina, Kabye, Dagomba	AU, UN
Tokelau	4.6	1,500	326	Island territory (New Zealand) ... C		English, Tokelauan	
Tonga	251	110,000	438	Constitutional monarchy ... A	Nuku'alofa	Tongan, English	UN
Trinidad and Tobago	1,980	1,100,000	556	Republic ... A	Port of Spain	English, Hindi, French, Spanish, Chinese	OAS, UN
Tristan da Cunha	40	300	7.5	Dependency (St. Helena) ... C	Edinburgh	English	
Tunisia	63,170	9,980,000	158	Republic ... A	Tunis	Arabic, French	AL, AU, UN
Turkey	302,541	68,505,000	226	Republic ... A	Ankara	Turkish, Kurdish, Arabic, Armenian, Greek	NATO, UN
Turkmenistan	188,457	4,820,000	26	Republic ... A	Ashgabat (Ashkhabad)	Turkmen, Russian, Uzbek	CIS, NATO/PP, UN
Turks and Caicos Islands	166	20,000	120	Overseas territory (U.K.) ... C	Grand Turk	English	
Tuvalu	10	11,000	1,100	Parliamentary state ... A	Funafuti	Tuvaluan, English, Samoan, I-Kiribati	UN
Uganda	93,065	26,010,000	279	Republic ... A	Kampala	English, Luganda, Swahili, indigenous, Arabic	AU, COMESA, UN
Ukraine	233,090	47,890,000	205	Republic ... A	Kiev (Kyïv)	Ukrainian, Russian, Romanian, Polish, Hungarian	CIS, NATO/PP, UN
United Arab Emirates	32,278	2,505,000	78	Federation of monarchs ... A	Abū Ẓaby (Abu Dhabi)	Arabic, Persian, English, Hindi, Urdu	AL, CAEU, OPEC, UN
United Kingdom	93,788	60,185,000	642	Constitutional monarchy ... A	London	English, Welsh, Scots Gaelic	EU, NATO, UN
United States	3,794,083	291,680,000	77	Federal republic ... A	Washington	English, Spanish	ANZUS, NAFTA, NATO, OAS, UN
Upper Volta see Burkina Faso							
Uruguay	67,574	3,425,000	51	Republic ... A	Montevideo	Spanish	MERCOSUR, OAS, UN
Utah	84,899	2,360,000	28	State (U.S.) ... D	Salt Lake City	English	
Uzbekistan	172,742	26,195,000	152	Republic ... A	Tashkent (Toshkent)	Uzbek, Russian, Tajik	CIS, NATO/PP, UN
Vanuatu	4,707	200,000	42	Republic ... A	Port Vila	Bislama, English, French	UN
Vatican City	0.2	900	4,500	Ecclesiastical state ... A	Vatican City	Italian, Latin, French, other	
Venezuela	352,145	24,835,000	71	Federal republic ... A	Caracas	Spanish, indigenous	OAS, OPEC, UN
Vermont	9,614	620,000	64	State (U.S.) ... D	Montpelier	English	
Victoria	87,807	4,905,000	56	State (Australia) ... D	Melbourne	English	
Vietnam	128,066	82,150,000	641	Socialist republic ... A	Hanoi	Vietnamese, English, French, Chinese, Khmer, indigenous	ASEAN, UN
Virginia	42,774	7,410,000	173	State (U.S.) ... D	Richmond	English	
Virgin Islands (U.S.)	134	110,000	821	Unincorporated territory (U.S.) ... C	Charlotte Amalie	English, Spanish, Creole	
Wake Island	3.0	(¹)		Unincorporated territory (U.S.) ... C		English	
Wales	8,023	2,965,000	370	Administrative division (U.K.) ... D	Cardiff	English, Welsh Gaelic	
Wallis and Futuna	99	16,000	162	Overseas territory (France) ... C	Mata-Utu	French, Wallisian	
Washington	71,300	6,150,000	86	State (U.S.) ... D	Olympia	English	
West Bank (incl. Jericho and East Jerusalem)	2,263	2,275,000	1,005	Israeli territory with limited self-government		Arabic, Hebrew	(⁴)
Western Australia	976,792	1,945,000	2.0	State (Australia) ... D	Perth	English	
Western Sahara	102,703	265,000	2.6	Occupied by Morocco ... C		Arabic	
West Virginia	24,230	1,815,000	75	State (U.S.) ... D	Charleston	English	
Wisconsin	65,498	5,490,000	84	State (U.S.) ... D	Madison	English	
Wyoming	97,814	505,000	5.2	State (U.S.) ... D	Cheyenne	English	
Xianggang (Hong Kong)	425	7,440,000	17,506	Special administrative region (China) ... D	Hong Kong (Xianggang)	Chinese (Cantonese), English	
Xinjiang Uygur (Sinkiang)	617,764	19,685,000	32	Autonomous region (China) ... D	Ürümqi	Turkish dialects, Mongolian, Tungus, English	
Xizang (Tibet)	471,045	2,680,000	5.7	Autonomous region (China) ... D	Lhasa	Tibetan dialects	
Yemen	203,850	19,680,000	97	Republic ... A	Şan'ā' (Sanaa)	Arabic	AL, CAEU, UN
Yugoslavia see Serbia and Montenegro							
Yukon Territory	186,272	32,000	0.2	Territory (Canada) ... D	Whitehorse	English, Inuktitut, indigenous	
Yunnan	152,124	43,850,000	288	Province (China) ... D	Kunming	Chinese (Mandarin), Tibetan dialects, Khmer, Miao-Yao	
Zaire see Congo, Democratic Republic of the							
Zambia	290,586	10,385,000	36	Republic ... A	Lusaka	English, indigenous	AU, COMESA, UN
Zhejiang	39,305	47,830,000	1,217	Province (China) ... D	Hangzhou	Chinese dialects	
Zimbabwe	150,873	12,630,000	84	Republic ... A	Harare (Salisbury)	English, indigenous	AU, COMESA, UN
WORLD	57,900,000	6,339,505,000	109				

... None, or not applicable
(1) No permanent population
(2) Claimed by Argentina
(3) Claimed by Spain
(4) The Palestinian Liberation Organization (PLO) is a member of AL and CAEU
(5) Future capital
(6) Claimed by Comoros
(7) Comprises Ceuta, Melilla, and several small islands

AL	Arab League (League of Arab States)
ANZUS	Australia-New Zealand-U.S. Security Treaty
ASEAN	Association of Southeast Asian Nations
AU	African Union
CAEU	Council of Arab Unity
CIS	Commonwealth of Independent States
COMESA	Common Market for Eastern and Southern Africa
EFTA	European Free Trade Association
EU	European Union
MERCOSUR	Southern Common Market
NAFTA	North American Free Trade Agreement
NATO	North Atlantic Treaty Organization
NATO/PP	NATO-Partnership for Peace Program
OAS	Organization of American States
OPEC	Organization of Petroleum Exporting Countries

WORLD DEMOGRAPHIC TABLE

CONTINENT/Country	Population Estimate 2004	Pop. Per Sq. Mile 2004	Percent Urban[1] 2001	Crude Birth Rate per 1,000[2] 2003	Crude Death Rate per 1,000[2] 2003	Natural Increase Percent[2] 2003	Fertility Rate (Children born/Woman)[3] 2003	Infant Mortality Rate per 1,000[3] 2003	Median Age[2] 2002	Life Expectancy Male[2] 2003	Life Expectancy Female[2] 2003
NORTH AMERICA											
Bahamas	300,000	56	64.7	19	9	1.0%	2	26	27	62	69
Belize	270,000	30	48.1	30	6	2.4%	4	27	19	65	70
Canada	32,360,000	8	78.9	11	8	0.3%	2	5	38	76	83
Costa Rica	3,925,000	199	59.5	19	4	1.5%	2	11	25	74	79
Cuba	11,290,000	264	75.5	12	7	0.5%	2	7	35	75	79
Dominica	69,000	238	71.4	17	7	1.0%	2	15	28	71	77
Dominican Republic	8,775,000	468	66.0	24	7	1.7%	3	34	24	66	70
El Salvador	6,530,000	804	61.5	28	6	2.2%	3	27	21	67	74
Guatemala	14,095,000	335	39.9	35	7	2.8%	5	38	18	64	66
Haiti	7,590,000	708	36.3	34	13	2.1%	5	76	18	50	53
Honduras	6,745,000	156	53.7	32	6	2.5%	4	30	19	65	68
Jamaica	2,705,000	637	56.6	17	5	1.2%	2	13	27	74	78
Mexico	104,340,000	138	74.6	22	5	1.7%	3	22	24	72	78
Nicaragua	5,180,000	103	56.5	26	5	2.2%	3	31	20	68	72
Panama	2,980,000	102	56.5	21	6	1.5%	3	21	26	70	75
St. Lucia	165,000	693	38.0	21	5	1.6%	2	14	24	70	77
Trinidad and Tobago	1,100,000	556	74.5	13	9	0.4%	2	25	30	67	72
United States	291,680,000	77	77.4	14	8	0.6%	2	7	36	74	80
SOUTH AMERICA											
Argentina	38,945,000	36	88.3	17	8	1.0%	2	16	29	72	79
Bolivia	8,655,000	20	62.9	26	8	1.8%	3	56	21	62	67
Brazil	183,080,000	55	81.7	18	6	1.2%	2	32	27	67	75
Chile	15,745,000	54	86.1	16	6	1.0%	2	9	30	73	80
Colombia	41,985,000	95	75.5	22	6	1.6%	3	22	26	67	75
Ecuador	13,840,000	126	63.4	25	5	2.0%	3	32	23	69	75
Guyana	705,000	9	36.7	18	9	0.9%	2	38	26	61	66
Paraguay	6,115,000	39	56.7	30	5	2.6%	4	28	21	72	77
Peru	28,640,000	58	73.1	23	6	1.7%	3	37	24	68	73
Suriname	435,000	7	74.8	19	7	1.3%	2	25	26	67	72
Uruguay	3,425,000	51	92.1	17	9	0.8%	2	14	32	73	79
Venezuela	24,835,000	71	87.2	20	5	1.5%	2	24	25	71	77
EUROPE											
Albania	3,535,000	318	42.9	15	5	1.0%	2	23	27	74	80
Austria	8,170,000	252	67.4	9	9	0%	1	5	39	76	82
Belarus	10,315,000	129	69.6	10	14	-0.4%	1	14	37	63	75
Belgium	10,340,000	877	97.4	11	10	0.1%	2	5	40	75	82
Bosnia and Herzegovina	4,000,000	202	43.4	13	8	0.4%	2	23	36	70	75
Bulgaria	7,550,000	176	67.4	10	14	-0.5%	1	22	41	68	75
Croatia	4,430,000	203	58.1	13	11	0.2%	2	7	39	71	78
Czech Republic	10,250,000	337	74.5	9	11	-0.1%	1	4	38	72	79
Denmark	5,405,000	325	85.1	12	11	0.1%	2	5	39	75	80
Estonia	1,405,000	80	69.4	9	13	-0.4%	1	12	38	64	77
Finland	5,210,000	40	58.5	11	10	0.1%	2	4	40	75	82
France	60,305,000	289	75.5	13	9	0.3%	2	4	38	76	83
Germany	82,415,000	598	87.7	9	10	-0.2%	1	4	41	75	82
Greece	10,635,000	209	60.3	10	10	0%	1	6	40	76	81
Hungary	10,045,000	280	64.8	10	13	-0.3%	1	9	38	68	77
Iceland	280,000	7	92.7	14	7	0.7%	2	4	34	78	82
Ireland	3,945,000	145	59.3	14	8	0.6%	2	6	33	75	80
Italy	58,030,000	499	67.1	9	10	-0.1%	1	6	41	76	83
Latvia	2,340,000	94	59.8	9	15	-0.6%	1	15	39	63	75
Lithuania	3,590,000	142	68.6	10	13	-0.2%	1	14	37	64	76
Luxembourg	460,000	460	91.9	12	8	0.4%	2	5	38	75	82
Macedonia	2,065,000	208	59.4	13	8	0.5%	2	12	33	72	77
Moldova	4,440,000	340	41.4	14	13	0.2%	2	42	32	61	69
Netherlands	16,270,000	1,007	89.6	12	9	0.3%	2	5	39	76	81
Norway	4,565,000	37	75.0	12	10	0.3%	2	4	38	77	82
Poland	38,625,000	320	62.5	10	10	0.1%	1	9	36	70	78
Portugal	10,110,000	285	65.8	11	10	0.1%	1	6	38	73	80
Romania	22,370,000	244	55.2	11	12	-0.1%	1	28	35	67	75
Serbia and Montenegro	10,660,000	270	51.7	13	11	0.2%	2	17	36	71	77
Slovakia	5,420,000	286	57.6	10	10	0.1%	1	8	35	70	78
Slovenia	1,935,000	247	49.1	9	10	-0.1%	1	4	39	72	80
Spain	40,250,000	207	77.8	10	9	0.1%	1	5	39	76	83
Sweden	8,980,000	52	83.3	11	10	0%	2	3	40	78	83
Switzerland	7,430,000	466	67.3	10	8	0.1%	1	4	40	77	83
Ukraine	47,890,000	205	68.0	10	16	-0.7%	1	21	38	61	72
United Kingdom	60,185,000	642	89.5	11	10	0.1%	2	5	38	76	81
Russia	144,310,000	22	72.9	10	14	-0.4%	1	20	38	62	73
ASIA											
Afghanistan	29,205,000	116	22.3	41	17	2.3%	6	142	19	48	46
Armenia	3,325,000	289	67.2	13	10	0.2%	2	41	32	62	71
Azerbaijan	7,850,000	235	51.8	19	10	1.0%	2	82	27	59	68
Bahrain	675,000	2,528	92.5	19	4	1.5%	3	19	29	71	76
Bangladesh	139,875,000	2,516	25.6	30	9	2.1%	3	66	21	61	61
Brunei	360,000	162	72.8	20	3	1.6%	2	14	26	72	77
Cambodia	13,245,000	189	17.5	27	9	1.8%	4	76	19	55	60
China	1,298,720,000	352	37.1	13	7	0.6%	2	25	32	70	74
Cyprus	775,000	217	70.2	13	8	0.5%	2	8	34	75	80
East Timor	1,010,000	176	7.5	28	6	2.1%	4	50	20	63	68
Georgia	4,920,000	183	56.5	12	15	-0.3%	2	51	35	61	68
India	1,057,415,000	865	27.9	23	8	1.5%	3	60	24	63	64
Indonesia	236,680,000	322	42.1	21	6	1.5%	3	38	26	67	71
Iran	68,650,000	108	64.7	17	6	1.2%	2	44	23	68	71
Iraq	25,025,000	148	67.4	34	6	2.8%	5	55	19	67	69
Israel	6,160,000	768	91.8	19	6	1.2%	3	7	29	77	81
Japan	127,285,000	873	78.8	10	9	0.1%	1	3	42	78	84
Jordan	5,535,000	160	78.7	24	3	2.1%	3	19	22	75	81
Kazakhstan	16,700,000	16	55.8	18	11	0.8%	2	59	28	58	69
Korea, North	22,585,000	485	60.5	18	7	1.1%	2	26	31	68	74
Korea, South	48,450,000	1,264	82.5	13	6	0.7%	2	7	33	72	79
Kuwait	2,220,000	323	96.1	22	2	1.9%	3	11	26	76	78

CONTINENT/Country	Population Estimate 2004	Pop. Per Sq. Mile 2004	Percent Urban[1] 2001	Crude Birth Rate per 1,000[2] 2003	Crude Death Rate per 1,000[2] 2003	Natural Increase Percent[2] 2003	Fertility Rate (Children born/Woman)[3] 2003	Infant Mortality Rate per 1,000[3] 2003	Median Age[2] 2002	Life Expectancy Male[2] 2003	Life Expectancy Female[2] 2003
Kyrgyzstan	4,930,000	64	34.3	26	9	1.7%	3	75	23	59	68
Laos	5,995,000	66	19.7	37	12	2.5%	5	89	19	52	56
Lebanon	3,755,000	935	90.1	20	6	1.3%	2	26	26	70	75
Malaysia	23,310,000	183	58.1	24	5	1.9%	3	19	24	69	75
Mongolia	2,730,000	5	56.6	21	7	1.4%	2	57	24	62	66
Myanmar	42,620,000	163	28.1	19	12	0.7%	2	70	25	54	58
Nepal	26,770,000	471	12.2	32	10	2.3%	4	71	20	59	59
Oman	2,855,000	24	76.5	37	4	3.4%	6	21	19	70	75
Pakistan	152,210,000	448	33.4	30	9	2.1%	4	77	20	61	63
Philippines	85,430,000	738	59.4	26	6	2.1%	3	25	22	66	72
Qatar	830,000	188	92.9	16	4	1.1%	3	20	31	71	76
Saudi Arabia	24,690,000	30	86.7	37	6	3.1%	6	48	19	67	71
Singapore	4,685,000	17,746	100.0	13	4	0.8%	1	4	35	77	84
Sri Lanka	19,825,000	783	23.1	16	6	1.0%	2	15	29	70	75
Syria	17,800,000	249	51.8	30	5	2.5%	4	32	20	68	71
Taiwan	22,675,000	1,631	[5]	13	6	0.7%	2	7	33	74	80
Tajikistan	6,935,000	126	27.7	33	8	2.4%	4	113	19	61	68
Thailand	64,570,000	326	20.0	16	7	1.0%	2	22	30	69	74
Turkey	68,505,000	226	66.2	18	6	1.2%	2	44	27	69	74
Turkmenistan	4,820,000	26	44.9	28	9	1.9%	4	73	21	58	65
United Arab Emirates	2,505,000	78	87.2	18	4	1.4%	3	16	28	72	77
Uzbekistan	26,195,000	152	36.6	26	8	1.8%	3	72	22	61	68
Vietnam	82,150,000	641	24.5	20	6	1.3%	2	31	25	68	73
Yemen	19,680,000	97	25.0	43	9	3.4%	7	65	16	59	63
AFRICA											
Algeria	33,090,000	36	57.7	22	5	1.7%	3	38	23	69	72
Angola	10,875,000	23	34.9	46	26	2.0%	6	194	18	36	38
Benin	7,145,000	164	43.0	43	14	3.0%	6	87	16	50	52
Botswana	1,570,000	7	49.4	26	31	-0.6%	3	67	19	32	32
Burkina Faso	13,400,000	127	16.9	45	19	2.6%	6	100	17	43	46
Burundi	6,165,000	574	9.3	40	18	2.2%	6	72	16	43	44
Cameroon	15,905,000	87	49.7	35	15	2.0%	5	70	18	47	49
Cape Verde	415,000	267	63.5	27	7	2.0%	4	51	19	67	73
Central African Republic	3,715,000	15	41.7	36	20	1.6%	5	93	18	40	43
Chad	9,395,000	19	24.1	47	16	3.1%	6	96	16	47	50
Comoros	640,000	742	33.8	39	9	3.0%	5	80	19	59	64
Congo	2,975,000	23	66.1	29	14	1.5%	4	95	20	49	51
Congo, Democratic Republic of the	57,445,000	63	30.7	45	15	3.0%	7	97	16	47	51
Cote d'Ivoire	17,145,000	138	44.0	40	18	2.2%	6	98	17	40	45
Djibouti	460,000	51	84.2	41	19	2.1%	6	107	18	42	44
Egypt	75,420,000	195	42.7	24	5	1.9%	3	35	23	68	73
Equatorial Guinea	515,000	48	49.3	37	13	2.4%	5	89	19	53	57
Eritrea	4,390,000	97	19.1	39	13	2.6%	6	76	18	51	55
Ethiopia	67,210,000	158	15.9	40	20	2.0%	6	103	17	40	42
Gabon	1,340,000	13	82.3	37	11	2.6%	5	55	19	55	59
Gambia, The	1,525,000	370	31.3	41	12	2.8%	6	75	17	52	56
Ghana	20,615,000	224	36.4	26	11	1.5%	3	53	20	56	57
Guinea	9,135,000	96	27.9	43	16	2.7%	6	93	18	48	51
Guinea-Bissau	1,375,000	99	32.3	38	17	2.2%	5	110	19	45	49
Kenya	31,840,000	142	34.4	29	16	1.3%	3	63	18	45	45
Lesotho	1,865,000	159	28.8	27	25	0.3%	4	86	20	37	37
Liberia	3,345,000	78	45.5	45	18	2.7%	6	132	18	47	49
Libya	5,565,000	8	88.0	27	3	2.4%	3	27	22	74	78
Madagascar	17,235,000	76	30.1	42	12	3.0%	6	80	17	54	59
Malawi	11,780,000	258	15.1	45	23	2.2%	6	105	16	38	38
Mali	11,790,000	25	30.9	48	19	2.9%	7	119	16	45	46
Mauritania	2,955,000	7	59.1	42	13	2.9%	6	74	17	50	54
Mauritius	1,215,000	1,542	41.6	16	7	0.9%	2	16	30	68	76
Morocco	31,950,000	185	56.1	23	6	1.7%	3	45	23	68	72
Mozambique	18,695,000	60	33.3	37	23	1.4%	5	138	19	39	37
Namibia	1,940,000	6	31.4	34	19	1.5%	5	68	18	44	41
Niger	11,210,000	23	21.1	50	22	2.8%	7	124	16	42	42
Nigeria	135,570,000	380	44.9	39	14	2.5%	5	71	18	51	51
Rwanda	7,880,000	775	6.3	40	22	1.8%	6	103	18	39	40
Sao Tome and Principe	180,000	484	47.7	42	7	3.5%	6	46	16	65	68
Senegal	10,715,000	141	48.2	36	11	2.5%	5	58	18	55	58
Sierra Leone	5,815,000	210	37.3	44	21	2.3%	6	147	18	40	45
Somalia	8,165,000	33	27.9	46	18	2.9%	7	120	18	46	49
South Africa	42,770,000	91	57.7	19	18	0%	2	61	25	47	47
Sudan	38,630,000	40	37.1	36	10	2.7%	5	66	18	57	59
Swaziland	1,165,000	174	26.7	29	21	0.8%	4	67	19	41	38
Tanzania	36,230,000	99	33.3	40	17	2.2%	5	104	18	43	46
Togo	5,495,000	251	33.9	35	12	2.4%	5	69	17	51	55
Tunisia	9,980,000	158	66.2	17	5	1.2%	2	27	26	73	76
Uganda	26,010,000	279	14.5	47	17	3.0%	7	88	15	43	46
Zambia	10,385,000	36	39.8	40	24	1.5%	5	99	17	35	35
Zimbabwe	12,630,000	84	36.0	30	22	0.8%	4	66	19	40	38
OCEANIA											
Australia	19,825,000	7	91.2	13	7	0.5%	2	5	36	77	83
Fiji	875,000	124	50.2	23	6	1.7%	3	13	24	66	71
Kiribati	100,000	319	38.6	31	9	2.3%	4	51	20	58	64
Micronesia, Federated States of	110,000	406	28.6	26	5	2.1%	4	32	19[4]	67	71
New Zealand	3,975,000	38	85.9	14	8	0.7%	2	6	33	75	81
Papua New Guinea	5,360,000	30	17.6	31	8	2.3%	4	55	21	62	66
Samoa	180,000	165	22.3	15	6	0.9%	3	30	24	67	73
Solomon Islands	515,000	47	20.2	32	4	2.8%	4	23	18	70	75
Tonga	110,000	438	33.0	25	6	1.9%	3	13	20	66	71
Vanuatu	200,000	42	22.1	24	8	1.6%	3	58	22	60	63

This table presents data for most independent nations having an area greater than 200 square miles
(1) Source: United Nations World Urbanization Prospects
(2) Source: United States Census Bureau International Database
(3) Source: United States Central Intelligence Agency World Factbook
(4) 2000 Census preliminary count from www.fsmgov.org/info/people.html
(5) Data for Taiwan is included with China

WORLD AGRICULTURE TABLE

CONTINENT/Country	Total Area Sq. Miles	Cropland Area[1] Sq. Miles	Cropland Area[1] %	Pasture Area[1] Sq. Miles	Pasture Area[1] %	Wheat[1] 1,000 metric tons	Rice[1] 1,000 metric tons	Corn[1] 1,000 metric tons	Cattle[1] 1,000	Pigs[1] 1,000	Sheep[1] 1,000
NORTH AMERICA											
Bahamas	5,382	46	0.9%	8	0.1%	-	-	-	1	5	6
Belize	8,867	402	4.5%	193	2.2%	-	12	36	52	25	4
Canada	3,855,103	177,144	4.6%	111,970	2.9%	24,676	-	8,168	13,340	12,970	819
Costa Rica	19,730	2,027	10.3%	9,035	45.8%	-	267	20	1,358	438	3
Cuba	42,804	17,239	40.3%	8,494	19.8%	-	342	207	4,305	2,600	310
Dominica	290	77	26.6%	8	2.7%	-	-	-	13	5	8
Dominican Republic	18,730	6,162	32.9%	8,108	43.3%	-	615	30	2,026	548	106
El Salvador	8,124	3,514	43.2%	3,066	37.7%	-	47	605	1,190	195	5
Guatemala	42,042	7,355	17.5%	10,046	23.9%	9	46	1,057	2,500	1,417	270
Haiti	10,714	4,247	39.6%	1,892	17.7%	-	111	211	1,390	934	147
Honduras	43,277	5,514	12.7%	5,822	13.5%	1	9	509	1,737	474	14
Jamaica	4,244	1,097	25.8%	884	20.8%	-	-	2	400	180	1
Mexico	758,452	105,406	13.9%	308,882	40.7%	3,263	324	18,466	30,428	16,112	6,048
Nicaragua	50,054	8,382	16.7%	18,591	37.1%	-	234	374	2,008	402	4
Panama	29,157	2,683	9.2%	5,927	20.3%	-	237	71	1,348	279	-
St. Lucia	238	69	29.2%	8	3.2%	-	-	-	12	15	13
Trinidad and Tobago	1,980	471	23.8%	42	2.1%	-	13	5	36	41	12
United States	3,794,083	684,401	18.0%	903,479	23.8%	58,862	9,222	244,296	98,197	60,229	7,071
SOUTH AMERICA											
Argentina	1,073,519	135,136	12.6%	548,265	51.1%	15,642	1,140	15,217	49,299	4,200	13,588
Bolivia	424,165	11,973	2.8%	130,618	30.8%	121	281	607	6,715	2,786	8,743
Brazil	3,300,172	256,623	7.8%	760,621	23.0%	2,461	10,998	35,119	170,295	30,608	14,728
Chile	291,930	8,880	3.0%	49,942	17.1%	1,490	113	685	4,117	2,395	4,153
Colombia	439,737	16,405	3.7%	161,391	36.7%	37	2,262	1,128	25,274	2,726	2,247
Ecuador	109,484	11,525	10.5%	19,653	18.0%	19	1,340	483	5,261	2,654	2,214
Guyana	83,000	1,969	2.4%	4,749	5.7%	-	560	3	220	20	130
Paraguay	157,048	12,008	7.6%	83,784	53.3%	256	112	804	9,758	2,633	402
Peru	496,225	16,255	3.3%	104,634	21.1%	180	1,963	1,205	4,936	2,795	14,414
Suriname	63,037	259	0.4%	81	0.1%	-	178	-	128	22	8
Uruguay	67,574	5,174	7.7%	52,290	77.4%	284	1,189	190	10,446	375	13,257
Venezuela	352,145	13,158	3.7%	70,425	20.0%	1	696	1,547	14,620	5,555	780
EUROPE											
Albania	11,100	2,699	24.3%	1,699	15.3%	298	-	203	719	96	1,929
Austria	32,378	5,676	17.5%	7,413	22.9%	1,412	-	1,774	2,166	3,556	357
Belarus	80,155	24,151	30.1%	11,564	14.4%	903	-	13	4,411	3,565	96
Belgium	11,787	3,344[2]	26.2%[2]	2,618[2]	20.5%[2]	1,535	-	420	3,165	7,462	150
Bosnia and Herzegovina	19,767	3,243	16.4%	4,633	23.4%	289	-	656	448	345	645
Bulgaria	42,855	17,900	41.8%	6,236	14.6%	3,071	8	1,137	664	1,459	2,536
Croatia	21,829	6,124	28.1%	6,035	27.6%	852	-	1,958	435	1,276	519
Czech Republic	30,450	12,788	42.0%	3,730	12.2%	4,196	-	324	1,604	3,761	87
Denmark	16,640	8,880	53.4%	1,452	8.7%	4,683	-	-	1,887	12,052	147
Estonia	17,462	2,691	15.4%	745	4.3%	123	-	-	276	304	29
Finland	130,559	8,490	6.5%	77	0.1%	427	-	-	1,060	1,303	101
France	208,482	75,618	36.3%	38,788	18.6%	35,327	110	15,928	20,377	14,693	9,754
Germany	137,847	46,409	33.7%	19,355	14.0%	21,358	-	3,362	14,723	26,021	2,746
Greece	50,949	14,873	29.2%	17,954	35.2%	2,111	153	2,007	584	925	8,977
Hungary	35,919	18,548	51.6%	4,097	11.4%	3,843	9	6,664	845	5,216	991
Iceland	39,769	27	0.1%	8,780	22.1%	-	-	-	72	44	477
Ireland	27,133	4,050	14.9%	12,934	47.7%	688	-	-	6,613	1,765	5,311
Italy	116,342	42,379	36.4%	16,907	14.5%	7,239	1,310	10,222	7,167	8,356	11,000
Latvia	24,942	7,220	28.9%	2,355	9.4%	410	-	-	393	407	28
Lithuania	25,213	11,541	45.8%	1,923	7.6%	1,062	-	-	856	984	14
Luxembourg	999	[3]	[3]	[3]	[3]	-	-	2	134	-	-
Macedonia	9,928	2,363	23.8%	2,432	24.5%	308	20	135	267	209	1,285
Moldova	13,070	8,398	64.3%	1,483	11.3%	902	-	1,096	423	646	929
Netherlands	16,164	3,622	22.4%	3,834	23.7%	995	-	148	4,108	13,253	1,335
Norway	125,050	3,398	2.7%	625	0.5%	265	-	-	1,017	414	2,342
Poland	120,728	55,267	45.8%	15,745	13.0%	8,946	-	962	6,124	17,588	366
Portugal	35,516	10,444	29.4%	5,548	15.6%	295	146	907	1,415	2,346	4,337
Romania	91,699	38,305	41.8%	19,039	20.8%	5,610	3	8,317	3,021	5,946	8,062
Serbia and Montenegro	39,449	14,394	36.5%	7,197	18.2%	2,207	-	5,013	1,550	4,012	1,853
Slovakia	18,924	6,085	32.2%	3,375	17.8%	1,445	-	612	671	1,548	344
Slovenia	7,821	784	10.0%	1,185	15.2%	153	-	283	473	585	80
Spain	194,885	69,298	35.6%	44,209	22.7%	5,785	844	4,208	6,140	22,079	24,185
Sweden	173,732	10,413	6.0%	1,726	1.0%	2,135	-	-	1,683	1,975	440
Switzerland	15,943	1,683	10.6%	4,417	27.7%	535	-	214	1,603	1,499	421
Ukraine	233,090	129,321	55.5%	30,541	13.1%	15,043	74	3,075	10,591	9,270	1,074
United Kingdom	93,788	22,019	23.5%	43,440	46.3%	14,380	-	-	11,052	6,537	41,205
Russia	6,592,849	485,400	7.4%	351,905	5.3%	37,455	509	1,133	27,936	17,076	12,954
ASIA											
Afghanistan	251,773	31,097	12.4%	115,831	46.0%	1,821	205	172	2,600	-	12,762
Armenia	11,506	2,162	18.8%	3,089	26.8%	211	-	9	478	75	515
Azerbaijan	33,437	7,471	22.3%	10,039	30.0%	1,172	19	107	1,965	21	5,321
Bahrain	267	23	8.7%	15	5.8%	-	-	-	12	-	17
Bangladesh	55,598	32,761	58.9%	2,317	4.2%	1,807	36,909	8	23,817	-	1,128
Brunei	2,226	27	1.2%	23	1.0%	-	-	-	2	6	2
Cambodia	69,898	14,699	21.0%	5,792	8.3%	-	4,035	146	2,896	2,079	-
China	3,690,045	599,520[4]	16.2%[4]	1,544,412[4]	41.9%[4]	102,463[4]	189,840[4]	116,240[4]	104,179[4]	440,384[4]	130,536[4]
Cyprus	3,572	436	12.2%	15	0.4%	12	-	-	55	419	240
East Timor	5,743	309	5.4%	579	10.1%	-	33	93	173	300	36
Georgia	26,911	4,104	15.3%	7,490	27.8%	207	-	358	1,117	433	541
India	1,222,510	655,987	53.7%	42,124	3.4%	72,140	132,818	12,285	217,773	17,000	57,900
Indonesia	735,310	129,730	17.6%	43,155	5.9%	-	50,953	9,409	11,370	6,098	7,316
Iran	636,372	63,892	10.0%	169,885	26.7%	8,740	2,103	1,113	8,273	-	53,900
Iraq	169,235	23,514	13.9%	15,444	9.1%	667	110	73	1,342	-	6,770
Israel	8,019	1,637	20.4%	548	6.8%	94	-	73	393	138	373
Japan	145,850	18,510	12.7%	1,564	1.1%	657	11,551	-	4,592	9,823	11
Jordan	34,495	1,544	4.5%	2,865	8.3%	18	-	13	66	-	1,900
Kazakhstan	1,049,156	83,672	8.0%	714,667	68.1%	10,938	225	256	4,021	984	8,785
Korea, North	46,540	10,811	23.2%	193	0.4%	88	2,031	1,253	575	3,076	186
Korea, South	38,328	7,293	19.0%	208	0.5%	4	7,204	67	2,191	8,266	1
Kuwait	6,880	58	0.8%	525	7.6%	-	-	-	19	-	543

CONTINENT/Country	Agricultural Area 2001 Total Area Sq. Miles	Cropland Area[1] Sq. Miles	Cropland Area[1] %	Pasture Area[1] Sq. Miles	Pasture Area[1] %	Average Production 1999-2001 Wheat[1] 1,000 metric tons	Rice[1] 1,000 metric tons	Corn[1] 1,000 metric tons	Average 1999-2001 Cattle[1] 1,000	Pigs[1] 1,000	Sheep[1] 1,000
Kyrgyzstan	77,182	5,664	7.3%	35,873	46.5%	1,113	17	363	942	98	3,101
Laos	91,429	3,699	4.0%	3,390	3.7%	-	2,213	108	1,106	1,390	-
Lebanon	4,016	1,208	30.1%	62	1.5%	60	-	4	76	63	354
Malaysia	127,320	29,286	23.0%	1,100	0.9%	-	2,170	63	744	1,943	167
Mongolia	604,829	4,633	0.8%	499,230	82.5%	148	-	-	2,997	17	14,587
Myanmar	261,228	41,023	15.7%	1,212	0.5%	105	20,683	413	10,974	3,923	390
Nepal	56,827	12,324	21.7%	6,784	11.9%	1,143	4,137	1,528	7,012	872	852
Oman	119,499	313	0.3%	3,861	3.2%	1	-	5	299	-	342
Pakistan	339,732	85,560	25.2%	19,305	5.7%	19,319	6,920	1,653	22,007	-	24,067
Philippines	115,831	41,120	35.5%	4,942	4.3%	-	12,377	4,540	2,467	10,724	30
Qatar	4,412	81	1.8%	193	4.4%	-	-	1	15	-	214
Saudi Arabia	830,000	14,649	1.8%	656,373	79.1%	1,871	-	5	304	-	7,848
Singapore	264	4	1.5%	-	0.0%	-	-	-	-	190	-
Sri Lanka	25,332	7,378	29.1%	1,699	6.7%	-	2,804	30	1,580	71	12
Syria	71,498	21,043	29.4%	31,942	44.7%	3,514	-	196	933	-	13,288
Taiwan	13,901	(5)	(5)	(5)	(5)	(5)	(5)	(5)	(5)	(5)	(5)
Tajikistan	55,251	4,093	7.4%	13,514	24.5%	375	67	38	1,045	1	1,481
Thailand	198,115	70,657	35.7%	3,089	1.6%	1	25,578	4,405	4,973	6,539	40
Turkey	302,541	101,757	33.6%	47,792	15.8%	19,341	350	2,266	10,949	4	29,394
Turkmenistan	188,457	7,008	3.7%	118,533	62.9%	1,472	33	9	863	46	5,750
United Arab Emirates	32,278	919	2.8%	1,178	3.6%	-	-	-	94	-	504
Uzbekistan	172,742	18,649	10.8%	88,031	51.0%	3,637	219	133	5,279	83	7,980
Vietnam	128,066	32,579	25.4%	2,479	1.9%	-	31,964	1,961	4,029	20,273	-
Yemen	203,850	6,158	3.0%	62,027	30.4%	145	-	48	1,320	-	4,758

AFRICA

CONTINENT/Country	Total Area Sq. Miles	Cropland Area Sq. Miles	Cropland Area %	Pasture Area Sq. Miles	Pasture Area %	Wheat	Rice	Corn	Cattle	Pigs	Sheep
Algeria	919,595	31,861	3.5%	122,780	13.4%	1,414	-	1	1,667	6	19,000
Angola	481,354	12,741	2.6%	208,495	43.3%	4	16	417	3,995	800	345
Benin	43,484	8,745	20.1%	2,124	4.9%	-	46	740	1,486	463	650
Botswana	224,607	1,440	0.6%	98,842	44.0%	1	-	8	2,035	6	347
Burkina Faso	105,869	15,444	14.6%	23,166	21.9%	-	102	500	4,767	621	6,722
Burundi	10,745	4,865	45.3%	3,610	33.6%	7	57	124	321	67	215
Cameroon	183,568	27,645	15.1%	7,722	4.2%	-	69	759	5,761	1,232	3,734
Cape Verde	1,557	158	10.2%	97	6.2%	-	-	27	22	195	9
Central African Republic	240,536	7,799	3.2%	12,066	5.0%	-	23	101	3,096	669	218
Chad	495,755	14,016	2.8%	173,746	35.0%	3	114	88	5,852	22	2,374
Comoros	863	510	59.1%	58	6.7%	-	17	4	51	-	21
Congo	132,047	849	0.6%	38,610	29.2%	-	1	6	87	46	102
Congo, Democratic Republic of the	905,446	30,425	3.4%	57,915	6.4%	9	338	1,184	823	1,050	925
Cote d'Ivoire	124,504	28,958	23.3%	50,193	40.3%	-	1,217	693	1,398	333	1,439
Djibouti	8,958	4	0.0%	5,019	56.0%	-	-	-	269	-	465
Egypt	386,662	12,888	3.3%	-	0.0%	6,388	5,681	6,487	3,583	29	4,510
Equatorial Guinea	10,831	888	8.2%	402	3.7%	-	-	-	5	6	37
Eritrea	45,406	1,942	4.3%	26,900	59.2%	32	-	13	2,150	-	1,570
Ethiopia	426,373	44,255	10.4%	77,220	18.1%	1,340	-	2,938	35,025	25	22,333
Gabon	103,347	1,911	1.8%	18,012	17.4%	-	1	26	36	213	197
Gambia, The	4,127	985	23.9%	1,772	42.9%	-	28	24	350	12	115
Ghana	92,098	22,780	24.7%	32,240	35.0%	-	244	988	1,297	327	2,715
Guinea	94,926	5,888	6.2%	41,313	43.5%	-	830	96	2,576	93	824
Guinea-Bissau	13,910	2,110	13.2%	4,170	29.9%	-	95	26	509	347	283
Kenya	224,961	19,923	8.9%	82,240	36.6%	184	58	2,419	13,229	311	7,000
Lesotho	11,720	1,290	11.0%	7,722	65.9%	39	-	128	547	63	839
Liberia	43,000	2,317	5.4%	7,722	18.0%	-	188	-	36	127	210
Libya	679,362	8,301	1.2%	51,352	7.6%	128	-	-	207	-	5,100
Madagascar	226,658	13,707	6.0%	92,664	40.9%	9	2,412	175	10,339	1,267	793
Malawi	45,747	9,035	19.7%	7,143	15.6%	2	86	2,190	741	450	110
Mali	478,841	18,147	3.8%	115,831	24.2%	8	801	378	6,594	72	6,282
Mauritania	397,956	1,931	0.5%	151,545	38.1%	-	65	7	1,470	-	7,437
Mauritius	788	409	51.9%	27	3.4%	-	-	-	27	12	10
Morocco	172,414	37,529	21.8%	81,081	47.0%	2,284	33	95	2,629	8	17,059
Mozambique	309,496	16,351	5.3%	169,885	54.9%	1	168	1,136	1,317	179	125
Namibia	317,818	3,166	1.0%	146,719	46.2%	4	-	26	2,436	21	2,330
Niger	489,192	17,375	3.6%	46,332	9.5%	10	66	5	2,217	39	4,386
Nigeria	356,669	120,464	33.8%	151,352	42.4%	75	3,109	4,734	19,677	5,000	20,833
Rwanda	10,169	5,019	49.4%	2,124	20.9%	6	13	66	766	172	264
Sao Tome and Principe	372	205	55.0%	4	1.0%	-	-	2	4	2	3
Senegal	75,951	9,653	12.7%	21,815	28.7%	-	229	84	3,076	263	4,619
Sierra Leone	27,699	2,178	7.9%	8,494	30.7%	-	215	9	413	52	365
Somalia	246,201	4,135	1.7%	166,024	67.4%	1	2	188	5,133	4	13,100
South Africa	470,693	60,664	12.9%	324,048	68.8%	2,200	3	9,147	13,594	1,542	28,677
Sudan	967,500	64,298	6.6%	452,434	46.8%	230	8	48	37,081	-	45,980
Swaziland	6,704	734	10.9%	4,633	69.1%	-	-	94	613	32	27
Tanzania	364,900	19,112	5.2%	135,136	37.0%	87	509	2,567	17,350	449	3,513
Togo	21,925	10,154	46.3%	3,861	17.6%	-	69	480	277	287	1,528
Tunisia	63,170	18,954	30.0%	15,792	25.0%	1,111	-	-	760	6	6,862
Uganda	93,065	27,799	29.9%	19,738	21.2%	12	106	1,108	5,977	1,540	1,065
Zambia	290,586	20,386	7.0%	115,831	39.9%	80	11	768	2,709	324	137
Zimbabwe	150,873	12,934	8.6%	66,410	44.0%	282	-	1,698	5,840	494	602

OCEANIA

CONTINENT/Country	Total Area Sq. Miles	Cropland Area Sq. Miles	Cropland Area %	Pasture Area Sq. Miles	Pasture Area %	Wheat	Rice	Corn	Cattle	Pigs	Sheep
Australia	2,969,910	195,368	6.6%	1,563,327	52.6%	23,654	1,417	363	27,645	2,607	116,736
Fiji	7,056	1,100	15.6%	676	9.6%	-	16	1	335	139	7
Kiribati	313	151	48.1%	-	0.0%	-	-	-	-	10	-
Micronesia, Federated States of	271	139	51.3%	42	15.7%	-	-	-	14	32	-
New Zealand	104,454	13,019	12.5%	53,525	51.2%	337	-	185	9,025	364	45,114
Papua New Guinea	178,704	3,320	1.9%	676	0.4%	-	1	7	87	1,583	6
Samoa	1,093	498	45.6%	8	0.7%	-	-	-	28	179	-
Solomon Islands	10,954	286	2.6%	154	1.4%	-	5	-	11	63	-
Tonga	251	185	73.8%	15	6.2%	-	-	-	11	81	-
Vanuatu	4,707	463	9.8%	162	3.4%	-	-	1	151	62	-

This table presents data for most independent nations having an area greater than 200 square miles
- Zero, insignificant, or not available
(1) Source: United Nations Food and Agriculture Organization
(2) Includes data for Luxembourg
(3) Data for Luxembourg is included with Belgium
(4) Includes data for Taiwan
(5) Data for Taiwan is included with China

WORLD ECONOMIC TABLE

CONTINENT/Country	GDP 2002		Trade		Commercial Energy Production Avg. 2000[2]					Average Production 1999-2001 in Metric Tons			
	Total GDP[1]	GDP Per Capita[1]	Value of Exports[1]	Value of Imports[1]	Total (1,000 Metric Tons of Coal Equiv.)	Solid %	Liquid %	Gas %	Hydro & Nuclear %	Coal[3]	Petroleum[3]	Iron Ore[4]	Bauxite[4]
NORTH AMERICA													
Bahamas	$4,590,000,000	$17,000	$560,700,000	$1,860,000,000	-	-	-	-	-	-	-	-	-
Belize	$1,280,000,000	$4,900	$290,000,000	$430,000,000	12	-	-	-	100%	-	-	-	-
Canada	$934,100,000,000	$29,400	$260,500,000,000	$229,000,000,000	507,218	10%	33%	43%	14%	70,711,084	97,834,913	20,527,000	-
Costa Rica	$32,000,000,000	$8,500	$5,100,000,000	$6,400,000,000	1,937	-	-	-	100%	-	-	-	-
Cuba	$30,690,000,000	$2,300	$1,800,000,000	$4,800,000,000	4,626	-	83%	17%	-	-	2,134,520	-	-
Dominica	$380,000,000	$5,400	$50,000,000	$135,000,000	4	-	-	-	100%	-	-	-	-
Dominican Republic	$53,780,000,000	$6,100	$5,300,000,000	$8,700,000,000	115	-	-	-	100%	-	-	-	-
El Salvador	$29,410,000,000	$4,700	$3,000,000,000	$4,900,000,000	1,110	-	-	-	100%	-	-	-	-
Guatemala	$53,200,000,000	$3,700	$2,700,000,000	$5,600,000,000	1,822	-	81%	1%	18%	-	1,076,526	9,000	-
Haiti	$10,600,000,000	$1,700	$298,000,000	$1,140,000,000	33	-	-	-	100%	-	-	-	-
Honduras	$16,290,000,000	$2,600	$1,300,000,000	$2,700,000,000	347	-	-	-	100%	-	-	-	-
Jamaica	$10,080,000,000	$3,900	$1,400,000,000	$3,100,000,000	18	-	-	-	100%	-	-	-	11,728,000
Mexico	$924,400,000,000	$9,000	$158,400,000,000	$168,400,000,000	340,594	1%	79%	16%	4%	11,097,943	150,165,451	6,860,000	-
Nicaragua	$11,160,000,000	$2,500	$637,000,000	$1,700,000,000	706	-	-	-	100%	-	-	-	-
Panama	$18,060,000,000	$6,000	$5,800,000,000	$6,700,000,000	418	-	-	-	100%	-	-	-	-
St. Lucia	$866,000,000	$5,400	$68,300,000	$319,400,000	-	-	-	-	-	-	-	-	-
Trinidad and Tobago	$11,070,000,000	$9,500	$4,200,000,000	$3,800,000,000	22,768	-	39%	61%	-	-	5,964,991	-	-
United States	$10,450,000,000,000	$37,600	$733,900,000,000	$1,194,100,000,000	2,342,228	33%	22%	30%	14%	996,498,186	289,640,487	35,178,000	-
SOUTH AMERICA													
Argentina	$403,800,000,000	$10,200	$25,300,000,000	$9,000,000,000	118,739	-	50%	45%	5%	260,299	38,783,798	-	-
Bolivia	$21,150,000,000	$2,500	$1,300,000,000	$1,600,000,000	7,732	-	33%	64%	3%	-	1,599,401	-	-
Brazil	$1,376,000,000,000	$7,600	$59,400,000,000	$46,200,000,000	143,640	3%	63%	6%	28%	4,446,477	61,155,586	124,667,000	13,654,000
Chile	$156,100,000,000	$10,000	$17,800,000,000	$15,600,000,000	6,180	6%	11%	45%	38%	475,484	349,201	5,523,000	-
Colombia	$251,600,000,000	$6,500	$12,900,000,000	$12,500,000,000	99,513	36%	52%	9%	4%	38,112,136	34,896,672	348,000	-
Ecuador	$42,650,000,000	$3,100	$4,900,000,000	$6,000,000,000	32,171	-	94%	3%	3%	-	19,520,185	-	-
Guyana	$2,628,000,000	$4,000	$500,000,000	$575,000,000	1	-	-	-	100%	-	-	-	2,272,000
Paraguay	$25,190,000,000	$4,200	$2,000,000,000	$2,400,000,000	6,577	-	-	-	100%	-	-	-	-
Peru	$138,800,000,000	$4,800	$7,600,000,000	$7,300,000,000	10,933	-	73%	9%	18%	52,297	4,932,561	2,701,000	-
Suriname	$1,469,000,000	$3,500	$445,000,000	$300,000,000	1,022	-	84%	-	16%	-	496,400	-	3,946,000
Uruguay	$26,820,000,000	$7,800	$2,100,000,000	$1,870,000,000	867	-	-	-	100%	-	-	-	-
Venezuela	$131,700,000,000	$5,500	$28,600,000,000	$18,800,000,000	311,899	3%	81%	14%	2%	7,482,998	146,621,238	10,497,000	4,309,000
EUROPE													
Albania	$15,690,000,000	$4,500	$340,000,000	$1,500,000,000	1,089	1%	42%	2%	55%	32,666	284,321	-	-
Austria	$227,700,000,000	$27,700	$70,000,000,000	$74,000,000,000	9,611	5%	15%	24%	56%	1,197,660	921,120	525,000	-
Belarus	$90,190,000,000	$8,200	$7,700,000,000	$8,800,000,000	3,644	18%	73%	9%	-	-	1,830,872	-	-
Belgium	$299,700,000,000	$29,000	$162,000,000,000	$152,000,000,000	18,451	2%	-	-	98%	318,998	-	-	-
Bosnia and Herzegovina	$7,300,000,000	$1,900	$1,150,000,000	$2,800,000,000	6,553	90%	-	-	10%	8,414,623	-	50,000	75,000
Bulgaria	$49,230,000,000	$6,600	$5,300,000,000	$6,900,000,000	13,500	46%	-	-	53%	28,841,963	37,048	310,000	-
Croatia	$43,120,000,000	$8,800	$4,900,000,000	$10,700,000,000	4,962	-	42%	43%	15%	5,104	1,191,360	-	-
Czech Republic	$157,100,000,000	$15,300	$40,800,000,000	$43,200,000,000	39,843	85%	1%	1%	14%	63,466,671	283,097	-	-
Denmark	$155,300,000,000	$29,000	$56,300,000,000	$47,900,000,000	36,502	-	70%	29%	2%	-	16,701,163	-	-
Estonia	$15,520,000,000	$10,900	$3,400,000,000	$4,400,000,000	3,892	100%	-	-	-	-	-	-	-
Finland	$133,800,000,000	$26,200	$40,100,000,000	$31,800,000,000	11,933	15%	-	-	85%	-	-	-	-
France	$1,558,000,000,000	$25,700	$307,800,000,000	$303,700,000,000	175,306	2%	4%	1%	93%	3,616,981	1,446,228	12,000	-
Germany	$2,160,000,000,000	$26,600	$608,000,000,000	$487,300,000,000	181,697	47%	2%	13%	38%	204,685,080	3,044,206	5,000	-
Greece	$203,300,000,000	$19,000	$12,600,000,000	$31,400,000,000	12,988	92%	3%	1%	4%	64,503,999	166,807	583,000	1,975,000
Hungary	$134,000,000,000	$13,300	$31,400,000,000	$33,900,000,000	16,319	25%	19%	24%	32%	14,796,257	1,301,710	-	994,000
Iceland	$8,444,000,000	$25,000	$2,300,000,000	$2,100,000,000	1,638	-	-	-	100%	-	-	-	-
Ireland	$113,700,000,000	$30,500	$86,600,000,000	$48,600,000,000	3,232	47%	-	47%	6%	-	-	-	-
Italy	$1,455,000,000,000	$25,000	$259,200,000,000	$238,200,000,000	40,332	-	16%	54%	30%	47,666	4,144,278	-	-
Latvia	$20,990,000,000	$8,300	$2,300,000,000	$3,900,000,000	369	6%	-	-	94%	-	-	-	-
Lithuania	$30,080,000,000	$8,400	$5,400,000,000	$6,800,000,000	3,677	-	12%	-	87%	-	251,824	-	-
Luxembourg	$21,940,000,000	$44,000	$10,100,000,000	$13,250,000,000	113	-	-	-	100%	-	-	-	-
Macedonia	$10,570,000,000	$5,000	$1,100,000,000	$1,900,000,000	3,038	95%	-	-	5%	7,463,628	-	9,000	-
Moldova	$11,510,000,000	$2,500	$590,000,000	$980,000,000	7	-	-	-	100%	-	-	-	-
Netherlands	$437,800,000,000	$26,900	$243,300,000,000	$201,100,000,000	87,974	-	4%	94%	2%	-	1,437,293	-	-
Norway	$149,100,000,000	$31,800	$68,200,000,000	$37,300,000,000	324,396	-	72%	22%	5%	847,996	154,419,533	355,000	-
Poland	$373,200,000,000	$9,500	$32,400,000,000	$43,400,000,000	108,277	94%	1%	5%	-	164,737,813	645,072	-	-
Portugal	$195,200,000,000	$18,000	$25,900,000,000	$39,000,000,000	1,560	-	-	-	100%	-	-	6,000	-
Romania	$169,300,000,000	$7,400	$13,700,000,000	$16,700,000,000	37,598	19%	24%	46%	10%	27,392,191	6,038,110	24,000	-
Serbia and Montenegro	$23,150,000,000	$2,370	$2,400,000,000	$6,300,000,000	14,188	74%	8%	8%	10%	34,480,488	810,787	10,000	580,000
Slovakia	$67,340,000,000	$12,200	$12,900,000,000	$15,400,000,000	8,813	17%	1%	2%	79%	3,606,648	48,134	200,000	-
Slovenia	$37,060,000,000	$18,000	$10,300,000,000	$11,100,000,000	3,644	38%	-	-	62%	4,391,644	991	-	-
Spain	$850,700,000,000	$20,700	$122,200,000,000	$156,600,000,000	40,444	28%	2%	1%	68%	23,479,212	296,665	-	-
Sweden	$230,700,000,000	$25,400	$80,600,000,000	$68,600,000,000	31,413	1%	-	-	99%	-	-	12,114,000	-
Switzerland	$233,400,000,000	$31,700	$100,300,000,000	$94,400,000,000	14,710	-	-	-	100%	-	-	-	-
Ukraine	$218,000,000,000	$4,500	$18,100,000,000	$18,000,000,000	118,973	50%	5%	20%	25%	81,998,575	3,747,936	28,933,000	-
United Kingdom	$1,528,000,000,000	$25,300	$286,300,000,000	$330,100,000,000	397,906	7%	47%	38%	8%	32,758,497	119,820,635	1,000	-
Russia	$1,409,000,000,000	$9,300	$104,600,000,000	$60,700,000,000	1,412,286	10%	33%	52%	5%	253,376,954	324,436,632	48,300,000	3,983,000
ASIA													
Afghanistan	$19,000,000,000	$700	$1,200,000,000	$1,300,000,000	195	1%	-	79%	20%	1,000	-	-	-
Armenia	$12,130,000,000	$3,800	$525,000,000	$991,000,000	901	-	-	-	100%	-	-	-	-
Azerbaijan	$28,610,000,000	$3,500	$2,000,000,000	$1,800,000,000	27,748	-	72%	27%	1%	-	14,183,985	-	-
Bahrain	$9,910,000,000	$14,000	$5,800,000,000	$4,200,000,000	14,442	-	22%	78%	-	-	1,827,397	-	-
Bangladesh	$238,200,000,000	$1,700	$6,200,000,000	$8,500,000,000	11,713	-	-	99%	1%	-	120,476	-	-
Brunei	$6,500,000,000	$18,600	$3,000,000,000	$1,400,000,000	27,922	-	49%	51%	-	-	9,435,323	-	-
Cambodia	$20,420,000,000	$1,500	$1,380,000,000	$1,730,000,000	10	-	-	-	100%	-	-	-	-
China	$5,989,000,000,000	$4,400	$658,260,000,000	$618,930,000,000	1,023,314[5]	70%[5]	23%[5]	4%[5]	3%[5]	1,251,423,183	161,226,848	72,967,000	9,000,000
Cyprus	$9,400,000,000	$15,000	$1,030,000,000	$3,900,000,000	-	-	-	-	-	-	-	-	-
East Timor	$440,000,000	$500	$8,000,000	$237,000,000	-	-	-	-	-	-	-	-	-
Georgia	$16,050,000,000	$3,100	$515,000,000	$750,000,000	963	1%	16%	8%	75%	10,000	102,258	-	-
India	$2,664,000,000,000	$2,540	$44,500,000,000	$53,800,000,000	367,807	73%	14%	8%	4%	304,842,421	32,123,682	48,080,000	7,554,000
Indonesia	$714,200,000,000	$3,100	$52,300,000,000	$32,100,000,000	279,695	27%	45%	26%	2%	79,664,587	70,565,213	282,000	1,168,000
Iran	$458,300,000,000	$7,000	$24,800,000,000	$21,800,000,000	350,729	-	77%	23%	-	1,376,993	181,632,777	5,367,000	136,000
Iraq	$58,000,000,000	$2,400	$13,000,000,000	$7,800,000,000	186,519	-	97%	3%	-	-	124,281,583	-	-
Israel	$117,400,000,000	$19,000	$28,100,000,000	$30,800,000,000	334	94%	2%	4%	1%	-	5,957	-	-
Japan	$3,651,000,000,000	$28,000	$383,800,000,000	$292,100,000,000	142,731	2%	1%	2%	95%	3,286,983	351,650	1,000	-
Jordan	$22,630,000,000	$4,300	$2,500,000,000	$4,400,000,000	316	-	1%	97%	2%	-	1,986	-	-
Kazakhstan	$120,000,000,000	$6,300	$10,300,000,000	$9,600,000,000	113,390	40%	45%	14%	1%	70,311,969	30,508,827	7,467,000	3,668,000
Korea, North	$22,260,000,000	$1,000	$842,000,000	$1,314,000,000	65,932	96%	-	-	4%	94,174,845	-	3,000,000	-
Korea, South	$941,500,000,000	$19,400	$162,600,000,000	$148,400,000,000	43,892	6%	-	-	94%	4,054,646	-	175,000	-
Kuwait	$36,850,000,000	$15,000	$16,000,000,000	$7,300,000,000	161,322	-	92%	8%	-	-	98,844,823	-	-

CONTINENT/Country	GDP 2002 Total GDP[1]	GDP Per Capita[1]	Trade Value of Exports[1]	Value of Imports[1]	Commercial Energy Production Avg. 2000[2] Total (1,000 Metric Tons of Coal Equiv.)	Solid %	Liquid %	Gas %	Hydro & Nuclear %	Average Production 1999-2001 in Metric Tons Coal[3]	Petroleum[3]	Iron Ore[4]	Bauxite[4]
Kyrgyzstan	$13,880,000,000	$2,800	$488,000,000	$587,000,000	2,026	9%	5%	2%	83%	423,664	91,503	-	-
Laos	$10,400,000,000	$1,700	$345,000,000	$555,000,000	146	1%	-	-	99%	1,000	-	-	-
Lebanon	$17,610,000,000	$5,400	$1,000,000,000	$6,000,000,000	55	-	-	-	100%	-	-	-	-
Malaysia	$198,400,000,000	$9,300	$95,200,000,000	$76,800,000,000	110,069	-	41%	58%	1%	314,332	33,792,132	208,000	137,000
Mongolia	$5,060,000,000	$1,840	$501,000,000	$659,000,000	2,212	100%	-	-	-	5,099,640	-	-	-
Myanmar	$73,690,000,000	$1,660	$2,700,000,000	$2,500,000,000	9,297	3%	6%	88%	2%	358,331	587,374	-	-
Nepal	$37,320,000,000	$1,400	$720,000,000	$1,600,000,000	172	10%	-	-	90%	9,667	-	-	-
Oman	$22,400,000,000	$8,300	$10,600,000,000	$5,500,000,000	74,376	-	92%	8%	-	-	46,989,489	-	-
Pakistan	$295,300,000,000	$2,100	$9,800,000,000	$11,100,000,000	33,773	6%	12%	74%	7%	3,247,391	2,768,108	-	10,000
Philippines	$379,700,000,000	$4,200	$35,100,000,000	$33,500,000,000	16,244	6%	-	-	94%	1,306,993	173,128	-	-
Qatar	$15,910,000,000	$21,500	$10,900,000,000	$3,900,000,000	92,237	-	57%	43%	-	-	35,018,538	-	-
Saudi Arabia	$268,900,000,000	$10,500	$71,000,000,000	$39,500,000,000	736,996	-	91%	9%	-	-	401,559,222	-	-
Singapore	$112,400,000,000	$24,000	$127,000,000,000	$113,000,000,000	-	-	-	-	-	-	-	-	-
Sri Lanka	$73,700,000,000	$3,700	$4,600,000,000	$5,400,000,000	394	-	-	-	100%	-	-	-	-
Syria	$63,480,000,000	$3,500	$6,200,000,000	$4,900,000,000	47,898	-	83%	15%	2%	-	26,119,029	-	-
Taiwan	$406,000,000,000	$18,000	$130,000,000,000	$113,000,000,000	[6]	[6]	[6]	[6]	[6]	58,284	38,686	-	-
Tajikistan	$8,476,000,000	$1,250	$710,000,000	$830,000,000	1,790	-	1%	3%	95%	20,667	16,613	-	-
Thailand	$445,800,000,000	$6,900	$67,700,000,000	$58,100,000,000	44,127	25%	24%	50%	2%	18,551,756	5,080,720	20,000	-
Turkey	$489,700,000,000	$7,000	$35,100,000,000	$50,800,000,000	28,167	69%	14%	3%	14%	65,334,995	2,642,106	2,300,000	303,000
Turkmenistan	$31,340,000,000	$5,500	$2,970,000,000	$2,250,000,000	71,764	-	15%	85%	-	-	7,139,688	-	-
United Arab Emirates	$53,970,000,000	$22,000	$44,900,000,000	$30,800,000,000	199,656	-	83%	17%	-	-	112,737,023	-	-
Uzbekistan	$66,060,000,000	$2,500	$2,800,000,000	$2,500,000,000	85,806	1%	13%	85%	1%	2,736,319	4,419,300	-	-
Vietnam	$183,800,000,000	$2,250	$16,500,000,000	$16,800,000,000	39,300	30%	59%	5%	7%	9,688,950	15,926,911	-	-
Yemen	$15,070,000,000	$840	$3,400,000,000	$2,900,000,000	30,622	-	100%	-	-	-	21,304,264	-	-
AFRICA													
Algeria	$173,800,000,000	$5,300	$19,500,000,000	$10,600,000,000	222,648	-	47%	53%	-	24,000	61,651,110	757,000	-
Angola	$18,360,000,000	$1,600	$8,600,000,000	$4,100,000,000	53,315	-	98%	1%	-	-	36,961,745	-	-
Benin	$7,380,000,000	$1,070	$207,000,000	$479,000,000	69	-	100%	-	-	-	39,547	-	-
Botswana	$13,480,000,000	$9,500	$2,400,000,000	$1,900,000,000	[7]	[7]	[7]	[7]	[7]	956,767	-	-	-
Burkina Faso	$14,510,000,000	$1,080	$250,000,000	$525,000,000	15	-	-	-	100%	-	-	-	-
Burundi	$3,146,000,000	$600	$26,000,000	$135,000,000	21	29%	-	-	71%	-	-	-	-
Cameroon	$26,840,000,000	$1,700	$1,900,000,000	$1,700,000,000	10,722	-	96%	-	4%	1,000	4,326,440	-	-
Cape Verde	$600,000,000	$1,400	$30,000,000	$220,000,000	-	-	-	-	-	-	-	-	-
Central African Republic	$4,296,000,000	$1,300	$134,000,000	$102,000,000	10	-	-	-	100%	-	-	-	-
Chad	$9,297,000,000	$1,100	$197,000,000	$570,000,000	-	-	-	-	-	-	-	-	-
Comoros	$441,000,000	$720	$16,300,000	$39,800,000	-	-	-	-	-	-	-	-	-
Congo	$2,500,000,000	$900	$2,400,000,000	$73,000,000	19,097	-	99%	1%	-	-	13,651,000	-	-
Congo, Democratic Republic of the	$34,000,000,000	$610	$1,200,000,000	$890,000,000	2,630	4%	71%	-	25%	96,000	1,194,669	-	-
Cote d'Ivoire	$24,030,000,000	$1,500	$4,400,000,000	$2,500,000,000	4,439	-	50%	45%	5%	-	620,450	-	-
Djibouti	$619,000,000	$1,300	$70,000,000	$255,000,000	-	-	-	-	-	-	-	-	-
Egypt	$289,800,000,000	$3,900	$7,000,000,000	$15,200,000,000	86,315	-	65%	32%	2%	-	38,024,058	1,283,000	-
Equatorial Guinea	$1,270,000,000	$2,700	$2,500,000,000	$562,000,000	7,531	-	100%	-	-	-	7,461,521	-	-
Eritrea	$3,300,000,000	$740	$20,000,000	$500,000,000	-	-	-	-	-	-	-	-	-
Ethiopia	$48,530,000,000	$750	$433,000,000	$1,630,000,000	211	-	-	-	100%	-	-	-	-
Gabon	$8,354,000,000	$5,700	$2,600,000,000	$1,100,000,000	23,273	-	95%	5%	-	-	15,674,359	-	-
Gambia, The	$2,582,000,000	$1,800	$138,000,000	$225,000,000	-	-	-	-	-	-	-	-	-
Ghana	$41,250,000,000	$2,100	$2,200,000,000	$2,800,000,000	830	-	2%	-	98%	-	330,933	-	525,000
Guinea	$18,690,000,000	$2,000	$835,000,000	$670,000,000	23	-	-	-	100%	-	-	-	15,663,000
Guinea-Bissau	$901,400,000	$800	$71,000,000	$59,000,000	-	-	-	-	-	-	-	-	-
Kenya	$32,890,000,000	$1,020	$2,100,000,000	$3,000,000,000	642	-	-	-	100%	-	-	-	-
Lesotho	$5,106,000,000	$2,700	$422,000,000	$738,000,000	[7]	[7]	[7]	[7]	[7]	-	-	-	-
Liberia	$3,116,000,000	$1,100	$110,000,000	$165,000,000	24	-	-	-	100%	-	-	-	-
Libya	$33,360,000,000	$7,600	$11,800,000,000	$6,300,000,000	103,205	-	92%	8%	-	-	67,767,436	-	-
Madagascar	$12,950,000,000	$760	$700,000,000	$985,000,000	64	-	-	-	100%	-	-	-	-
Malawi	$6,811,000,000	$670	$435,000,000	$505,000,000	107	-	-	-	100%	-	-	-	-
Mali	$9,775,000,000	$860	$680,000,000	$630,000,000	29	-	-	-	100%	-	-	-	-
Mauritania	$4,891,000,000	$1,900	$355,000,000	$360,000,000	4	-	-	-	100%	-	-	7,492,000	-
Mauritius	$12,150,000,000	$11,000	$1,600,000,000	$1,800,000,000	12	-	-	-	100%	-	-	-	-
Morocco	$121,800,000,000	$3,900	$7,500,000,000	$10,400,000,000	201	14%	9%	33%	43%	61,000	15,223	4,000	-
Mozambique	$19,520,000,000	$1,000	$680,000,000	$1,180,000,000	874	2%	-	-	98%	18,667	-	-	8,000
Namibia	$13,150,000,000	$6,900	$1,210,000,000	$1,380,000,000	[7]	[7]	[7]	[7]	[7]	-	-	-	-
Niger	$8,713,000,000	$830	$293,000,000	$368,000,000	175	100%	-	-	-	151,666	-	-	-
Nigeria	$112,500,000,000	$875	$17,300,000,000	$13,600,000,000	172,641	-	90%	10%	-	61,000	108,397,478	-	-
Rwanda	$8,920,000,000	$1,200	$68,000,000	$253,000,000	20	-	-	-	100%	-	-	-	-
Sao Tome and Principe	$200,000,000	$1,200	$5,500,000	$24,800,000	1	-	-	-	100%	-	-	-	-
Senegal	$15,640,000,000	$1,500	$1,150,000,000	$1,460,000,000	1	-	-	100%	-	-	-	-	-
Sierra Leone	$2,826,000,000	$580	$35,000,000	$190,000,000	[7]	[7]	[7]	[7]	[7]	-	-	-	-
Somalia	$4,270,000,000	$550	$126,000,000	$343,000,000	-	-	-	-	-	-	-	-	-
South Africa	$427,700,000,000	$10,000	$31,800,000,000	$26,600,000,000	245,195[8]	92%[8]	5%[8]	1%[8]	2%[8]	224,286,505	1,277,485	20,751,000	-
Sudan	$52,900,000,000	$1,420	$1,800,000,000	$1,500,000,000	13,436	-	99%	-	1%	-	7,679,837	-	-
Swaziland	$5,542,000,000	$4,400	$820,000,000	$938,000,000	[7]	[7]	[7]	[7]	[7]	288,665	-	-	-
Tanzania	$20,420,000,000	$630	$863,000,000	$1,670,000,000	343	23%	-	-	77%	5,000	-	-	-
Togo	$7,594,000,000	$1,500	$449,000,000	$561,000,000	-	-	-	-	-	-	-	-	-
Tunisia	$67,130,000,000	$6,500	$6,800,000,000	$8,700,000,000	8,065	-	66%	34%	-	-	3,826,400	105,000	-
Uganda	$30,490,000,000	$1,260	$476,000,000	$1,140,000,000	193	-	-	-	100%	-	-	3,000	-
Zambia	$8,240,000,000	$890	$709,000,000	$1,123,000,000	1,117	15%	-	-	85%	192,358	-	-	-
Zimbabwe	$26,070,000,000	$2,400	$1,570,000,000	$1,739,000,000	4,801	92%	-	-	8%	4,508,643	-	237,000	-
OCEANIA													
Australia	$525,500,000,000	$27,000	$66,300,000,000	$68,000,000,000	331,923	71%	14%	14%	1%	307,176,075	31,728,994	104,014,000	51,834,000
Fiji	$4,822,000,000	$5,500	$442,000,000	$642,000,000	53	-	-	-	100%	-	-	-	-
Kiribati	$79,000,000	$840	$6,000,000	$44,000,000	-	-	-	-	-	-	-	-	-
Micronesia, Federated States of	$277,000,000	$2,000	$22,000,000	$149,000,000	-	-	-	-	-	-	-	-	-
New Zealand	$78,400,000,000	$20,200	$15,000,000,000	$12,500,000,000	19,812	14%	13%	40%	33%	3,452,315	1,839,394	660,000	-
Papua New Guinea	$10,860,000,000	$2,300	$1,800,000,000	$1,100,000,000	5,864	-	96%	2%	2%	-	3,874,601	-	-
Samoa	$1,000,000,000	$5,600	$15,500,000	$130,100,000	3	-	-	-	100%	-	-	-	-
Solomon Islands	$800,000,000	$1,700	$47,000,000	$82,000,000	-	-	-	-	-	-	-	-	-
Tonga	$236,000,000	$2,200	$8,900,000	$70,000,000	-	-	-	-	-	-	-	-	-
Vanuatu	$563,000,000	$2,900	$22,000,000	$93,000,000	-	-	-	-	-	-	-	-	-

This table presents data for most independent nations having an area greater than 200 square miles
- Zero, insignificant, or not available
(1) Source: United States Central Intelligence Agency World Factbook
(2) Source: United Nations Energy Statistics Yearbook
(3) Source: United States Energy Information Administration International Energy Annual
(4) Source: United States Geological Survey Minerals Yearbook
(5) Includes data for Taiwan
(6) Data for Taiwan is included with China
(7) Data for countries in the South Africa Customs Union are included with South Africa
(8) Includes data for countries in the South Africa Customs Union

WORLD ENVIRONMENT TABLE

CONTINENT/Country	Total Area Sq. Miles	Protected Area 2002[1,2] Sq. Miles	%	Endangered Species 2003[3] Mammal	Bird	Reptile	Amphib.	Fish	Invrt.	Forest Cover[4] Sq. Miles 2000	Percent Change 1990-2000
NORTH AMERICA											
Bahamas	5,382	-	-	5	4	6	0	15	1	3,251	
Belize	8,867	3,999	45.1%	5	2	4	0	17	1	5,205	-20.9%
Canada	3,855,103	427,916	11.1%	16	8	2	1	25	11	944,294	
Costa Rica	19,730	4,538	23.0%	13	13	7	1	13	9	7,598	-7.4%
Cuba	42,804	29,578	69.1%	11	18	7	0	23	3	9,066	13.4%
Dominica	290	-	-	1	3	4	0	11	0	178	-8.0%
Dominican Republic	18,730	9,721	51.9%	5	15	10	1	10	2	5,313	
El Salvador	8,124	33	0.4%	2	0	4	0	5	1	467	-37.3%
Guatemala	42,042	8,408	20.0%	7	6	8	0	14	8	11,004	-15.9%
Haiti	10,714	43	0.4%	4	14	8	1	12	2	340	-44.3%
Honduras	43,277	2,770	6.4%	10	5	6	0	14	2	20,784	-9.9%
Jamaica	4,244	3,590	84.6%	5	12	8	4	12	5	1,255	-14.2%
Mexico	758,452	77,362	10.2%	72	40	18	4	106	41	213,148	-10.3%
Nicaragua	50,054	8,910	17.8%	6	5	7	0	17	2	12,656	-26.3%
Panama	29,157	6,327	21.7%	17	16	7	0	17	2	11,104	-15.3%
St. Lucia	238	-	-	2	5	6	0	10	0	35	-35.7%
Trinidad and Tobago	1,980	119	6.0%	1	1	5	0	15	0	1,000	-7.8%
United States	3,794,083	982,668	25.9%	39	56	27	25	155	557	872,563	1.7%
SOUTH AMERICA											
Argentina	1,073,519	70,852	6.6%	32	39	5	5	9	10	133,777	-7.6%
Bolivia	424,165	56,838	13.4%	25	28	2	1	0	1	204,897	-2.9%
Brazil	3,300,172	221,112	6.7%	74	113	22	6	33	34	2,100,028	-4.1%
Chile	291,930	55,175	18.9%	21	22	0	3	9	0	59,985	-1.3%
Colombia	439,737	44,853	10.2%	39	78	14	0	23	0	191,510	-3.7%
Ecuador	109,484	20,036	18.3%	34	62	10	0	11	48	40,761	-11.5%
Guyana	83,000	249	0.3%	13	2	6	0	13	1	65,170	-2.8%
Paraguay	157,048	5,497	3.5%	10	26	2	0	0	0	90,240	-5.0%
Peru	496,225	30,270	6.1%	46	76	6	1	8	2	251,796	-4.0%
Suriname	63,037	3,089	4.9%	12	1	6	0	12	0	54,491	-
Uruguay	67,574	203	0.3%	6	11	3	0	8	1	4,988	63.3%
Venezuela	352,145	224,669	63.8%	26	24	13	0	19	1	191,144	-4.2%
EUROPE											
Albania	11,100	422	3.8%	3	3	4	0	16	4	3,826	-7.3%
Austria	32,378	10,685	33.0%	7	3	0	0	7	44	15,004	2.0%
Belarus	80,155	5,050	6.3%	7	3	0	0	0	5	36,301	37.5%
Belgium	11,787	-	-	11	2	0	0	7	11	2,811	-1.8%
Bosnia and Herzegovina	19,767	99	0.5%	10	3	1	1	10	10	8,776	
Bulgaria	42,855	1,928	4.5%	14	10	2	0	10	9	14,247	5.9%
Croatia	21,829	1,637	7.5%	9	4	1	1	26	11	6,884	1.1%
Czech Republic	30,450	4,902	16.1%	8	2	0	0	7	19	10,162	0.2%
Denmark	16,640	5,658	34.0%	5	1	0	0	7	11	1,757	2.2%
Estonia	17,462	2,061	11.8%	5	3	0	0	1	4	7,954	6.5%
Finland	130,559	12,142	9.3%	4	3	0	0	1	10	84,691	0.4%
France	208,482	27,728	13.3%	18	5	3	2	15	65	59,232	4.2%
Germany	137,847	43,973	31.9%	11	5	0	0	12	31	41,467	-
Greece	50,949	1,834	3.6%	13	7	6	1	26	11	13,896	9.1%
Hungary	35,919	2,514	7.0%	9	8	1	0	8	25	7,104	4.1%
Iceland	39,769	3,897	9.8%	7	0	0	0	8	0	120	24.0%
Ireland	27,133	461	1.7%	6	1	0	0	6	3	2,544	34.8%
Italy	116,342	9,191	7.9%	14	5	4	4	16	58	38,622	3.0%
Latvia	24,942	3,342	13.4%	5	3	0	0	3	8	11,286	4.5%
Lithuania	25,213	2,597	10.3%	6	4	0	0	3	5	7,699	2.5%
Luxembourg	999	-	-	3	1	0	0	0	4	-	-
Macedonia	9,928	705	7.1%	11	3	2	0	4	5	3,498	-
Moldova	13,070	183	1.4%	6	5	1	0	9	5	1,255	2.2%
Netherlands	16,164	2,295	14.2%	10	4	0	0	7	7	1,448	2.7%
Norway	125,050	8,503	6.8%	10	2	0	0	7	9	34,240	3.6%
Poland	120,728	14,970	12.4%	14	4	0	0	3	15	34,931	2.0%
Portugal	35,516	2,344	6.6%	17	7	0	1	19	82	14,154	18.4%
Romania	91,699	4,310	4.7%	17	8	2	0	10	22	24,896	2.3%
Serbia and Montenegro	39,449	1,302	3.3%	12	5	1	0	19	19	11,147	-0.5%
Slovakia	18,924	4,315	22.8%	9	4	1	0	8	19	8,405	9.0%
Slovenia	7,821	469	6.0%	9	1	0	1	15	42	4,274	2.0%
Spain	194,885	16,565	8.5%	24	7	7	3	23	63	55,483	6.4%
Sweden	173,732	15,810	9.1%	6	2	0	0	6	13	104,765	-
Switzerland	15,943	4,783	30.0%	5	2	0	0	4	30	4,629	3.7%
Ukraine	233,090	9,091	3.9%	16	8	2	0	11	14	37,004	3.3%
United Kingdom	93,788	19,602	20.9%	12	2	0	0	11	10	10,788	6.5%
Russia	6,592,849	514,242	7.8%	45	38	6	0	18	30	3,287,242	0.2%
ASIA											
Afghanistan	251,773	755	0.3%	13	11	1	1	0	1	5,216	
Armenia	11,506	874	7.6%	11	4	5	0	1	7	1,355	13.6%
Azerbaijan	33,437	2,040	6.1%	13	8	5	0	5	6	4,224	13.5%
Bahrain	267	-	-	1	6	0	0	6	0	-	
Bangladesh	55,598	445	0.8%	22	23	20	0	8	0	5,151	14.1%
Brunei	2,226	-	-	11	14	4	0	6	0	1,707	-2.2%
Cambodia	69,898	12,931	18.5%	24	19	10	0	11	0	36,043	-5.7%
China	3,690,045	287,824	7.8%	81	75	31	1	46	4	631,200	12.4%
Cyprus	3,572	-	-	3	3	3	0	6	0	664	44.5%
East Timor	5,743	-	-	0	6	0	0	2	0	1,958	-6.3%
Georgia	26,911	619	2.3%	13	3	7	1	6	10	11,537	
India	1,222,510	63,571	5.2%	86	72	25	3	27	23	247,542	0.6%
Indonesia	735,310	151,474	20.6%	147	114	28	0	91	31	405,353	-11.1%
Iran	636,372	30,546	4.8%	22	13	8	2	14	3	28,182	
Iraq	169,235	-	-	11	11	2	0	3	2	3,085	
Israel	8,019	1,267	15.8%	15	12	4	0	10	10	510	61.0%
Japan	145,850	9,918	6.8%	37	35	11	10	27	45	92,977	0.1%
Jordan	34,495	1,173	3.4%	9	8	1	0	5	3	332	
Kazakhstan	1,049,156	28,327	2.7%	17	15	2	1	7	4	46,904	24.5%
Korea, North	46,540	1,210	2.6%	13	19	0	0	5	1	31,699	
Korea, South	38,328	2,645	6.9%	13	25	0	0	7	1	24,124	-0.8%

CONTINENT/Country	Total Area Sq. Miles	Protected Area 2002[1,2] Sq. Miles	%	Endangered Species 2003[3] Mammal	Bird	Reptile	Amphib.	Fish	Invrt.	Forest Cover[4] Sq. Miles 2000	Percent Change 1990-2000
Kuwait	6,880	103	1.5%	1	7	1	0	6	0	19	66.7%
Kyrgyzstan	77,182	2,779	3.6%	7	4	2	0	0	3	3,873	29.4%
Laos	91,429	11,429	12.5%	31	20	11	0	6	0	48,498	-4.0%
Lebanon	4,016	20	0.5%	6	7	1	0	8	1	139	-2.7%
Malaysia	127,320	7,257	5.7%	50	37	21	0	34	3	74,487	52.4%
Mongolia	604,829	69,555	11.5%	14	16	0	0	1	3	41,101	-5.3%
Myanmar	261,228	784	0.3%	39	35	20	0	7	2	132,892	-13.1%
Nepal	56,827	5,058	8.9%	29	25	6	0	0	1	15,058	-16.7%
Oman	119,499	16,730	14.0%	11	10	4	0	17	1	4	-
Pakistan	339,732	16,647	4.9%	17	17	9	0	14	0	9,116	-14.3%
Philippines	115,831	6,602	5.7%	50	67	8	23	48	19	22,351	-13.3%
Qatar	4,412	-	-	0	6	1	0	4	0	-	-
Saudi Arabia	830,000	317,890	38.3%	9	15	2	0	8	1	5,807	-
Singapore	264	13	4.9%	3	7	3	0	12	1	8	-
Sri Lanka	25,332	3,420	13.5%	22	14	8	0	22	2	7,490	-15.2%
Syria	71,498	-	-	4	8	3	0	8	3	1,780	-
Taiwan	13,901	-	-	12	21	8	0	23	0	-	-
Tajikistan	55,251	2,321	4.2%	9	7	1	0	3	2	1,544	5.3%
Thailand	198,115	27,538	13.9%	37	37	19	0	35	1	56,996	-7.1%
Turkey	302,541	4,841	1.6%	17	11	12	3	29	13	39,479	2.2%
Turkmenistan	188,457	7,915	4.2%	13	6	2	0	8	5	14,498	-
United Arab Emirates	32,278	-	-	4	8	1	0	6	0	1,239	32.1%
Uzbekistan	172,742	3,455	2.0%	9	9	2	0	4	1	7,602	2.4%
Vietnam	128,066	4,738	3.7%	42	37	24	1	22	0	37,911	5.5%
Yemen	203,850	-	-	6	12	2	0	10	2	1,734	-17.0%
AFRICA											
Algeria	919,595	45,980	5.0%	13	6	2	0	9	12	8,282	14.2%
Angola	481,354	31,769	6.6%	19	15	4	0	8	6	269,329	-1.7%
Benin	43,484	4,957	11.4%	9	2	1	0	7	0	10,232	-20.9%
Botswana	224,607	41,552	18.5%	7	7	0	0	0	0	47,981	-8.7%
Burkina Faso	105,869	12,175	11.5%	7	2	1	0	0	0	27,371	-2.1%
Burundi	10,745	612	5.7%	6	7	0	0	0	3	363	-61.0%
Cameroon	183,568	8,261	4.5%	38	15	1	1	34	4	92,116	-8.5%
Cape Verde	1,557	-	-	3	2	0	0	13	0	328	142.9%
Central African Republic	240,536	20,927	8.7%	14	3	1	0	0	0	88,444	-1.3%
Chad	495,755	45,114	9.1%	15	5	1	0	0	1	49,004	-6.0%
Comoros	863	-	-	2	9	2	0	3	4	31	-33.3%
Congo	132,047	6,602	5.0%	15	3	1	0	9	1	85,174	-0.8%
Congo, Democratic Republic of the	905,446	58,854	6.5%	40	28	2	0	9	45	522,037	-3.8%
Cote d'Ivoire	124,504	7,470	6.0%	19	12	2	1	10	1	27,479	-27.1%
Djibouti	8,958	-	-	5	5	0	0	9	0	23	-
Egypt	386,662	37,506	9.7%	13	7	6	0	13	1	278	38.5%
Equatorial Guinea	10,831	-	-	16	5	2	1	7	2	6,765	-5.7%
Eritrea	45,406	1,952	4.3%	12	7	6	0	8	0	6,120	3.3%
Ethiopia	426,373	72,057	16.9%	35	16	1	0	0	4	17,734	-8.1%
Gabon	103,347	723	0.7%	14	5	1	0	11	1	84,271	-0.5%
Gambia, The	4,127	95	2.3%	3	2	1	0	10	0	1,857	10.3%
Ghana	92,098	5,157	5.6%	14	8	2	0	7	0	24,460	-15.9%
Guinea	94,926	664	0.7%	12	10	1	1	7	3	26,753	-4.8%
Guinea-Bissau	13,948	-	-	3	0	1	0	9	1	8,444	-9.0%
Kenya	224,961	17,997	8.0%	50	24	5	0	27	15	66,008	-5.2%
Lesotho	11,720	23	0.2%	6	7	0	0	1	1	54	-
Liberia	43,000	731	1.7%	16	11	2	0	7	2	13,440	-17.9%
Libya	679,362	679	0.1%	8	1	3	0	8	0	1,382	15.1%
Madagascar	226,658	9,746	4.3%	50	27	18	2	25	32	45,278	-9.1%
Malawi	45,747	5,124	11.2%	8	11	0	0	0	8	9,892	-21.6%
Mali	478,841	17,717	3.7%	13	4	1	0	1	0	50,911	-7.0%
Mauritania	397,956	6,765	1.7%	10	2	2	0	10	1	1,224	-23.6%
Mauritius	788	-	-	3	9	4	0	7	32	62	-5.9%
Morocco	172,414	1,207	0.7%	16	9	2	0	10	8	11,680	-0.4%
Mozambique	309,496	25,998	8.4%	15	16	5	0	19	7	118,151	2.0%
Namibia	317,818	43,223	13.6%	14	11	3	1	11	1	31,043	-8.4%
Niger	489,192	37,668	7.7%	11	3	0	0	0	1	5,127	-31.7%
Nigeria	356,669	11,770	3.3%	27	9	2	0	11	1	52,189	-22.8%
Rwanda	10,169	630	6.2%	8	9	0	0	0	2	1,185	-32.8%
Sao Tome and Principe	372	-	-	3	9	1	0	6	2	104	-
Senegal	75,951	8,810	11.6%	12	4	6	0	17	0	23,958	-6.8%
Sierra Leone	27,699	582	2.1%	12	10	3	0	7	4	4,073	-25.5%
Somalia	246,201	1,970	0.8%	19	10	2	0	16	1	29,016	-9.3%
South Africa	470,693	25,888	5.5%	36	28	19	9	47	113	34,429	-0.9%
Sudan	967,500	50,310	5.2%	22	6	2	0	7	1	237,943	-13.5%
Swaziland	6,704	-	-	5	5	0	0	0	0	2,015	12.5%
Tanzania	364,900	108,740	29.8%	41	33	5	0	26	47	149,850	-2.3%
Togo	21,925	1,732	7.9%	9	0	2	0	7	0	1,969	-29.1%
Tunisia	63,170	190	0.3%	11	5	3	0	8	5	1,969	2.2%
Uganda	93,065	22,894	24.6%	20	13	0	0	27	10	16,178	-17.9%
Zambia	290,586	92,697	31.9%	11	11	0	0	0	6	120,641	-21.4%
Zimbabwe	150,873	18,256	12.1%	11	10	0	0	0	2	73,514	-14.4%
OCEANIA											
Australia	2,969,910	397,968	13.4%	63	35	38	35	74	282	596,678	-1.8%
Fiji	7,056	78	1.1%	5	13	6	1	8	2	3,147	-2.0%
Kiribati	313	-	-	0	4	1	0	4	1	108	-
Micronesia, Federated States of	271	-	-	6	5	2	0	6	4	58	-37.5%
New Zealand	104,454	30,918	29.6%	8	63	11	1	16	13	30,680	5.2%
Papua New Guinea	178,704	4,110	2.3%	58	32	9	0	31	12	118,151	-3.6%
Samoa	1,093	-	-	3	8	1	0	4	1	405	-19.2%
Solomon Islands	10,954	33	0.3%	20	23	4	0	4	6	9,792	-1.7%
Tonga	251	-	-	2	3	2	0	3	2	15	-
Vanuatu	4,707	-	-	5	8	2	0	4	0	1,726	1.4%

This table presents data for most independent nations having an area greater than 200 square miles
- Zero, insignificant, or not available
(1) Source: World Resources Institute, 2003. Earth Trends: The Environmental Information Portal. Available at http://earthtrends.wri.org. Washington D. C. World Resources Institute
(2) Source: United Nations Environment Programme - World Conservation Monitoring Centre (UNEP-WCMC); World Database on Protected Areas
(3) Source: International Union of Conservation of Nature and Natural Resources; IUCN 2003 Red List of Threatened Species <www.redlist.org>
(4) Source: United Nations Food and Agriculture Organization; Global Forest Resources Assessment 2000

WORLD COMPARISONS

General Information

Equatorial diameter of the earth, 7,926.38 miles.
Polar diameter of the earth, 7,899.80 miles.
Mean diameter of the earth, 7,917.52 miles.
Equatorial circumference of the earth, 24,901.46 miles.
Polar circumference of the earth, 24,855.34 miles.
Mean distance from the earth to the sun, 93,020,000 miles.
Mean distance from the earth to the moon, 238,857 miles.
Total area of the earth, 197,000,000 sq. miles.
Highest elevation on the earth's surface, Mt. Everest, Asia, 29,028 ft.
Lowest elevation on the earth's land surface, shores of the Dead Sea, Asia, 1,339 ft. below sea level.
Greatest known depth of the ocean, southwest of Guam, Pacific Ocean, 35,810 ft.
Total land area of the earth (incl. inland water and Antarctica), 57,900,000 sq. miles.
Area of Africa, 11,700,000 sq. miles.
Area of Antarctica, 5,400,000 sq. miles.
Area of Asia, 17,300,000 sq. miles.
Area of Europe, 3,800,000 sq. miles.
Area of North America, 9,500,000 sq. miles.
Area of Oceania (incl. Australia) 3,300,000 sq. miles.
Area of South America, 6,900,000 sq. miles.
Population of the earth (est. 1/1/04), 6,339,505,000.

Principal Islands and Their Areas

ISLAND	Area (Sq. Mi.)
Baffin I., Canada	195,928
Banks I., Canada	27,038
Borneo (Kalimantan), Asia	287,300
Bougainville, Papua New Guinea	3,591
Cape Breton I., Canada	3,981
Celebes (Sulawesi), Indonesia	73,057
Ceram (Seram), Indonesia	7,191
Corsica, France	3,367
Crete, Greece	3,189
Cuba, N. America	42,780
Cyprus, Asia	3,572
Devon I., Canada	21,331
Ellesmere I., Canada	75,767
Flores, Indonesia	5,502
Great Britain, U.K.	88,795
Greenland, N. America	840,000
Guadalcanal, Solomon Is.	2,060
Hainan Dao, China	13,127
Hawaii, U.S.	4,028
Hispaniola, N. America	29,300
Hokkaidō, Japan	32,245
Honshū, Japan	89,176
Iceland, Europe	39,769
Ireland, Europe	32,587
Jamaica, N. America	4,247
Java (Jawa), Indonesia	51,038
Kodiak I., U.S.	3,670
Kyūshū, Japan	17,129
Lyete, Philippines	2,785
Long Island, U.S.	1,377
Luzon, Philippines	40,420
Madagascar, Africa	226,642
Melville I., Canada	16,274
Mindanao, Philippines	36,537
Mindoro, Philippines	3,759
Negros, Philippines	4,907
New Britain, Papua New Guinea	14,093
New Caledonia, Oceania	6,252
Newfoundland, Canada	42,031
New Guinea, Asia-Oceania	308,882
New Ireland, Papua New Guinea	3,475
North East Land, Norway	6,350
North I., New Zealand	44,333
Novaya Zemlya, Russia	31,892
Palawan, Philippines	4,550
Panay, Philippines	4,446
Prince of Wales I., Canada	12,872
Puerto Rico, N. America	3,514
Sakhalin, Russia	29,498
Samar, Philippines	5,050
Sardinia, Italy	9,301
Shikoku, Japan	7,258
Sicily, Italy	9,926
Somerset I., Canada	9,570
Southampton I., Canada	15,913
South I., New Zealand	57,708
Spitsbergen, Norway	15,260
Sri Lanka, Asia	24,942
Sumatra (Sumatera), Indonesia	182,860
Taiwan, Asia	13,900
Tasmania, Australia	26,178
Tierra del Fuego, S. America	18,600
Timor, Asia	5,743
Vancouver I., Canada	12,079
Victoria I., Canada	83,897
Vrangelya (Wrangel), Russia	2,819

Principal Lakes, Oceans, Seas, and Their Areas

LAKE Country	Area (Sq. Mi.)
Arabian Sea	1,492,000
Aral Sea, Kazakhstan-Uzbekistan	13,000
Arctic Ocean	5,400,000
Athabasca, L., Canada	3,064
Atlantic Ocean	29,600,000
Balqash köli (L. Balkhash), Kazakhstan	7,027
Baltic Sea, Europe	163,000
Baykal, Ozero (L. Baikal), Russia	12,162
Bering Sea, Asia-N.A.	876,000
Black Sea, Europe-Asia	178,000
Caribbean Sea, N.A.-S.A.	1,063,000
Caspian Sea, Asia-Europe	144,402
Chad, L., Cameroon-Chad-Nigeria	595
Erie, L., Canada-U.S.	9,910
Eyre, L., Australia	3,668
Gairdner, L., Australia	1,076
Great Bear Lake, Canada	12,096
Great Salt Lake, U.S.	1,700
Great Slave Lake, Canada	11,030
Hudson Bay, Canada	475,000
Huron, L., Canada-U.S.	23,000
Indian Ocean	26,500,000
Japan, Sea of, Asia	389,000
Koko Nor (Qinghai Hu), China	1,722
Ladozhskoye Ozero (L. Ladoga), Russia	7,002
Manitoba, L., Canada	1,785
Mediterranean Sea, Europe-Africa-Asia	967,000
Mexico, Gulf of, N. America	596,000
Michigan, L., U.S.	22,300
Nicaragua, Lago de, Nicaragua	3,147
North Sea, Europe	222,000
Nyasa, L., Malawi-Mozambique-Tanzania	11,120
Onezhskoye Ozero (L. Onega), Russia	3,819
Ontario, L., Canada-U.S.	7,340
Pacific Ocean	60,100,000
Red Sea, Africa-Asia	169,000
Rudolf, L., Ethiopia-Kenya	2,471
Southern Ocean	7,800,000
Superior, L., Canada-U.S.	31,700
Tanganyika, L., Africa	12,355
Titicaca, Lago, Bolivia-Peru	3,232
Torrens, L., Australia	1,076
Vänern (L.), Sweden	2,181
Van Gölü (L.), Turkey	1,434
Victoria, L., Kenya-Tanzania-Uganda	26,564
Winnipeg, L., Canada	9,416
Winnipegosis, L., Canada	2,075
Yellow Sea, China-Korea	480,000

Principal Mountains and Their Heights

MOUNTAIN Country	Elev. (Ft.)
Aconcagua, Cerro, Argentina	22,831
Annapurna, Nepal	26,504
Aoraki, New Zealand	12,316
Api, Nepal	23,399
Apo, Philippines	9,692
Ararat, Mt., Turkey	16,854
Barú, Volcán, Panama	11,401
Bangueta, Mt., Papua New Guinea	13,520
Belukha, Mt., Kazakhstan-Russia	14,783
Bia, Phou, Laos	9,249
Blanc, Mont (Monte Bianco), France-Italy	15,771
Blanca Pk., Colorado, U.S.	14,345
Bolívar, Pico, Venezuela	16,427
Bonete, Cerro, Argentina	22,546
Borah Pk., Idaho, U.S.	12,662
Boundary Pk., Nevada, U.S.	13,140
Cameroon Mtn., Cameroon	13,451
Carrauntoohil, Ireland	3,406
Chaltel, Cerro (Monte Fitzroy), Argentina-Chile	10,958
Chimborazo, Ecuador	20,702
Chirripó, Cerro, Costa Rica	12,530
Colima, Nevado de, Mexico	13,911
Cotopaxi, Ecuador	19,347
Cristóbal Colón, Pico, Colombia	19,029
Damāvand, Qolleh-ye, Iran	18,386
Dhawalāgiri, Nepal	26,810
Duarte, Pico, Dominican Rep.	10,417
Dufourspitze (Monte Rosa), Italy-Switzerland	15,203
Elbert, Mt., Colorado, U.S.	14,433
El'brus, Gora, Russia	18,510
Elgon, Mt., Kenya-Uganda	14,178
Erciyeş, Dağı, Turkey	12,848
Etna, Mt., Italy	10,902
Everest, Mt., China-Nepal	29,028
Fairweather, Mt., Alaska-Canada	15,300
Folādī, Koh-e, Afghanistan	16,847
Foraker, Mt., Alaska, U.S.	17,400
Fuji San, Japan	12,388
Galdhøpiggen, Norway	8,100
Gannett Pk., Wyoming, U.S.	13,804
Gasherbrum, China-Pakistan	26,470
Gerlachovský štít, Slovakia	8,711
Giluwe, Mt., Papua New Guinea	14,331
Gongga Shan, China	24,790
Grand Teton, Wyoming, U.S.	13,770
Grossglockner, Austria	12,457
Hadūr Shu'ayb, Yemen	12,008
Haleakalā Crater, Hawaii, U.S.	10,023
Hekla, Iceland	4,892
Hood, Mt., Oregon, U.S.	11,239
Huascarán, Nevado, Peru	22,133
Huila, Nevado de, Colombia	18,865
Hvannadalshnúkur, Iceland	6,952
Illampu, Nevado, Bolivia	21,066
Illimani, Nevado, Bolivia	20,741
Ismail Samani, pik, Tajikistan	24,590
Iztaccíhuatl, Mexico	17,159
Jaya, Puncak, Indonesia	16,503
Jungfrau, Switzerland	13,642
K2 (Qogir Feng), China-Pakistan	28,250
Kāmet, China-India	25,447
Kānchenjunga, India-Nepal	28,208
Kātrīnā, Jabal, Egypt	8,668
Kebnekaise, Sweden	6,926
Kenya, Mt. (Kirinyaga), Kenya	17,058
Kerinci, Gunung, Indonesia	12,467
Kilimanjaro, Tanzania	19,340
Kinabalu, Gunong, Malaysia	13,455
Klyuchevskaya, Russia	15,584
Kosciuszko, Mt., Australia	7,313
Koussi, Emi, Chad	11,204
Kula Kangri, Bhutan	24,784
La Selle, Massif de, Haiti	8,793
Lassen Pk., California, U.S.	10,457
Llullaillaco, Volcán, Argentina-Chile	22,110
Logan, Mt., Canada	19,551
Longs Pk., Colorado, U.S.	14,255
Makālu, China-Nepal	27,825
Margherita Peak, Dem. Rep. of the Congo-Uganda	16,763
Markham, Mt., Antarctica	14,049
Maromokotro, Madagascar	9,436
Massive, Mt., Colorado, U.S.	14,421
Matterhorn, Italy-Switzerland	14,692
Mauna Kea, Hawaii, U.S.	13,796
Mauna Loa, Hawaii, U.S.	13,679
Mayon Volcano, Philippines	8,077
McKinley, Mt., Alaska, U.S.	20,320
Meron, Hare, Israel	3,963
Meru, Mt., Tanzania	14,978
Misti, Volcán, Peru	19,101
Mitchell, Mt., North Carolina, U.S.	6,684
Môco, Serra do, Angola	8,596
Moldoveanu, Romania	8,346
Mulhacén, Spain	11,424
Musala, Bulgaria	9,596
Muztag, China	25,338
Muztagata, China	24,757
Namjagbarwa Feng, China	25,446
Nanda Devi, India	25,645
Nanga Parbat, Pakistan	26,660
Narodnaya, Gora, Russia	6,217
Nevis, Ben, United Kingdom	4,406
Ojos del Salado, Nevado, Argentina-Chile	22,615
Ólimbos, Cyprus	6,401
Ólympos, Greece	9,570
Olympus, Mt., Washington, U.S.	7,965
Orizaba, Pico de, Mexico	18,406
Paektu San, North Korea-China	9,003
Paricutín, Mexico	9,186
Parnassós, Greece	8,061
Pelée, Montagne, Martinique	4,583
Pidurutalagala, Sri Lanka	8,281
Pikes Pk., Colorado, U.S.	14,110
Pobedy, pik, China-Kyrgyzstan	24,406
Popocatépetl, Volcán, Mexico	17,930
Pulog, Mt., Philippines	9,626
Rainier, Mt., Washington, U.S.	14,410
Ramm, Jabal, Jordan	5,755
Ras Dashen Terara, Ethiopia	15,158
Rinjani, Gunung, Indonesia	12,224
Robson, Mt., Canada	12,972
Roraima, Mt., Brazil-Guyana-Venezuela	9,432
Ruapehu, Mt., New Zealand	9,177
St. Elias, Mt., Alaska, U.S.-Canada	18,008
Sajama, Nevado, Bolivia	21,391
Semeru, Gunung, Indonesia	12,060
Shām, Jabal ash, Oman	9,957
Shasta, Mt., California, U.S.	14,162
Snowdon, United Kingdom	3,560
Tahat, Algeria	9,541
Tajumulco, Guatemala	13,845
Taranaki, Mt., New Zealand	8,260
Tirich Mīr, Pakistan	25,230
Tomanivi (Victoria), Fiji	4,341
Toubkal, Jebel, Morocco	13,665
Triglav, Slovenia	9,396
Trikora, Puncak, Indonesia	15,584
Tupungato, Cerro, Argentina-Chile	21,555
Turquino, Pico, Cuba	6,470
Uluru (Ayers Rock), Australia	2,844
Uncompahgre Pk, Colorado, U.S.	14,309
Vesuvio (Vesuvius), Italy	4,190
Victoria, Mt., Papua New Guinea	13,238
Vinson Massif, Antarctica	16,066
Waddington, Mt., Canada	13,163
Washington, Mt., New Hampshire, U.S.	6,288
Whitney, Mt., California, U.S.	14,494
Wilhelm, Mt., Papua New Guinea	14,793
Wrangell, Mt., Alaska, U.S.	14,163
Xixabangma Feng (Gosainthan), China	26,286
Yü Shan, Taiwan	13,114
Zugspitze, Austria-Germany	9,718

Principal Rivers and Their Lengths

RIVER Continent	Length (Mi.)
Albany, N. America	610
Aldan, Asia	1,412
Amazonas-Ucayali, S. America	4,000
Amu Darya, Asia	1,578
Amur, Asia	1,752
Araguaia, S. America	1,367
Arkansas, N. America	1,460
Atchafalaya, N. America	1,420
Athabasca, N. America	765
Brahmaputra, Asia	1,770
Brazos, N. America	1,280
Canadian, N. America	906
Churchill, N. America	1,000
Colorado, N. America (U.S.-Mexico)	1,450
Colorado, N. America (Texas)	862
Columbia, N. America	1,240
Congo (Zaïre), Africa	2,715
Danube, Europe	1,777
Darling, Australia	864
Dnieper (Dnipro), Europe	1,367
Don, Europe	1,162
Elbe, Europe	690
Essequibo, S. America	603
Euphrates, Asia	1,510
Fraser, N. America	851
Ganges, Asia	1,864
Gila, N. America	649
Godāvari, Asia	932
Huang (Yellow), Asia	2,902
Indigirka, Asia	1,072
Indus, Asia	1,118
Irrawaddy, Asia	1,300
Juruá, S. America	1,250
Kama, Europe	1,122
Kasai, Africa	1,338
Kolyma, Asia	1,323
Lena, Asia	2,734
Limpopo, Africa	1,100
Loire, Europe	690
Mackenzie, N. America	2,635
Madeira, S. America	2,013
Magdalena, S. America	951
Marañón, S. America	1,000
Mekong, Asia	2,796
Meuse, Europe	575
Mississippi, N. America	2,340
Mississippi-Missouri, N. America	3,710
Missouri, N. America	2,540
Murray-Darling, Australia	2,169
Negro, S. America	1,305
Nelson, N. America	1,600
Niger, Africa	2,585
Nile, Africa	4,132
Ob', Asia	2,268
Oder, Europe	565
Ohio, N. America	1,310
Oka, Europe	932
Orange, Africa	1,300
Orinoco, S. America	1,703
Ottawa, N. America	790
Paraguay, S. America	1,610
Parnaíba, S. America	901
Peace, N. America	1,195
Pechora, Europe	1,125
Pecos, N. America	926
Pilcomayo, S. America	1,550
Plata-Paraná, S. America	2,920
Platte, N. America	990
Purús, S. America	1,860
Red, N. America	1,290
Rhine, Europe	820
Rhône, Europe	503
Rio Grande, N. America	1,900
Roosevelt, S. America	950
St. Lawrence, N. America	1,900
Salado, S. America	870
Salween (Nu), Asia	1,750
São Francisco, S. America	1,740
Saskatchewan-Bow, N. America	1,205
Severnaya Dvina (Northern Dvina), Europe	462
Snake, N. America	1,040
Sungari (Songhua), Asia	1,140
Syr Darya, Asia	1,370
Tagus, Europe	625
Tarim, Asia	1,328
Tennessee, N. America	886
Tigris, Asia	1,180
Tisa, Europe	607
Tocantins, S. America	1,640
Ucayali, S. America	1,220
Ural, Asia	1,509
Uruguay, S. America	1,025
Verkhnyaya Tunguska (Angara), Asia	1,105
Vilyuy, Asia	1,647
Volga, Europe	2,082
Volta, Africa	994
Wisła (Vistula), Europe	630
Xiang, Asia	930
Xingu, S. America	1,230
Yangtze (Chang), Asia	3,915
Yellowstone, N. America	692
Yenisey, Asia	2,169
Yukon, N. America	1,980
Zambezi, Africa	1,653

PRINCIPAL CITIES OF THE WORLD

Abidjan, Cote d'Ivoire 1,929,079
Abū Ẓaby (Abu Dhabi), United Arab Emirates 242,975
Accra, Ghana (1,390,000) 949,113
Addis Ababa, Ethiopia 2,424,000
Ahmadābād, India (4,519,278) 3,515,361
Aleppo (Ḩalab), Syria (1,640,000) . . .1,591,400
Alexandria (Al Iskandarīyah), Egypt (3,350,000) .3,339,076
Algiers (El Djazaïr), Algeria (2,547,983) .1,507,241
Al Jīzah (Giza), Egypt (*Al Qāhirah) 2,221,817
Almaty, Kazakhstan (1,190,000) 1,129,356
'Ammān, Jordan (1,500,000) 1,147,447
Amsterdam, Netherlands (1,121,303) 727,053
Ankara, Turkey (3,294,220) 2,984,099
Antananarivo, Madagascar 1,250,000
Antwerp (Antwerpen), Belgium (1,135,000) .453,030
Ashgabat (Ashkhabad), Turkmenistan 557,600
Asmera, Eritrea 358,100
Astana (Aqmola), Kazakhstan (319,324) 312,965
Asunción, Paraguay (700,000) 546,637
Athens (Athína), Greece (3,150,000) . .772,072
Atlanta, Georgia, U.S. (4,112,198) 416,474
Auckland, New Zealand (1,074,510) . .367,737
Baghdād, Iraq 3,841,268
Baku (Bakı), Azerbaijan (2,020,000) .1,792,300
Bamako, Mali 658,275
Bandung, Indonesia 5,919,400
Bangalore, India (5,686,844) 4,292,223
Banghāzī, Libya 800,000
Bangkok (Krung Thep), Thailand (7,060,000) .5,620,591
Bangui, Central African Republic 451,690
Barcelona, Spain (4,000,000) 1,496,266
Beijing, China (7,320,000) 6,690,000
Beirut (Bayrūt), Lebanon (1,675,000) . .509,000
Belfast, N. Ireland, U.K. (730,000) 297,300
Belgrade (Beograd), Serbia and Montenegro 1,594,483
Belo Horizonte, Brazil (4,055,000) . .1,366,301
Berlin, Germany (4,220,000) 3,386,667
Birmingham, England, U.K. (2,705,000) .1,020,589
Bishkek, Kyrgyzstan 753,400
Bogotá, Colombia 6,422,198
Bonn, Germany (600,000) 301,048
Boston, Massachusetts, U.S. (5,819,100) .589,141
Brasília, Brazil 1,947,133
Bratislava, Slovakia 451,395
Brazzaville, Congo 693,712
Brisbane, Australia (1,627,535) 888,449
Brussels (Bruxelles), Belgium (2,390,000) .133,845
Bucharest (București), Romania (2,300,000) .2,016,131
Budapest, Hungary (2,450,000) 1,825,153
Buenos Aires, Argentina (11,000,000) 2,960,976
Cairo (Al Qāhirah), Egypt (9,300,000) .6,800,992
Calgary, Alberta, Canada (951,395) . . .878,866
Cali, Colombia 2,128,920
Canberra, Australia (342,798) 311,518
Cape Town, South Africa (1,900,000) .854,616
Caracas, Venezuela (4,000,000) 1,822,465
Cardiff, Wales, U.K. (645,000) 315,040
Casablanca, Morocco (3,400,000) . . .3,022,000
Changchun, China 2,470,000
Chelyabinsk, Russia (1,320,000) 1,086,300
Chengdu, China 2,760,000
Chennai (Madras), India (6,424,624) .4,216,268
Chicago, Illinois, U.S. (9,157,540) . . .2,896,016
Chișinău (Kishinev), Moldova (746,500) 658,300
Chittagong, Bangladesh (2,342,662) .1,566,070
Chongqing, China 3,870,000
Cincinnati, Ohio, U.S. (1,979,202) 331,285
Cleveland, Ohio, U.S. (2,945,831) 478,403
Cologne (Köln), Germany (1,830,000) .962,507
Colombo, Sri Lanka (2,050,000) 615,000
Conakry, Guinea 950,000
Copenhagen (København), Denmark (2,030,000) 499,148
Córdoba, Argentina (1,260,000) 1,179,067

Cotonou, Benin 650,660
Curitiba, Brazil (2,595,000) 1,586,848
Dakar, Senegal (1,976,533) 879,703
Dalian, China 2,400,000
Dallas, Texas, U.S. (5,221,801) 1,188,580
Damascus (Dimashq), Syria (2,230,000) .1,549,932
Dar es Salaam, Tanzania 1,360,850
Delhi, India (12,791,458) 9,817,439
Denver, Colorado, U.S. (2,581,506) 554,636
Detroit, Michigan, U.S. (5,456,428) 951,270
Dhaka (Dacca), Bangladesh (6,537,308) .3,637,892
Djibouti, Djibouti 329,337
Dnipropetrovs'k, Ukraine (1,590,000) .1,108,682
Donets'k, Ukraine (2,090,000) 1,050,369
Douala, Cameroon 712,251
Dublin (Baile Átha Cliath), Ireland (1,175,000) 481,854
Durban, South Africa (1,740,000) 669,242
Dushanbe, Tajikistan (700,000) 528,600
Düsseldorf, Germany (1,200,000) 568,855
Edinburgh, Scotland, U.K. (640,000) . .448,850
Edmonton, Alberta, Canada (937,845) 666,104
Eşfahān, Iran (1,525,000) 1,266,072
Essen, Germany (5,040,000) 599,515
Fortaleza, Brazil (2,780,000) 788,956
Frankfurt am Main, Germany (1,960,000) .643,821
Fukuoka, Japan (2,000,000) 1,341,489
Geneva (Génève), Switzerland (450,592) 172,598
Glasgow, Scotland, U.K. (1,870,000) . .616,430
Goiânia, Brazil 1,075,761
Guadalajara, Mexico (3,669,021) . . .1,646,183
Guangzhou (Canton), China 3,750,000
Guatemala, Guatemala (1,500,000) .1,006,954
Guayaquil, Ecuador 2,117,553
Halifax, Nova Scotia, Canada (359,183) 119,300
Hamburg, Germany (2,460,000) 1,704,735
Hannover, Germany (1,015,000) 514,718
Hanoi, Vietnam (1,275,000) 1,073,760
Harare, Zimbabwe (1,470,000) 1,189,103
Harbin, China 3,120,000
Havana (La Habana), Cuba (2,285,000) .2,189,716
Helsinki, Finland (939,697) 548,720
Hiroshima, Japan (1,600,000) 1,126,282
Hô Chi Minh City (Saigon), Vietnam (3,300,000) .3,015,743
Hong Kong (Xianggang), China (4,770,000) .1,250,993
Honolulu, Hawaii, U.S. (876,156) 371,657
Houston, Texas, U.S. (4,669,571) 1,953,631
Hyderābād, India (5,533,640) 3,449,878
Ibadan, Nigeria 1,144,000
Islāmābād, Pakistan (*Rawalpindi) . . .529,180
İstanbul, Turkey (8,506,026) 8,260,438
İzmir, Turkey (2,554,363) 2,081,556
Jaipur, India 2,324,319
Jakarta, Indonesia (10,200,000) 9,373,900
Jerusalem (Yerushalayim), Israel (685,000) 633,700
Jiddah, Saudi Arabia 1,450,000
Jinan, China 2,150,000
Johannesburg, South Africa (4,000,000) .752,349
Kābul, Afghanistan 1,424,400
Kampala, Uganda 773,463
Kānpur, India (2,690,486) 2,540,069
Kaohsiung, Taiwan (1,845,000) 1,468,586
Karāchi, Pakistan 9,339,023
Katowice, Poland (2,755,000) 343,158
Kharkiv, Ukraine (1,950,000) 1,494,235
Khartoum (Al Kharṭūm), Sudan (1,450,000) .947,483
Kiev (Kyïv), Ukraine (3,250,000) 2,589,541
Kingston, Jamaica (830,000) 516,500
Kinshasa, Dem. Rep. of the Congo 3,000,000
Kitakyūshū, Japan (1,550,000) 1,011,491
Kolkata (Calcutta), India (13,216,546) 4,580,544
Kuala Lumpur, Malaysia (2,500,000) .1,297,526
Kuwait (Al Kuwayt), Kuwait (1,126,000) 28,859
Lagos, Nigeria (3,800,000) 1,213,000
Lahore, Pakistan 5,143,495
La Paz, Bolivia (1,487,854) 792,611
Libreville, Gabon (418,616) 362,386
Lilongwe, Malawi 435,964

Lima, Peru (6,321,173) 340,422
Lisbon (Lisboa), Portugal (2,350,000) .563,210
Liverpool, England, U.K. (1,515,000) . .467,995
Ljubljana, Slovenia 263,832
Lomé, Togo .450,000
London, England, U.K. (12,000,000) 7,074,265
Los Angeles, California, U.S. (16,373,645) 3,694,820
Luanda, Angola 1,459,900
Lucknow, India (2,266,933) 2,207,340
Lusaka, Zambia 1,269,848
Lyon, France (1,648,216) 445,452
Madrid, Spain (4,690,000) 2,882,860
Managua, Nicaragua 864,201
Manaus, Brazil 1,394,724
Manchester, England, U.K. (2,760,000) 430,818
Manila, Philippines (11,200,000) 1,654,761
Mannheim, Germany (1,525,000) 307,730
Maputo, Mozambique 966,837
Maracaibo, Venezuela 1,249,670
Marseille, France (1,516,340) 798,430
Mashhad, Iran 1,887,405
Mecca (Makkah), Saudi Arabia 630,000
Medan, Indonesia 1,988,200
Medellín, Colombia (2,290,000) 1,885,001
Melbourne, Australia (3,366,542) 67,784
Mexico City (Ciudad de México), Mexico (17,786,983) 8,605,239
Miami, Florida, U.S. (3,876,380) 362,470
Milan (Milano), Italy (3,790,000) 1,305,591
Milwaukee, Wisconsin, U.S. (1,689,572) 596,974
Minneapolis, Minnesota, U.S. (2,968,806) 382,618
Minsk, Belarus (1,680,567) 1,677,137
Mogadishu (Muqdisho), Somalia 600,000
Monrovia, Liberia 465,000
Monterrey, Mexico (3,236,604) 1,110,909
Montevideo, Uruguay (1,650,000) . .1,303,182
Montréal, Quebec, Canada (3,426,350) .1,039,534
Moscow (Moskva), Russia (12,850,000) 8,389,700
Mumbai (Bombay), India (16,368,084) 11,914,398
Munich (München), Germany (1,930,000) .1,194,560
Nagoya, Japan (5,250,000) 2,171,378
Nāgpur, India (2,122,965) 2,051,320
Nairobi, Kenya 2,143,254
Nanjing, China 2,490,000
Naples (Napoli), Italy (3,150,000) . . .1,046,987
N'Djamena, Chad 546,572
Newcastle upon Tyne, England, U.K. (1,350,000) 282,338
New Delhi, India (*Delhi) 294,783
New York, New York, U.S. (21,199,865) 8,008,278
Niamey, Niger 392,165
Nizhniy Novgorod, Russia (1,950,000) .1,364,900
Nouakchott, Mauritania 393,325
Novosibirsk, Russia (1,505,000) 1,402,400
Nürnberg, Germany (1,065,000) 486,628
Odesa, Ukraine (1,150,000) 1,002,246
Omsk, Russia (1,190,000) 1,157,600
Ōsaka, Japan (17,050,000) 2,598,589
Oslo, Norway (773,498) 504,040
Ottawa, Ontario, Canada (1,063,664) 774,072
Ouagadougou, Burkina Faso 634,479
Palembang, Indonesia 1,415,500
Panamá, Panama (995,000) 415,964
Paris, France (11,174,743) 2,125,246
Patna, India (1,707,429) 1,376,950
Perm', Russia (1,110,000) 1,017,100
Perth, Australia (1,244,320) 10,195
Philadelphia, Pennsylvania, U.S. (6,188,463) .1,517,550
Phnom Penh (Phnum Pénh), Cambodia .570,155
Phoenix, Arizona, U.S. (3,251,876) . . .1,321,045
Port Moresby, Papua New Guinea . . .246,664
Port-au-Prince, Haiti (1,425,594) 990,558
Portland, Oregon, U.S. (2,265,223) 529,121
Porto, Portugal (1,230,000) 273,060
Porto Alegre, Brazil (3,375,000) 1,304,998
Prague (Praha), Czech Republic (1,328,000) .1,193,270
Pretoria, South Africa (1,100,000) 692,348
Pune, India (3,755,525) 2,540,069
Pusan, South Korea 3,814,325
P'yŏngyang, North Korea 2,741,260
Qingdao, China 2,300,000

Québec, Quebec, Canada (682,757) . . .169,076
Quezon City, Philippines (*Manila) .1,989,419
Quito, Ecuador 1,615,809
Rabat, Morocco (1,200,000) 717,000
Rangoon (Yangon), Myanmar (2,800,000) .2,705,039
Recife, Brazil (3,160,000) 1,421,993
Regina, Saskatchewan, Canada (192,800) 178,225
Reykjavík, Iceland (166,015) 107,684
Rīga, Latvia (1,000,000) 792,508
Rio de Janeiro, Brazil (10,465,000) . .5,851,914
Riyadh (Ar Riyāḑ), Saudi Arabia . . .1,800,000
Rome (Roma), Italy (3,235,000) 2,649,765
Rosario, Argentina (1,190,000) 894,645
Rostov-na-Donu, Russia (1,160,000) .1,017,300
Rotterdam, Netherlands (1,089,979) . .539,000
Sacramento, California, U.S. (1,796,857) .407,018
St. Louis, Missouri, U.S. (2,603,607) . . .348,189
St. Petersburg (Leningrad), Russia (6,000,000) .4,728,200
Salvador, Brazil (2,855,000) 2,439,823
Samara, Russia (1,450,000) 1,168,000
San Diego, California, U.S. (2,813,833) .1,223,400
San Francisco, California, U.S. (7,039,362) 776,733
San José, Costa Rica (996,194) 309,672
San Juan, Puerto Rico (1,967,627) 421,958
San Salvador, El Salvador (1,908,921) 473,372
Santiago, Chile 4,788,543
Santo Domingo, Dominican Republic .2,677,056
São Paulo, Brazil (17,380,000) 9,713,692
Sapporo, Japan (2,000,000) 1,822,300
Sarajevo, Bosnia and Herzegovina . . .367,703
Saratov, Russia (1,135,000) 881,000
Seattle, Washington, U.S. (3,554,760) 563,374
Seoul (Sŏul), South Korea (15,850,000) 10,231,217
Shanghai, China (11,010,000) 8,930,000
Shenyang (Mukden), China 4,050,000
Singapore, Singapore (4,400,000) . . .4,017,700
Skopje, Macedonia 440,577
Sofia (Sofiya), Bulgaria (1,189,794) .1,138,629
Stockholm, Sweden (1,643,366) 743,703
Stuttgart, Germany (2,020,000) 582,443
Surabaya, Indonesia 2,801,300
Sūrat, India (2,811,466) 2,433,787
Sydney, Australia (3,741,290) 11,115
T'aipei, Taiwan (6,200,000) 2,640,322
Tallinn, Estonia 403,981
Tashkent (Toshkent), Uzbekistan (2,325,000) .2,142,700
Tbilisi, Georgia (1,460,000) 1,279,000
Tegucigalpa, Honduras 576,661
Tehrān, Iran (8,800,000) 6,758,845
Tel Aviv-Yafo, Israel (1,890,000) 348,100
Tianjin (Tientsin), China 5,000,000
Tiranë, Albania 244,153
Tōkyō, Japan (30,300,000) 8,130,408
Toronto, Ontario, Canada (4,682,897) .2,481,494
Tripoli (Ṭarābulus), Libya 1,500,000
Tunis, Tunisia (1,300,000) 702,330
Turin (Torino), Italy (1,550,000) 921,485
Ufa, Russia (1,110,000) 1,088,900
Ulan Bator (Ulaanbaatar), Mongolia 672,882
Ürümqi, China 1,130,000
València, Spain (1,340,000) 739,014
Vancouver, British Columbia, Canada (1,986,965) 545,671
Viangchan (Vientiane), Laos 464,000
Vienna (Wien), Austria (1,950,000) . .1,609,631
Vilnius, Lithuania 578,334
Volgograd (Stalingrad), Russia (1,358,000) .1,000,000
Warsaw (Warszawa), Poland (2,300,000) .1,615,369
Washington, D.C., U.S. (7,608,070) . . .572,059
Wellington, New Zealand (346,500) . . .167,400
Winnipeg, Manitoba, Canada (671,274) 619,544
Wuhan, China 3,870,000
Xi'an, China 2,410,000
Yekaterinburg, Russia (1,530,000) . .1,272,900
Yerevan, Armenia (1,315,000) 1,249,202
Yokohama, Japan (*Tōkyō) 3,426,506
Zagreb, Croatia 867,865
Zürich, Switzerland (932,681) 337,553

Metropolitan area populations are shown in parentheses.
* City is located within the metropolitan area of another city; for example, Yokohama, Japan is located in the Tōkyō metropolitan area.

GLOSSARY OF FOREIGN GEOGRAPHICAL TERMS

Annam — Annamese
Arab — Arabic
Bantu — Bantu
Bur — Burmese
Camb — Cambodian
Celt — Celtic
Chn — Chinese
Czech — Czech
Dan — Danish
Du — Dutch
Fin — Finnish
Fr — French
Ger — German
Gr — Greek
Hung — Hungarian
Ice — Icelandic
India — India
Indian — American Indian
Indon — Indonesian
It — Italian
Jap — Japanese
Kor — Korean
Mal — Malayan
Mong — Mongolian
Nor — Norwegian
Per — Persian
Pol — Polish
Port — Portuguese
Rom — Romanian
Rus — Russian
Siam — Siamese
So. Slav — Southern Slavonic
Sp — Spanish
Swe — Swedish
Tib — Tibetan
Tur — Turkish
Yugo — Yugoslav

å, Nor., Swe — brook, river
aa, Dan., Nor — brook
aas, Dan., Nor — ridge
āb, Per — water, river
abad, India, Per — town, city
ada, Tur — island
adrar, Berber — mountain
air, Indon — stream
akrotírion, Gr — cape
älf, Swe — river
alp, Ger — mountain
altipiano, It — plateau
alto, Sp — height
archipel, Fr — archipelago
archipiélago, Sp — archipelago
arquipélago, Port — archipelago
arroyo, Sp — brook, stream
ås, Nor., Swe — ridge
austral, Sp — southern
baai, Du — bay
bab, Arab — gate, port
bach, Ger — brook, stream
backe, Swe — hill
bad, Ger — bath, spa
bahía, Sp — bay, gulf
bahr, Arab — river, sea, lake
baia, It — bay, gulf
baía, Port — bay
baie, Fr — bay, gulf
bajo, Sp — depression
bak, Indon — stream
bakke, Dan., Nor — hill
balkan, Tur — mountain range
bana, Jap — point, cape
banco, Sp — bank
bandar, Mal., Per. — town, port, harbor
bang, Siam — village
bassin, Fr — basin
batang, Indon., Mal — river
ben, Celt — mountain, summit
bender, Arab — harbor, port
bereg, Rus — coast, shore
berg, Du., Ger., Nor., Swe. — mountain, hill
bir, Arab — well
birkat, Arab — lake, pond, pool
bit, Arab — house
bjaerg, Dan., Nor — mountain
bocche, It — mouth
boğazı, Tur — strait
bois, Fr — forest, wood
boloto, Rus — marsh
bolsón, Sp. — flat-floored desert valley
boreal, Sp — northern
borg, Dan., Nor., Swe — castle, town
borgo, It — town, suburb
bosch, Du — forest, wood
bouche, Fr — river mouth
bourg, Fr — town, borough
bro, Dan., Nor., Swe — bridge
brücke, Ger — bridge
bucht, Ger — bay, bight
bugt, Dan., Nor., Swe — bay, gulf
bulu, Indon — mountain
burg, Du., Ger — castle, town
buri, Siam — town
burun, burnu, Tur — cape
by, Dan., Nor., Swe — village
caatinga, Port. (Brazil) — open brushland
cabezo, Sp — summit
cabo, Port., Sp — cape
campo, It., Port., Sp — plain, field
campos, Port. (Brazil) — plains
cañón, Sp — canyon
cap, Fr — cape
capo, It — cape
casa, It., Port., Sp — house
castello, It., Port — castle, fort
castillo, Sp — castle
càte, Fr — hill
çay, Tur — stream, river
cayo, Sp — rock, shoal, islet
cerro, Sp — mountain, hill
champ, Fr — field
chang, Chn — village, middle
château, Fr — castle
chen, Chn — market town
chiang, Chn — river
chott, Arab — salt lake
chou, Chn. — capital of district; island
chu, Tib — water, stream
cidade, Port — town, city
cima, Sp — summit, peak
città, It — town, city
ciudad, Sp — town, city
cochilha, Port — ridge
col, Fr — pass
colina, Sp — hill
cordillera, Sp — mountain chain
costa, It., Port., Sp — coast
côte, Fr — coast
cuchilla, Sp — mountain ridge
dağ, Tur — mountain(s)
dake, Jap — peak, summit
dal, Dan., Du., Nor., Swe — valley
dan, Kor — point, cape
danau, Indon — lake
dar, Arab — house, abode, country
darya, Per — river, sea
dasht, Per — plain, desert
deniz, Tur — sea
désert, Fr — desert
deserto, It — desert
desierto, Sp — desert
détroit, Fr — strait
dijk, Du — dam, dike
djebel, Arab — mountain
do, Kor — island
dorf, Ger — village
dorp, Du — village
duin, Du — dune
dzong, Tib. — fort, administrative capital
eau, Fr — water
ecuador, Sp — equator
eiland, Du — island
elv, Dan., Nor — river, stream
embalse, Sp — reservoir
erg, Arab — dune, sandy desert
est, Fr., It — east
estado, Sp — state
este, Port., Sp — east
estrecho, Sp — strait
étang, Fr — pond, lake
état, Fr — state
eyjar, Ice — islands
feld, Ger — field, plain
festung, Ger — fortress
fiume, It — river
fjäll, Swe — mountain
fjärd, Swe — bay, inlet
fjeld, Nor — mountain, hill
fjord, Dan., Nor — fiord, inlet
fjördur, Ice — fiord, inlet
fleuve, Fr — river
flod, Dan., Swe — river
flói, Ice — bay, marshland
fluss, Ger — river
foce, It — river mouth
fontein, Du — a spring
forêt, Fr — forest
fors, Swe — waterfall
forst, Ger — forest
fos, Dan., Nor — waterfall
fu, Chn — town, residence
fuente, Sp — spring, fountain
fuerte, Sp — fort
furt, Ger — ford
gang, Kor — stream, river
gangri, Tib — mountain
gat, Dan., Nor — channel
gàve, Fr — stream
gawa, Jap — river
gebergte, Du — mountain range
gebiet, Ger — district, territory
gebirge, Ger — mountains
ghat, India — pass, mountain range
gobi, Mong — desert
gol, Mong — river
göl, gölü, Tur — lake
golf, Du., Ger — gulf, bay
golfe, Fr — gulf, bay
golfo, It., Port., Sp — gulf, bay
gomba, gompa, Tib — monastery
gora, Rus., So. Slav — mountain
góra, Pol — mountain
gorod, Rus — town
grad, Rus., So. Slav — town
guba, Rus — bay, gulf
gundung, Indon — mountain
guntō, Jap — archipelago
gunung, Mal — mountain
haf, Swe — sea, ocean
hafen, Ger — port, harbor
haff, Ger — gulf, inland sea
hai, Chn — sea, lake
hama, Jap — beach, shore
hamada, Arab — rocky plateau
hamn, Swe — harbor
hāmūn, Per — swampy lake, plain
hantō, Jap — peninsula
hassi, Arab — well, spring
haus, Ger — house
haut, Fr — summit, top
hav, Dan., Nor — sea, ocean
havn, Dan., Nor — harbor, port
havre, Fr — harbor, port
háza, Hung — house, dwelling of
heim, Ger — hamlet, home
hem, Swe — hamlet, home
higashi, Jap — east
hisar, Tur — fortress
hissar, Arab — fort
ho, Chn — river
hoek, Du — cape
hof, Ger — court, farmhouse
höfn, Ice — harbor
hoku, Jap — north
holm, Dan., Nor., Swe — island
hora, Czech — mountain
horn, Ger — peak
hoved, Dan., Nor — cape
hsien, Chn — district, district capital
hu, Chn — lake
hügel, Ger — hill
huk, Dan., Swe — point
hus, Dan., Nor., Swe — house
île, Fr — island
ilha, Port — island
indsö, Dan., Nor — lake
insel, Ger — island
insjö, Swe — lake
irmak, irmagi, Tur — river
isla, Sp — island
isola, It — island
istmo, It., Sp — isthmus
järvi, jaur, Fin — lake
jebel, Arab — mountain
jima, Jap — island
jökel, Nor — glacier
joki, Fin — river
jökull, Ice — glacier
kaap, Du — cape
kai, Jap — bay, gulf, sea
kaikyō, Jap — channel, strait
kalat, Per — castle, fortress
kale, Tur — fort
kali, Mal — creek, river
kand, Per — village
kang, Chn — mountain ridge; village
kap, Dan., Ger — cape
kapp, Nor., Swe — cape
kasr, Arab — fort, castle
kawa, Jap — river
kefr, Arab — village
kei, Jap — creek, river
ken, Jap — prefecture
khor, Arab — bay, inlet
khrebet, Rus — mountain range
kiang, Chn — large river
king, Chn — capital city, town
kita, Jap — north
ko, Jap — lake
köbstad, Dan — market-town
kol, Mong — lake
kólpos, Gr — gulf
kong, Chn — river
kopf, Ger — head, summit, peak
köpstad, Swe — market-town
körfezi, Tur — gulf
kosa, Rus — spit
kou, Chn — river mouth
köy, Tur — village
kraal, Du. (Africa) — native village
ksar, Arab — fortified village
kuala, Mal — bay, river mouth
kuh, Per — mountain
kum, Tur — sand
kuppe, Ger — summit
küste, Ger — coast
kyo, Jap — town, capital
la, Tib — mountain pass
labuan, Mal — anchorage, port
lac, Fr — lake
lago, It., Port., Sp — lake
lagoa, Port — lake, marsh
laguna, It., Port., Sp — lagoon, lake
lahti, Fin — bay, gulf
län, Swe — county
landsby, Dan., Nor — village
liehtao, Chn — archipelago
liman, Tur — bay, port
ling, Chn — pass, ridge, mountain
llanos, Sp — plains
loch, Celt. (Scotland) — lake, bay
loma, Sp — long, low hill
lough, Celt. (Ireland) — lake, bay
machi, Jap — town
man, Kor — bay
mar, Port., Sp — sea
mare, It., Rom — sea
marisma, Sp — marsh, swamp
mark, Ger — boundary, limit
massif, Fr — block of mountains
mato, Port — forest, thicket
me, Siam — river
meer, Du., Ger — lake, sea
mer, Fr — sea
mesa, Sp — flat-topped mountain
meseta, Sp — plateau
mina, Port., Sp — mine
minami, Jap — south
minato, Jap — harbor, haven
misaki, Jap — cape, headland
mont, Fr — mount, mountain
montagna, It — mountain
montagne, Fr — mountain
montaña, Sp — mountain
monte, It., Port., Sp. — mount, mountain
more, Rus., So. Slav — sea
morro, Port., Sp — hill, bluff
mühle, Ger — mill
mund, Ger — mouth, opening
mündung, Ger — river mouth
mura, Jap — township
myit, Bur — river
mys, Rus — cape
nada, Jap — sea
nadi, India — river, creek
naes, Dan., Nor — cape
nafud, Arab — desert of sand dunes
nagar, India — town, city
nahr, Arab — river
nam, Siam — river, water
nan, Chn., Jap — south
näs, Nor., Swe — cape
nez, Fr — point, cape
nishi, nisi, Jap — west
njarga, Fin — peninsula
nong, Siam — marsh
noord, Du — north
nor, Mong — lake
nord, Dan., Fr., Ger., It., Nor., Swe — north
norte, Port., Sp — north
nos, Rus — cape
nyasa, Bantu — lake
ö, Dan., Nor., Swe — island
occidental, Sp — western
ocna, Rom — salt mine
odde, Dan., Nor — point, cape
oeste, Port., Sp — west
oka, Jap — hill
oost, Du — east
oriental, Sp — eastern
óros, Gr — mountain
ost, Ger., Swe — east
öster, Dan., Nor., Swe — eastern
ostrov, Rus — island
oued, Arab — river, stream
ouest, Fr — west
ozero, Rus — lake
pää, Fin — mountain
padang, Mal — plain, field
pampas, Sp. (Argentina) — grassy plains
pará, Indian (Brazil) — river
pas, Fr — channel, passage
paso, Sp — mountain pass, passage
passo, It., Port. — mountain pass, passage, strait
patam, India — city, town
pei, Chn — north
pélagos, Gr — open sea
pegunungan, Indon — mountains
peña, Sp — rock
peresheyek, Rus — isthmus
pertuis, Fr — strait
peski, Rus — desert
pic, Fr — mountain peak
pico, Port., Sp — mountain peak
piedra, Sp — stone, rock
ping, Chn — plain, flat
planalto, Port — plateau
planina, Yugo — mountains
playa, Sp — shore, beach
pnom, Camb — mountain
pointe, Fr — point
polder, Du., Ger — reclaimed marsh
polje, So. Slav — plain, field
poluostrov, Rus — peninsula
pont, Fr — bridge
ponta, Port — point, headland
ponte, It., Port — bridge
pore, India — city, town
porthmós, Gr — strait
porto, It., Port — port, harbor
potamós, Gr — river
p'ov, Rus — peninsula
prado, Sp — field, meadow
presqu'ile, Fr — peninsula
proliv, Rus — strait
pu, Chn — commercial village
pueblo, Sp — town, village
puerto, Sp — port, harbor
pulau, Indon — island
punkt, Ger — point
punt, Du — point
punta, It., Sp — point
pur, India — city, town
puy, Fr — peak
qal'a, qal'at, Arab — fort, village
qasr, Arab — fort, castle
rann, India — wasteland
ra's, Arab — cape, head
reka, Rus., So. Slav — river
represa, Port — reservoir
rettō, Jap — island chain
ria, Sp — estuary
ribeira, Port — stream
riberão, Port — river
rio, It., Port — stream, river
río, Sp — river
rivière, Fr — river
roca, Sp — rock
rt, Yugo — cape
rūd, Per — river
saari, Fin — island
sable, Fr — sand
sahara, Arab — desert, plain
saki, Jap — cape
sal, Sp — salt
salar, Sp — salt flat, salt lake
salto, Sp — waterfall
san, Jap., Kor — mountain, hill
sat, satul, Rom — village
schloss, Ger — castle
sebkha, Arab — salt marsh
see, Ger — lake, sea
şehir, Tur — town, city
selat, Indon — stream
selvas, Port. (Brazil) — tropical rain forests
seno, Sp — bay
serra, Port — mountain chain
serranía, Sp — mountain ridge
seto, Jap — strait
severnaya, Rus — northern
shahr, Per — town, city
shan, Chn — mountain, hill, island
shatt, Arab — river
shi, Jap — city
shima, Jap — island
shōtō, Jap — archipelago
si, Chn — west, western
sierra, Sp — mountain range
sjö, Nor., Swe — lake, sea
sö, Dan., Nor — lake, sea
söder, södra, Swe — south
song, Annam — river
sopka, Rus — peak, volcano
source, Fr — a spring
spitze, Ger — summit, point
staat, Ger — state
stad, Dan., Du., Nor., Swe. — city, town
stadt, Ger — city, town
stato, It — state
step', Rus — treeless plain, steppe
straat, Du — strait
strand, Dan., Du., Ger., Nor., Swe — shore, beach
stretto, It — strait
strom, Ger — river, stream
ström, Dan., Nor., Swe. — stream, river
stroom, Du — stream, river
su, suyu, Tur — water, river
sud, Fr., Sp — south
süd, Ger — south
suidō, Jap — channel
sul, Port — south
sund, Dan., Nor., Swe — sound
sungai, sungei, Indon., Mal — river
sur, Sp — south
syd, Dan., Nor., Swe — south
tafelland, Ger — plateau
take, Jap — peak, summit
tal, Ger — valley
tanjung, tanjong, Mal — cape
tao, Chn — island
târg, târgul, Rom — market, town
tell, Arab — hill
teluk, Indon — bay, gulf
terra, It — land
terre, Fr — earth, land
thal, Ger — valley
tierra, Sp — earth, land
tō, Jap — east; island
tonle, Camb — river, lake
top, Du — peak
torp, Swe — hamlet, cottage
tsangpo, Tib — river
tsi, Chn — village, borough
tso, Tib — lake
tsu, Jap — harbor, port
tundra, Rus — treeless arctic plains
tung, Chn — east
tuz, Tur — salt
udde, Swe — cape
ufer, Ger — shore, riverbank
ujung, Indon — point, cape
umi, Jap — sea, gulf
ura, Jap — bay, coast, creek
ust'ye, Rus — river mouth
valle, It., Port., Sp — valley
vallée, Fr — valley
valli, It — lake
vár, Hung — fortress
város, Hung — town
varoš, So. Slav — town
veld, Du — open plain, field
verkh, Rus — top, summit
ves, Czech — village
vest, Dan., Nor., Swe — west
vik, Swe — cove, bay
vila, Port — town
villa, Sp — town
villar, Sp — village, hamlet
ville, Fr — town, city
vostok, Rus — east
wad, wādī, Arab. — intermittent stream
wald, Ger — forest, woodland
wan, Chn., Jap — bay, gulf
weiler, Ger — hamlet, village
westersch, Du — western
wüste, Ger — desert
yama, Jap — mountain
yarimada, Tur — peninsula
yug, Rus — south
zaki, Jap — cape
zaliv, Rus — bay, gulf
zapad, Rus — west
zee, Du — sea
zemlya, Rus — land
zuid, Du — south

ABBREVIATIONS OF GEOGRAPHICAL NAMES AND TERMS

Afg.	Afghanistan
Afr.	Africa
Ak., U.S.	Alaska, U.S.
Al., U.S.	Alabama, U.S.
Alb.	Albania
Alg.	Algeria
Am. Sam.	American Samoa
And.	Andorra
Ang.	Angola
Ant.	Antarctica
Antig.	Antigua and Barbuda
aq.	Aqueduct
Ar., U.S.	Arkansas, U.S.
Arg.	Argentina
Arm.	Armenia
arpt.	Airport
Aus.	Austria
Austl.	Australia
Az., U.S.	Arizona, U.S.
Azer.	Azerbaijan
b.	Bay, Gulf, Inlet, Lagoon
Bah.	Bahamas
Bahr.	Bahrain
Barb.	Barbados
Bdi.	Burundi
Bel.	Belgium
Bela.	Belarus
Ber.	Bermuda
Bhu.	Bhutan
bk.	Undersea Bank
bldg.	Building
Blg.	Bulgaria
Bngl.	Bangladesh
Bol.	Bolivia
Bos.	Bosnia and Herzegovina
Bots.	Botswana
Braz.	Brazil
Bru.	Brunei
Br. Vir. Is.	British Virgin Islands
bt.	Dight
Burkina	Burkina Faso
c.	Cape, Point
Ca., U.S.	California, U.S.
Cam.	Cameroon
Camb.	Cambodia
can.	Canal
Can.	Canada
C.A.R.	Central African Republic
Cay. Is.	Cayman Islands
C. Iv.	Cote d'Ivoire
clf.	Cliff, Escarpment
co.	County, Parish
Co., U.S.	Colorado, U.S.
Col.	Colombia
Com.	Comoros
cont.	Continent
Cook Is.	Cook Islands
C.R.	Costa Rica
Cro.	Croatia
cst.	Coast, Beach
Ct., U.S.	Connecticut, U.S.
C.V.	Cape Verde
Cyp.	Cyprus
Czech Rep.	Czech Republic
d.	Delta
D.C., U.S.	District of Columbia, U.S.
Den.	Denmark
dep.	Dependency, Colony
depr.	Depression
dept.	Department, District
des.	Desert
Dji.	Djibouti
Dom.	Dominica
Dom. Rep.	Dominican Republic
D.R.C.	Democratic Republic of the Congo
Ec.	Ecuador
educ.	Educational Facility
El Sal.	El Salvador
Eng., U.K.	England, U.K.
Eq. Gui.	Equatorial Guinea
Erit.	Eritrea
Est.	Estonia
est.	Estuary
Eth.	Ethiopia
E. Timor	East Timor
Eur.	Europe
Falk. Is.	Falkland Islands
Far. Is.	Faroe Islands
Fin.	Finland
fj.	Fjord
Fl., U.S.	Florida, U.S.
for.	Forest, Moor
Fr.	France
Fr. Gu.	French Guiana
Fr. Poly.	French Polynesia
Ga., U.S.	Georgia, U.S.
Gam.	The Gambia
Gaza	Gaza Strip
Geor.	Georgia
Ger.	Germany
Grc.	Greece
Gren.	Grenada
Grnld.	Greenland
Guad.	Guadeloupe
Guat.	Guatemala
Guern.	Guernsey
Gui.	Guinea
Gui.-B.	Guinea-Bissau
Guy.	Guyana
Hi., U.S.	Hawaii, U.S.
hist.	Historic Site, Ruins
hist. reg.	Historic Region
Hond.	Honduras
Hung.	Hungary
i.	Island
Ia., U.S.	Iowa, U.S.
ice	Ice Feature, Glacier
Ice.	Iceland
Id., U.S.	Idaho, U.S.
Il., U.S.	Illinois, U.S.
In., U.S.	Indiana, U.S.
Indon.	Indonesia
I. of Man	Isle of Man
I.R.	Indian Reservation
Ire.	Ireland
is.	Islands
Isr.	Israel
isth.	Isthmus
Jam.	Jamaica
Jord.	Jordan
Kaz.	Kazakhstan
Kir.	Kiribati
Kor., N.	Korea, North
Kor., S.	Korea, South
Ks., U.S.	Kansas, U.S.
Kuw.	Kuwait
Ky., U.S.	Kentucky, U.S.
Kyrg.	Kyrgyzstan
l.	Lake, Pond
La., U.S.	Louisiana, U.S.
Lat.	Latvia
Leb.	Lebanon
Leso.	Lesotho
Lib.	Liberia
Liech.	Liechtenstein
Lith.	Lithuania
Lux.	Luxembourg
Ma., U.S.	Massachusetts, U.S.
Mac.	Macedonia
Madag.	Madagascar
Malay.	Malaysia
Mald.	Maldives
Marsh. Is.	Marshall Islands
Mart.	Martinique
Maur.	Mauritania
May.	Mayotte
Md., U.S.	Maryland, U.S.
Me., U.S.	Maine, U.S.
Mex.	Mexico
Mi., U.S.	Michigan, U.S.
Micron.	Micronesia, Federated States of
Mn., U.S.	Minnesota, U.S.
Mo., U.S.	Missouri, U.S.
Mol.	Moldova
Mong.	Mongolia
Monts.	Montserrat
Mor.	Morocco
Moz.	Mozambique
Ms., U.S.	Mississippi, U.S.
Mt., U.S.	Montana, U.S.
mth.	River Mouth or Channel
mtn.	Mountain
mts.	Mountains
Mwi.	Malawi
Mya.	Myanmar
N.A.	North America
N.C., U.S.	North Carolina, U.S.
N. Cal.	New Caledonia
N.D., U.S.	North Dakota, U.S.
Ne., U.S.	Nebraska, U.S.
neigh.	Neighborhood
Neth.	Netherlands
Neth. Ant.	Netherlands Antilles
N.H., U.S.	New Hampshire, U.S.
Nic.	Nicaragua
Nig.	Nigeria
N. Ire., U.K.	Northern Ireland, U.K.
N.J., U.S.	New Jersey, U.S.
N.M., U.S.	New Mexico, U.S.
N. Mar. Is.	Northern Mariana Islands
Nmb.	Namibia
Nor.	Norway
Nv., U.S.	Nevada, U.S.
N.Y., U.S.	New York, U.S.
N.Z.	New Zealand
o.	Ocean
Oc.	Oceania
Oh., U.S.	Ohio, U.S.
Ok., U.S.	Oklahoma, U.S.
Or., U.S.	Oregon, U.S.
p.	Pass
Pa., U.S.	Pennsylvania, U.S.
Pak.	Pakistan
Pan.	Panama
Pap. N. Gui.	Papua New Guinea
Para.	Paraguay
pen.	Peninsula
Phil.	Philippines
Pit.	Pitcairn
pl.	Plain, Flat
plat.	Plateau, Highland
Pol.	Poland
Port.	Portugal
P.R.	Puerto Rico
prov.	Province, Region
pt. of i.	Point of Interest
r.	River, Creek
Reu.	Reunion
rec.	Recreational Site, Park
reg.	Physical Region
rel.	Religious Institution
res.	Reservoir
rf.	Reef, Shoal
R.I., U.S.	Rhode Island, U.S.
Rom.	Romania
Rw.	Rwanda
S.A.	South America
S. Afr.	South Africa
Sau. Ar.	Saudi Arabia
S.C., U.S.	South Carolina, U.S.
sci.	Scientific Station
Scot., U.K.	Scotland, U.K.
S.D., U.S.	South Dakota, U.S.
sea feat.	Undersea Feature
Sen.	Senegal
Serb.	Serbia and Montenegro
Sey.	Seychelles
S. Geor.	South Georgia
Sing.	Singapore
S.L.	Sierra Leone
Slvk.	Slovakia
Slvn.	Slovenia
S. Mar.	San Marino
Sol. Is.	Solomon Islands
Som.	Somalia
Sp. N. Afr.	Spanish North Africa
Sri L	Sri Lanka
St. Hel.	St. Helena
St. K./N.	St. Kitts and Nevis
St. Luc.	St. Lucia
St. P./M.	St. Pierre and Miquelon
strt.	Strait, Channel, Sound
S. Tom./P.	Sao Tome and Principe
St. Vin.	St. Vincent and the Grenadines
Sur.	Suriname
Sval.	Svalbard
sw.	Swamp, Marsh
Swaz.	Swaziland
Swe.	Sweden
Switz.	Switzerland
Tai.	Taiwan
Taj.	Tajikistan
Tan.	Tanzania
T./C. Is.	Turks and Caicos Islands
ter.	Territory
Thai.	Thailand
Tn., U.S.	Tennessee, U.S.
trans.	Transportation Facility
Trin.	Trinidad and Tobago
Tun.	Tunisia
Tur.	Turkey
Turkmen.	Turkmenistan
Tx., U.S.	Texas, U.S.
U.A.E.	United Arab Emirates
Ug.	Uganda
U.K.	United Kingdom
Ukr.	Ukraine
Ur.	Uruguay
U.S.	United States
Ut., U.S.	Utah, U.S.
Uzb.	Uzbekistan
Va., U.S.	Virginia, U.S.
val.	Valley, Watercourse
Ven.	Venezuela
Viet.	Vietnam
V.I.U.S.	Virgin Islands (U.S.)
vol.	Volcano
Vt., U.S.	Vermont, U.S.
Wa., U.S.	Washington, U.S.
W.B.	West Bank
Wi., U.S.	Wisconsin, U.S.
W. Sah.	Western Sahara
wtfl.	Waterfall
W.V., U.S.	West Virginia, U.S.
Wy., U.S.	Wyoming, U.S.
Zam.	Zambia
Zimb.	Zimbabwe

PRONUNCIATION OF GEOGRAPHICAL NAMES

Key to the Sound Values of Letters and Symbols Used in the Index to Indicate Pronunciation

ă-ăt; băttle
ā-fīnăl; appeăl
ā-rāte; elāte
å-senăte; inanimåte
ä-ärm; cälm
à-àsk; båth
a-sofă; mărine (short neutral or indeterminate sound)
â-fâre; prepâre
ch-choose; church
dh-as th in other; either
ē-bē; ēve
ė-ėvent; crėate
ĕ-bĕt; ĕnd
ĕ-recĕnt (short neutral or indeterminate sound)
ẽ-cratẽr; cindẽr
g-gō; gāme
gh-guttural g
ĭ-bĭt; wĭll
ĭ-(short neutral or indeterminate sound)
ī-rīde; bīte
ᴋ-gutteral k as ch in German ich
ng-sing
ŋ-bank; linger
ɴ-indicates nasalized
ŏ-nŏd; ŏdd
ŏ-cŏmmit; cŏnnect
ō-ōld; bōld
ô-ôbey; hôtel
ô-ôrder; nôrth
oi-boil
ōō-tōōd; rōōt
o-as oo in foot; wood
ou-out; thou
s-soft; so; sane
sh-dish; finish
th-thin; thick
ū-pūre; cūre
ů-ůnite; ůsŭrp
û-ûrn; fûr
ŭ-stŭd; ŭp
ŭ-circŭs; sŭbmit
ü-as in French tu
zh-as z in azure
'-indeterminate vowel sound

In many cases the spelling of foreign geographical names does not even remotely indicate the pronunciation to an American, i.e., Słupsk in Poland is pronounced swôpsk; Jujuy in Argentina is pronounced hōōhwē'; La Spezia in Italy is lä-spē'zyä.

This condition is hardly surprising, however, when we consider that in our own language Worcester, Massachusetts, is pronounced wôs'tẽr; Sioux City, Iowa, sōō sī'ti; Schuylkill Haven, Pennsylvania, skōōl'kĭl hā-vĕn; Poughkeepsie, New York, pŏ-kĭp'sĕ.

The indication of pronunciation of geographic names presents several peculiar problems:

1. Many foreign tongues use sounds that are not present in the English language and which an American cannot normally articulate. Thus, though the nearest English equivalent sound has been indicated, only approximate results are possible.

2. There are several dialects in each foreign tongue which cause variation in the local pronunciation of names. This also occurs in identical names in the various divisions of a great language group, as the Slavic or the Latin.

3. Within the United States there are marked differences in pronunciation, not only of local geographic names, but also of common words, indicating that the sound and tone values for letters as well as the placing of the emphasis vary considerably from one part of the country to another.

4. A number of different letters and diacritical combinations could be used to indicate essentially the same or approximate pronunciations.

Some variation in pronunciation other than that indicated in this index may be encountered, but such a difference does not necessarily indicate that either is in error, and in many cases it is a matter of individual choice as to which is preferred. In fact, an exact indication of pronunciation of many foreign names using English letters and diacritical marks is extremely difficult and sometimes impossible.

PRONOUNCING INDEX

This universal index includes in a single alphabetical list approximately 4,000 names of features that appear on the reference maps. Each name is followed by a page reference and geographical coordinates.

Abbreviation and Capitalization Abbreviations of names on the maps have been standardized as much as possible. Names that are abbreviated on the maps are generally spelled out in full in the index. Periods are used after all abbreviations regardless of local practice. The abbreviation "St." is used only for "Saint". "Sankt" and other forms of this term are spelled out.

Most initial letters of names are capitalized, except for a few Dutch names, such as "s-Gravenhage". Capitalization of noninitial words in a name generally follows local practice.

Alphabetization Names are alphabetized in the order of the letters of the English alphabet. Spanish *ll* and *ch*, for example, are not treated as direct letters. Furthermore, diacritical marks are disregarded in alphabetization — German or Scandinavian *ä* or *ö* are treated as *a* or *o*.

The names of physical features may appear inverted, since they are always alphabetized under the proper, not the generic, part of the name, thus: "Gibraltar, Strait of". Otherwise every entry, whether consisting of one word or more, is alphabetized as a single continuous entity. "Lakeland", for example, appears after "La Crosse" and before "La Salle". Names beginning with articles (Le Harve, Den Helder, Al Manāmah, Ad Dawhah) are not inverted.

In the case of identical names, towns are listed first, then political divisions, then physical features.

Generic Terms Except for cities, the names of all features are followed by terms that represent broad classes of features, for example, Mississippi, r. or Alabama, state. A list of all abbreviations used in the index is on page 37.

Country names and the names of features that extend beyond the boundaries of one county are followed by the name of the continent in which each is located. Country designations follow the names of all other places in the index. The locations of places in the United States and the United Kingdom are further defined by abbreviations that include the state or political division in which each is located.

Pronunciations Pronunciations are included for most names listed. An explanation of the pronunciation system used appears on page 37.

Page References and Geographical Coordinates The geographical coordinates and page references are found in the last columns of each entry.

If a page contains several maps or insets, a lowercase letter identifies the specific map or inset.

Latitude and longitude coordinates for point features, such as cities and mountain peaks, indicate the location of the symbols. For extensive areal features, such as countries or mountain ranges, or linear features, such as canals and rivers, locations are given for the position of the type as it appears on the map.

PLACE (Pronunciation)	PAGE	LAT.	LONG.

A

PLACE (Pronunciation)	PAGE	LAT.	LONG.
Abaetetuba, Braz. (ä´bȧĕ-tē-tōō´bä)	19	1°44´s	48°45´w
Abancay, Peru (ä-bän-kä´ē)	18	13°44´s	72°46´w
Abasolo, Mex. (ä-bä-sō´lō)	6	24°05´n	98°24´w
Abasolo, Mex. (ä-bä-sō´lō)	2	27°13´n	101°25´w
Abraão, Braz. (äbrä-oun´)	17a	23°10´s	44°10´w
Abraham's Bay, b., Bah.	11	22°20´n	73°50´w
Abrolhos, Arquipélago dos, is., Braz.	19	17°58´s	38°40´w
Abunã, r., S.A. (ä-bōō-nä´)	18	10°25´s	67°00´w
Acaclas, Col. (a-ka´sēäs)	18a	3°59´n	73°44´w
Acajutla, El Sal. (ä-kä-hōōt´lä)	8	13°37´n	89°50´w
Acala, Mex. (ä-kä´lä)	7	16°38´n	92°49´w
Acámbaro, Mex. (ä-käm´bä-rō)	6	20°03´n	100°42´w
Acancéh, Mex. (ä-kän-sě´)	8a	20°50´n	89°27´w
Acapetlahuaya, Mex. (ä-kä-pĕt´lä-hwä´yä)	6	18°24´n	100°04´w
Acaponeta, Mex. (ä-kä-pō-nā´tä)	6	22°31´n	105°25´w
Acaponeta, r., Mex. (ä-kä-pō-nä´tä)	6	22°47´n	105°23´w
Acapulco, Mex. (ä-kä-pōōl´kō)	4	16°49´n	99°57´w
Acaraí Mountains, mts., S.A.	19	1°30´n	57°40´w
Acarigua, Ven. (äkä-rē´gwä)	18	9°29´n	69°11´w
Acatlán de Osorio, Mex. (ä-kät-län´dä ō-sō´rē-ō)	6	18°11´n	98°04´w
Acatzingo de Hidalgo, Mex.	7	18°58´n	97°4l´w
Acayucan, Mex. (ä-kä-yōō´kän)	7	17°56´n	94°55´w
Achacachi, Bol. (ä-chä-kä´chē)	18	16°11´s	68°32´w
Acklins, i., Bah. (äk´lĭns)	5	22°30´n	73°55´w
Acklins, The Bight of, b., Bah. (äk´lĭns)	11	22°35´n	74°20´w
Acolmán, Mex. (ä-kôl-má´n)	7a	19°38´n	98°56´w
Aconcagua, prov., Chile (ä-kōn-kä´gwä)	17b	32°20´s	71°00´w
Aconcagua, r., Chile (ä-kōn-kä´gwä)	17b	32°43´s	70°53´w
Aconcagua, Cerro, mtn., Arg. (ä-kōn-kä´gwä)	20	32°30´s	70°00´w
Acoyapa, Nic. (ä-kō-yä´pä)	8	11°54´n	85°11´w
Acre, state, Braz. (ä´krä)	18	8°40´s	70°45´w
Acre, r., S.A.	18	10°33´s	68°34´w
Actopán, Mex. (äk-tō-pän´)	6	20°16´n	98°57´w
Actópan, r., Mex. (äk-tō´pän)	7	19°25´n	96°37´w
Acuitzio del Canje, Mex. (ä-kwēt´zē-ō dĕl kän´há)	6	19°28´n	101°21´w
Acul, Baie de l', b., Haiti (ä-kōōl´)	11	19°55´n	72°20´w
Afonso Claudio, Braz. (äl-fōn´sō-klou´dēō)	17a	20°05´s	41°05´w
Agalta, Cordillera de, mts., Hond. (kôr´dĕl yě´rä dě ä gäl´tä)	8	15°15´n	85°42´w
Agua, vol., Guat. (ä´gwä)	8	14°28´n	90°43´w
Agua Blanca, Río, r., Mex. (rě´ō-ä-gwä-blä´n-kä)	6	21°46´n	102°54´w
Agua Brava, Laguna de, l., Mex.	6	22°04´n	105°40´w
Aguada, Cuba (ä gwä´dȧ)	10	22°25´n	80°50´w
Aguada, I., Mex. (ä gwä´dȧ)	8a	10°40´n	89°40´w
Aguadas, Col. (ä-gwä´däs)	18	5°37´n	75°27´w
Aguadilla, P.R. (ä-gwä-dēl´yä)	5b	18°27´n	67°10´w
Aguadulce, Pan. (ä-gwä-dōōl´sä)	9	8°15´n	80°33´w
Agua Escondida, Meseta de, plat., Mex.	7	16°54´n	91°35´w
Aguai, Braz. (ägwa-e´)	17a	22°04´s	46°57´w
Agualeguas, Mex. (ä-gwä-lā´gwäs)	2	26°19´n	99°33´w
Aguán, r., Hond. (ä-gwä´n)	8	15°22´n	87°00´w
Aguanaval, r., Mex. (ä-guä-nä-väl´)	2	25°12´n	103°28´w
Aguascalientes, Mex. (ä´gwäs-käl-yěn´tás)	4	21°52´n	102°17´w
Aguascalientes, state, Mex. (ä´gwäs-käl-yěn´tás)	6	22°00´n	102°18´w
Aguililla, Mex. (ä-gē-lēl-yä)	6	18°44´n	102°44´w
Aguililla, r., Mex. (ä-gē-lēl-yä)	6	18°30´n	102°48´w
Aguja, Punta, c., Peru (pūn´tä ä-gōō´ hä)	18	6°00´s	81°15´w
Ahuacatlán, Mex. (ä-wä-kät-län´)	6	21°05´n	104°28´w
Ahuachapán, El Sal. (ä-wä-chä-pän´)	8	13°57´n	89°53´w
Ahualulco, Mex. (ä-wä-lōōl´kō)	6	20°43´n	103°57´w
Ahuatempan, Mex. (ä-wä-těm-pän)	6	18°11´n	98°02´w
Aimores, Serra dos, mts., Braz. (sě´r-rä-dôs-ī-mō-rě´s)	19	17°40´s	42°38´w
Aipe, Col. (ī´pě)	18a	3°13´n	75°15´w
Aiuruoca, Braz. (äē´ōō-rōōō´-kä)	17a	21°57´s	44°36´w
Aiuruoca, r., Braz.	17a	22°11´s	44°35´w
Ajalpan, Mex. (ä-häl´pän)	7	18°21´n	97°14´w
Ajuchitlán del Progreso, Mex. (ä-hōō-chet-län)	6	18°11´n	100°32´w
Ajusco, Mex. (ä-hōō´s-kō)	7a	19°13´n	99°12´w
Ajusco, Cerro, mtn., Mex. (sě´r-rō-ä-hōō´s-kō)	7a	19°12´n	99°16´w
Alacranes, Cuba (ä-lä-krä´nás)	10	22°45´n	81°35´w
Alagôas, state, Braz. (ä-lä-gō´äzh)	19	9°50´s	36°33´w
Alagoinhas, Braz. (a-lä-gō-ēn´yäzh)	19	12°13´s	38°12´w
Alahuatán, r., Mex. (ä-lä-wä-tá´n)	6	18°30´n	100°00´w
Alajuela, C.R. (ä-lä-hwa´lä)	9	10°01´n	84°14´w
Alajuela, Lago, l., Pan. (ä-lä-hwa´lä)	4a	9°15´n	79°34´w
Álamo, Mex. (ä´lä-mô)	7	20°55´n	97°41´w
Álamo, r., Mex. (ä´lä-mô)	2	26°33´n	99°35´w
Alaquines, Mex. (ä-lä-kē´nás)	6	22°07´n	99°35´w
Alberti, Arg. (äl-bě´r-tē)	17c	35°01´s	60°16´w
Albina, Sur. (äl-bē´nä)	19	5°30´n	54°33´w
Albuquerque, Cayos de, is., Col.	9	12°12´n	81°24´w
Alcântara, Braz. (äl-kän´tä-rä)	19	2°17´s	44°29´w
Alcorta, Arg. (äl-kōr´tä)	17c	33°32´s	61°08´w
Aldama, Mex. (al-dä´mä)	6	22°54´n	98°04´w
Aldama, Mex. (äl-dä´mä)	2	28°50´n	105°54´w
Alegre, Braz. (älě´grě)	17a	20°41´s	41°32´w
Alegre, r., Braz. (älě´grě)	20b	22°22´s	43°34´w
Alegrete, Braz. (ä-lä-grä´tä)	20	29°46´s	55°44´w
Alemán, Presa, res., Mex. (prä sä-lě-mä´ń)	7	18°20´n	96°35´w
Alem Paraíba, Braz. (ä-lě´m-pá-räē´bä)	17a	21°54´s	42°40´w
Alenquer, Braz. (ä-lěn-kěr´)	19	1°58´s	54°44´w
Alfenas, Braz. (äl-fē´näs)	17a	21°26´s	45°55´w
Allende, Mex. (äl-yěn´dä)	7	18°23´n	92°49´w
Allende, Mex.	2	28°20´n	100°50´w
All Pines, Belize (ól pĭnz)	8a	16°55´n	88°15´w
Almagre, Laguna, l., Mex. (lä-gó´nä-äl-mä´grě)	7	23°48´n	97°45´w
Almirante, Pan. (äl-mē-rän´tä)	9	9°18´n	82°24´w
Almirante, Bahía de, b., Pan.	9	9°22´n	82°07´w
Almoloya, Mex. (äl-mō-lō´yä)	6	19°32´n	99°44´w
Almoloya, Mex. (äl-mō-lō´yä)	7a	19°11´n	99°28´w
Alpujarra, Col. (äl-pōō-kä´rä)	18a	3°23´n	74°56´w
Alta Gracia, Arg. (äl´tä grä´sě-a)	20	31°41´s	64°19´w
Altagracia, Ven.	18	10°42´n	71°34´w
Altagracia de Orituco, Ven.	19b	9°53´n	66°22´w
Altamira, Braz. (äl-tä-mē´rä)	19	3°13´s	52°14´w
Altamira, Mex.	7	22°25´n	97°55´w
Altamirano, Arg. (äl-tä-mē-rä´nō)	20	35°26´s	58°12´w
Altiplano, pl., Bol. (äl-tē-plá´nō)	18	18°38´s	68°20´w
Alto Marañón, r., Peru (äl´tō-mä-rän-yō´n)	18	8°18´s	77°13´w
Alto Rio Doce, Braz. (äl´tō-rē´ō-dō´sě)	17a	21°02´s	43°23´w
Alto Songo, Cuba (äl-tō-sôn´gō)	11	20°10´n	75°45´w
Altotonga, Mex. (äl-tō-tôn´gä)	7	19°44´n	97°13´w
Alto Velo, i., Dom. Rep. (äl-tō-vě´lō)	11	17°30´s	71°35´w
Alvarado, Mex. (äl-vä-rä´dhō)	7	18°48´n	95°45´w
Alvarado, Laguna de, l., Mex. (lä-gó´nä-dě-äl-vä-rä´dō)	7	18°44´n	95°45´w
Alvinópolis, Braz. (äl-vēnō´pō-lēs)	17a	20°07´s	43°03´w
Amacuzac, r., Mex. (a-ma-kōō-zäk´)	6	18°00´n	99°03´w
Amalfi, Col. (ä´mä´l-fē)	18a	6°55´n	75°04´w
Amambai, Serra de, mts., S.A.	19	20°06´s	57°08´w
Amapala, Hond. (ä-mä-pä´lä)	8	13°16´n	87°39´w
Amarante, Braz. (ä-mä-rän´tä)	19	6°17´s	42°43´w
Amatenango, Mex. (ä-mä-tä-nan´gō)	7	16°30´n	92°29´w
Amatique, Bahía de, b., N.A. (bä-ē´ä-dě-ä-mä-tē´kä)	8	15°58´n	88°50´w
Amatitlán, Guat. (ä-mä-tē-tlän´)	8	14°27´n	90°39´w
Amatlán de Cañas, Mex. (ä-mät-län´dä kän-yäs)	6	20°50´n	104°22´w
Amazon (Amazonas) (Solimões), r., S.A.	19	2°03´s	53°18´w
Amazonas, state, Braz. (ä-mä-thō´näs)	18	4°15´s	64°30´w
Ambalema, Col. (äm-bä-lā´mä)	18	4°47´n	74°45´w
Ambato, Ec. (äm-bä´tō)	18	1°15´s	70°30´w
Ambergris Cay, i., Belize (äm´běr-grēs kāz)	8a	18°04´n	87°43´w
Ambergris Cays, is., T./C. Is.	11	21°20´n	71°40´w
Amealco, Mex. (ä-mā-äl´kō)	6	20°12´n	100°08´w
Ameca, Mex. (ä-mē´kä)	4	20°34´n	104°02´w

ät; finäl; räte; senäte; ärm; äsk; sofä; färe; ch-choose; dh-as th in other; bē; ěvent; bět; recěnt; cratěr; g-gō; gh-guttural g; bǐt; ī-short neutral; rīde; к-guttural k as ch in German ich;

ăt; fĭnȧl; rāte; senâte; ärm; ȧsk; sofȧ; fâre; ch-choose; dh-as th in other; bē; ĕvent; bĕt; recĕnt; cratèr; g-gō; gh-guttural g; bĭt; ĭ-short neutral; rīde; ᴋ-guttural k as ch in German ich;

PLACE (Pronunciation)	PAGE	LAT.	LONG.
Bluefields, Nic. (blo͞o′fēldz)	5	12°03′N	83°45′W
Blue Mountains, mts., Jam.	10	18°05′N	76°35′W
Blumenau, Braz. (blo͞o′mĕn-ou)	20	26°53′S	48°58′W
Boaco, Nic. (bô-ä′kō)	8	12°24′N	85°41′W
Boa Vista do Rio Branco, Braz.	19	2°46′N	60°45′W
Boca del Pozo, Ven. (bô-ka-del-pô′zô)	19b	11°00′N	64°21′W
Boca de Uchire, Ven. (bô-kä-dĕ-o͞o-chē′rĕ)	19b	10°09′N	65°27′W
Bocaina, Serra da, mtn., Braz. (sĕ′r-rä-dȧ-bô-kä′ē-nä)	17a	22°47′S	44°39′W
Bocas, Mex. (bō′käs)	6	22°29′N	101°03′W
Bocas del Toro, Pan. (bō′käs dĕl tō′rō)	9	9°24′N	82°15′W
Boggy Peak, mtn., Antig. (bŏg′ĭ-pēk)	9b	17°03′N	61°50′W
Bogotá, Col. (Santa Fé de Bogotá)	18	4°36′N	74°05′W
Boguete, Pan. (bô-gĕ′tĕ)	9	8°54′N	82°29′W
Bohom, Mex. (bŏ-ô′m)	7	16°47′N	92°42′W
Bolaños, Mex. (bō-län′yŏs)	6	21°40′N	103°48′W
Bolaños, r., Mex.	6	21°26′N	103°54′W
Bolívar, Arg. (bô-lē′vär)	20	36°15′S	61°05′W
Bolívar, Col.	18	1°46′N	76°58′W
Bolívar, Pico, mtn., Ven.	18	8°44′N	70°54′W
Bolivia, nation, S.A. (bô-lĭv′ĭ-ȧ)	18	17°00′S	64°00′W
Bolonchenticul, Mex. (bô-lôn-chĕn-tē-ko͞o′l)	8a	20°03′N	89°47′W
Bolondrón, Cuba (bô-lôn-drōn′)	10	22°45′N	81°25′W
Bom Jardim, Braz. (bôn zhär-dēɴ′)	17a	22°10′S	42°25′W
Bom Jesus do Itabapoana, Braz.	17a	21°08′S	41°51′W
Bom Sucesso, Braz. (bôn-so͞o-sĕ′sŏ)	17a	21°02′S	44°44′W
Bonaire, i., Neth. Ant. (bô-nâr′)	18	12°10′N	68°15′W
Bonds Cay, i., Bah. (bŏnds kē)	10	25°30′N	77°45′W
Bonete, Cerro, mtn., Arg. (bô′nĕtĕh çĕrrô)	20	27°50′S	68°35′W
Bonfim, Braz. (bôɴ-fē′ɴ)	17a	20°20′S	44°15′W
Bonhomme, Pic, mtn., Haiti	11	19°10′N	72°20′W
Booby Rocks, is., Bah. (bo͞o′bĭ rŏks)	10	23°55′N	77°00′W
Borba, Braz. (bôr′bä)	19	4°23′S	59°31′W
Borborema, Planalto da, plat., Braz. (plä-nȧl′tô-dä-bôr-bō-rĕ′mä)	19	7°35′S	36°40′W
Borracha, Isla la, i., Ven. (ē′s-lä-lȧ-bôr-rȧ′chä)	19b	10°18′N	64°44′W
Braço Maior, mth., Braz.	19	11°00′S	51°00′W
Braço Menor, mth., Braz. (brä′zô-mĕ-nō′r)	19	11°38′S	50°00′W
Bragado, Arg. (brä-gä′dō)	20	35°07′S	60°28′W
Bragança, Braz. (brä-gän′sä)	19	1°02′S	46°50′W
Bragança Paulista, Braz. (brä-gän′sä-pä′o͞o-lē′s-tȧ)	20	22°58′S	46°31′W
Branca, Pedra, mtn., Braz. (pĕ′drä brȧ′ɴ-kä)	20b	22°55′S	43°28′W
Branco, r., Braz. (bräŋ′kō)	19	2°21′N	60°38′W
Brasília, Braz. (brä-sē′lvä)	19	15°49′S	47°39′W
Brasilia Legal, Braz.	19	3°45′S	55°46′W
Brasópolis, Braz. (brä-sô′pô-lēs)	17a	22°30′S	45°36′W
Brazil, nation, S.A.	19	9°00′S	53°00′W
Brazilian Highlands, mts., Braz. (brȧ zĭl yăn hī-lăndz)	15	14°00′S	48°00′W
Brejo, Braz. (brå′zhô)	19	3°33′S	42°46′W
Breves, Braz. (brā′vĕzh)	19	1°32′S	50°13′W
Brewster, Cerro, mtn., Pan. (sĕ′r-rô-bro͞o′stĕr)	9	9°19′N	79°15′W
Bridge Point, c., Bah. (brĭj)	10	25°35′N	76°40′W
Bridgetown, Barb. (brĭj′ toun)	5	13°08′N	59°37′W
Broa, Ensenada de la, b., Cuba	10	22°30′N	82°00′W
Brothers, is., Bah. (brŭd′hērs)	10	26°05′N	79°00′W
Brown Bank, bk.	11	21°30′N	74°35′W
Brunswick, Península de, pen., Chile	20	53°25′S	71°15′W
Brusque, Braz. (bro͞o′s-ko͞oĕ)	20	27°15′S	48°45′W
Bucaramanga, Col. (bo͞o-kä′rä-mäŋ′gä)	18	7°12′N	73°14′W
Buenaventura, Col. (bwā′nä-vĕn-to͞o′rȧ)	18	3°46′N	77°09′W
Buenaventura, Cuba	11a	22°53′N	82°22′W
Buenaventura, Bahía de, b., Col.	18	3°45′N	79°23′W
Buena Vista, Bahía, b., Cuba (bä-ē′ä-bwĕ-nä-vē′s-tä)	10	22°30′N	79°10′W
Buenos Aires, Arg. (bwā′nōs ī′rās)	20	34°20′S	58°30′W
Buenos Aires, Col.	18a	3°01′N	76°34′W
Buenos Aires, C.R.	9	9°10′N	83°21′W
Buenos Aires, prov., Arg.	20	30°15′S	61°45′W
Buenos Aires, l., S.A.	20	46°30′S	72°15′W
Buga, Col. (bo͞o′gä)	18	3°54′N	76°17′W
Buin, Chile (bô-ēn′)	17b	33°44′S	70°44′W
Bull Head, mtn., Jam.	10	18°10′N	77°15′W
Burgos, Mex. (bôr′gōs)	2	24°57′N	98°47′W
Burica, Punta, c., N.A. (po͞o′n-tä-bo͞o′rē-kä)	9	8°02′N	83°12′W
Burro, Serranías del, mts., Mex. (sĕr-rä-nē′äs dĕl bo͞o′r-rô)	2	29°39′N	102°07′W
Burzaco, Arg. (bo͞or-zȧ′kô)	20a	34°50′S	58°23′W
Bustamante, Mex. (bo͞os-tȧ-män′tȧ)	2	26°34′N	100°30′W

C

PLACE (Pronunciation)	PAGE	LAT.	LONG.
Caballones, Canal de, strt., Cuba (kä-nä′l-dĕ-kä-bäl-yô′nĕs)	10	20°45′N	79°20′W
Cabedelo, Braz. (kä-bē-dā′lô)	19	6°58′S	34°49′W
Cabeza, Arrecife, i., Mex.	7	19°07′N	95°52′W
Cabimas, Ven. (kä-bē′mäs)	18	10°21′N	71°27′W
Cabo Frio, Braz. (kä′bô-frē′ô)	17a	22°53′S	42°02′W
Cabo Frio, Ilha do, Braz. (ē′lä-dô-kä′bô frē′ô)	17a	23°01′S	42°00′W
Cabo Gracias a Dios, Hond. (kȧ′bô-grä-syäs-ä-dyô′s)	9	15°00′N	83°13′W
Cabuçu, r., Braz. (kä-bo͞o′-so͞o)	20b	22°57′S	43°36′W
Caçapava, Braz. (kä′sä-pȧ′vä)	17a	23°05′S	45°52′W
Cáceres, Braz. (kȧ′sĕ-rĕs)	19	16°11′S	57°32′W
Cachapoal, r., Chile (kä-chä-pô-ȧ′l)	17b	34°23′S	70°19′W
Cachi, Nevados de, mtn., Arg. (nĕ-vȧ′dôs-dĕ-kȧ′chē)	20	25°05′S	66°40′W
Cochinal, Chile (kä-chē-näl′)	20	24°57′S	69°33′W
Cachoeira, Braz. (kä-shô-ā′rä)	19	12°32′S	38°47′W
Cachoeirá do Sul, Braz. (kä-shô-ā′rä-dô-so͞o′l)	20	30°02′S	52°49′W
Cachoeiras de Macacu, Braz. (kä-shô-ā′räs-dĕ-mä-kȧ′ko͞o)	17a	22°28′S	42°39′W
Cachoeiro de Itapemirim, Braz.	19	20°51′S	41°06′W
Cadereyta, Mex. (kä-då-rā′tä)	6	20°42′N	99°47′W
Cadereyta Jiménez, Mex. (kä-dä-rā′tä hē-mā′nāz)	2	25°36′N	99°59′W
Cagua, Ven. (kä′gwä)	19b	10°12′N	67°27′W
Caguas, P.R. (kä′gwäs)	5b	18°12′N	66°01′W
Cahuacán, Mex. (kä-wä-kä′n)	7a	19°38′N	99°25′W
Cahuita, Punta, c., C.R. (po͞o′n-tä-kä-wē′tȧ)	9	9°47′N	82°41′W
Caibarién, Cuba (kī-bä-rē-ĕn′)	10	22°35′N	79°30′W
Caicedonia, Col. (kī-sĕ-dô-nēä)	18a	4°21′N	75°48′W
Caicos Bank, bk., (kī′kōs)	11	21°35′N	72°00′W
Caicos Islands, is., T./C. Is.	5	21°45′N	71°50′W
Caicos Passage, strt., N.A.	11	21°55′N	72°45′W
Caimanera, Cuba (kī-mä-nā′rä)	11	20°00′N	75°10′W
Caimito, r., Pan. (kä-ē-mē′tô)	4a	8°50′N	79°45′W
Caimito del Guayabal, Cuba (kä-ē-mê′tō-dĕl-gwä-yä-bä′l)	11a	22°57′N	82°36′W
Cairo, C.R. (kī′rô)	9	10°06′N	83°47′W
Cajamarca, Col. (kä-kä-mȧ′r-kä)	18a	4°25′N	75°25′W
Cajamarca, Peru (kä-hä-mär′kä)	18	7°16′S	78°30′W
Cajuru, Braz. (kȧ-zho͞o′ro͞o)	17a	21°17′S	47°17′W
Calabazar, Cuba (kä-lä-bä-zä′r)	11a	23°02′N	82°25′W
Calabozo, Ven. (kä-lä-bō′zō)	18	8°48′N	67°27′W
Calama, Chile (kä-lä′mä)	20	22°17′S	68°58′W
Calamar, Col. (kä-lä-mär′)	18	10°24′N	75°00′W
Calamar, Col.	18	1°55′N	72°33′W
Caldas, Col. (kȧ′l-däs)	18a	6°06′N	75°38′W
Caldas, dept., Col.	18a	5°20′N	75°38′W
Caldera, Chile (käl-dā′rä)	20	27°02′S	70°53′W
Calera Victor Rosales, Mex. (kä-lā′rä-vē′k-tôr-rô-sȧ′lĕs)	6	22°57′N	102°42′W
Cali, Col. (kä′lē)	18	3°26′N	76°30′W
California, Golfo de, b., Mex. (gôl-fô-dĕ-kä-lē-fôr-nyä)	4	30°30′N	113°45′W
Calkini, Mex. (käl-kĕ-nē′)	7	20°21′N	90°06′W
Callao, Peru (käl-yä′ô)	18	12°02′S	77°07′W
Calotmul, Mex. (kä-lôt-mo͞ol)	8a	20°58′N	88°11′W
Calpulalpan, Mex. (käl-po͞o-lȧl′pän)	6	19°35′N	98°33′W
Calvillo, Mex. (käl-vēl′yō)	7	21°51′N	102°44′F
Camaguey, Cuba (kä-mä-gwā′)	5	21°25′N	78°00′W
Camagüey, prov., Cuba	10	21°30′N	78°10′W
Camajuani, Cuba (kä-mä-hwä′nē)	10	22°25′N	79°50′W
Camargo, Mex. (kä-mär gō)	2	26°19′N	98°49′W
Camarón, Cabo, c., Hond. (kȧ′bô-kä-mä-rōn′)	8	16°06′N	85°05′W
Camatagua, Ven. (kä-mä-tȧ′gwä)	19b	9°49′N	66°55′W
Cambuci, Braz. (käm-bo͞o′sē)	17a	21°35′S	41°54′W
Camoçim, Braz. (kä-mô-sēɴ′)	19	2°56′S	40°55′W
Campana, Arg. (käm-pä′nä)	17c	34°10′S	58°58′W
Campana, i., Chile (käm-pän′yä)	20	48°20′S	75°15′W
Campanha, Braz. (käm-pän-yäɴ′)	17a	21°51′S	45°24′W
Campeche, Mex. (käm-pā′chå)	4	19°51′N	90°32′W
Campeche, state, Mex.	4	18°55′N	90°20′W
Campeche, Bahía de, b., Mex. (bä-ē′ä-dĕ-käm-pā′chā)	4	19°30′N	93°40′W
Campechuela, Cuba (käm-på-chwā′lä)	10	20°15′N	77°15′W
Campina Grande, Braz. (käm pē′nä grän′dĕ)	19	7°15′S	35°49′W
Campinas, Braz. (käm-pē′näzh)	19	22°53′S	47°03′W
Campoalegre, Col. (kä′m-pô-ȧlĕ′grĕ)	18	2°34′N	75°20′W
Campo Belo, Braz.	17a	20°52′S	45°15′W
Campo Florido, Cuba (kä′m-pō flô-rē′dō)	11a	23°07′N	82°07′W
Campo Grande, Braz. (käm-pô gran′de)	19	20°28′S	54°32′W
Campo Grande, Braz.	20b	22°54′S	43°33′W
Campo Maior, Braz. (käm-pô mä-yôr′)	19	4°48′S	42°12′W
Campos, Braz. (kä′m-pôs)	19	21°46′S	41°19′W
Campos do Jordão, Braz. (kä′m-pôs-dô-zhôr-dou′ɴ)	17a	22°45′S	45°35′W
Campos Gerais, Braz. (kä′m-pôs-zhĕ-rȧēs)	17a	21°17′S	45°43′W
Camu, r., Dom. Rep. (kä′mo͞o)	11	19°05′N	70°15′W
Cañada de Gómez, Arg. (kä-nyä′dä-dĕ-gô′mĕz)	20	32°49′S	61°24′W
Cananea, Mex. (kä-nä-nĕ′ä)	4	31°00′N	110°20′W
Canarreos, Archipiélago de los, is., Cuba	10	21°35′N	82°20′W
Cañas, C.R. (kä′-nyäs)	8	10°26′N	85°06′W
Cañas, r., C.R.	8	10°20′N	85°21′W
Cañasgordas, Col. (kä′nyäs-gô′r-däs)	18a	6°44′N	76°01′W
Canastra, Serra de, mts., Braz. (sĕ′r-rä-dĕ-kä-nä′s-trä)	19	19°53′S	46°57′W
Canatlán, Mex. (kä-nät-län′)	2	24°30′N	104°45′W
Canavieiras, Braz. (kä-nȧ-vē-ā′räs)	19	15°40′S	38°49′W
Canchyuaya, Cerros de, mts., Peru (sĕ′r-rôs-dĕ-kän-cho͞o-ȧ′ïä)	18	7°30′S	74°30′W
Cancuc, Mex. (kän-ko͞ok)	7	16°58′N	92°17′W
Cancún, Mex.	8a	21°25′N	86°50′W
Candelaria, Cuba (kän-dĕ-lä′ryä)	10	22°45′N	82°55′W
Candelaria, r., Mex. (kän-dĕ-lä-ryä)	7	18°25′N	91°21′W
Canelones, Ur. (kä-nĕ-lô-nĕs)	17c	34°32′S	56°19′W
Canelones, dept., Ur.	17c	34°34′S	56°15′W
Cañete, Peru (kän-yā′tå)	18	13°06′S	76°17′W
Caney, Cuba (kä-nā′) (kā′nĭ)	11	20°05′N	75°45′W
Cañitas, Mex. (kän-yē′täs)	6	23°38′N	102°44′W
Caño, Isla do, i., C.R. (ē′s-lä-dĕ-kä′nō)	9	8°30′N	84°00′W
Canouan, i., St. Vin.	9b	12°44′N	61°10′W
Cansahcab, Mex.	8a	21°11′N	89°05′W
Cantagalo, Braz. (kän-tä-gȧ′lo)	17a	21°59′S	42°22′W
Cantiles, Cayo, i., Cuba (ky-ō-kän-tē′lås)	10	21°40′N	82°00′W
Cañuelas, Arg. (kä-nyôĕ′-läs)	17c	35°03′S	58°45′W
Capaya, r., Ven. (kä-pä-īä)	19b	10°28′N	66°15′W
Capesterre, Guad.	9b	16°02′N	61°37′W
Cap-Haïtien, Haiti (kȧp ȧ-ē-syăn′)	5	19°45′N	72°15′W
Capilla de Señor, Arg. (kä-pēl′yä da san-yör′)	17c	34°18′S	59°07′W
Capivari, Braz. (kä-pē-vȧ′rē)	17a	22°59′S	47°29′W
Capivari, r., Braz.	20b	22°39′S	43°19′W
Capulhuac, Mex. (kä-pôl-hwäk′)	6	19°33′N	99°43′W
Capultitlán, Mex. (kä-pô′l-tē-tlȧ′n)	7a	19°15′N	99°40′W
Caquetá (Japurá), r., S.A.	18	0°20′S	73°00′W
Carabobo, dept., Ven. (kä-rä-bô′-bô)	19b	10°07′N	68°06′W
Caracas, Ven. (kä-rä′käs)	18	10°30′N	66°58′W
Carácuaro de Morelos, Mex. (kä-rä′kwä-rō-dĕ-mô-rĕ-lôs)	6	18°44′N	101°04′W
Caraguatatuba, Braz. (kä-rä-gwä-tȧ-to͞o′bä)	17a	23°37′S	45°26′W
Carajás, Serra dos, mts., Braz. (sĕ′r-rä-dôs-kä-rä-zhȧ′s)	19	5°58′S	51°45′W
Caramanta, Cerro, mtn., Col. (sĕ′r-rô-kä-rä-mȧ′n-tä)	18a	5°29′N	76°01′W
Carangola, Braz. (kä-rȧn′gŏ′lä)	17a	20°46′S	42°02′W
Carata, Laguna, l., Nic. (lä-gô′nä-kä-rä′tä)	9	13°59′N	83°41′W
Caratasca, Laguna, l., Hond. (lä-gô′nä-kä-rä-täs′kä)	9	15°20′N	83°45′W
Caravelas, Braz. (kȧ-rä-vĕl′äzh)	19	17°46′S	39°06′W
Carayaca, Ven. (kȧ-rä-īȧ′kä)	19b	10°32′N	67°07′W
Caràzinho, Braz. (kä-rȧ′zē-nyŏ)	20	28°22′S	52°33′W
Carbet, Pitons du, mtn., Mart.	9b	14°40′N	61°05′W
Cárdenas, Cuba (kär′då näs)	5	23°00′N	81°10′W
Cárdenas, Mex. (kȧ′r-dĕ-näs)	7	17°59′N	93°23′W
Cárdenas, Mex.	6	22°01′N	99°38′W
Cárdenas, Bahía de, b., Cuba (bä-ē′ä-dĕ-kär′då-näs)	10	23°10′N	81°10′W
Caribbean Sea, sea, (kăr-ĭ-bē′ăn)	5	14°30′N	75°30′W
Caribe, Arroyo, r., Mex. (är-ro′i-kä-rē′bĕ)	7	18°18′N	90°38′W
Carinhanha, Braz. (kä-rĭ-nyän′yä)	19	14°14′S	43°44′W
Carlos Casares, Arg. (kär-lôs-kä-sȧ′rĕs)	20	35°38′S	61°17′W
Carmelo, Ur. (kär-mĕ′lo)	17c	33°59′S	58°15′W
Carmen, Isla del, i., Mex. (ē′s-lä-dĕl-kȧ′r-mĕn)	7	18°43′N	91°40′W
Carmen, Laguna del, l., Mex. (lä gô′nä dĕl kȧ′r mĕn)	7	18°15′N	93°26′W
Carmen de Areco, Arg. (kär′mĕn′ dā ä-rā′kô)	17c	34°21′S	59°50′W
Carmen de Patagones, Arg. (kȧ′r-mĕn-dĕ-pä-tä-gô′nĕs)	20	41°00′S	63°00′W
Carmo, Braz. (kȧ′r-mô)	17a	21°57′S	42°45′W
Carmo do Rio Clara, Braz. (kȧ′r-mô-dô-rē′ô-klä′rä)	17a	20°57′S	46°04′W
Carolina, Braz. (kä-rô-lē′nä)	19	7°26′S	47°16′W
Carolina, l., Mex. (kä-rō-lē′nä)	8a	18°41′N	89°40′W
Caroni, r., Ven. (kä-rō′nē)	18	5°49′N	62°57′W
Carora, Ven. (kä-rô′rä)	18	10°09′N	70°12′W
Carretas, Punta, c., Peru (po͞o′n-tä-kär-rĕ′tĕ′rȧs)	18	14°15′S	76°25′W
Carriacou, i., Gren.	9b	12°28′N	61°20′W
Carrion Crow Harbor, b., Bah. (kăr′ĭŭn krō)	10	26°35′N	77°55′W
Cartagena, Col. (kär-tä-hā′nä)	18	10°30′N	75°40′W
Cartago, Col. (kär-tä′gō)	18a	4°44′N	75°54′W
Cartago, C.R.	5	9°52′N	83°56′W
Caruaru, Braz. (kä-rô-ȧ-ro͞o′)	19	8°19′S	35°52′W
Carúpano, Ven. (kä-ro͞o′pä-nō)	18	10°45′N	63°21′W
Casablanca, Chile (kä-sä-bläŋ′kä)	17b	33°19′S	71°24′W
Casa Branca, Braz. (kȧ′sä-brȧ′ɴ-kä)	17a	21°47′S	47°04′W
Caseros, Arg. (kä-sā′rôs)	20a	34°35′S	58°34′W
Casilda, Arg. (kä-sē′l-dä)	20	33°02′S	61°11′W
Casilda, Cuba	10	21°50′N	80°00′W
Casimiro de Abreu, Braz. (kä′sē-mē′ro-dĕ-ȧ-brĕ′o͞o)	17a	22°30′S	42°11′W
Casiquiare, r., Ven. (kä-sē-kyä′rā)	18	2°11′N	66°15′W
Cássia, Braz. (kȧ′syä)	17a	20°36′S	46°53′W
Castelli, Arg. (kȧs-tĕ′zhē)	17c	36°07′S	57°48′W
Castelo, Braz. (käs-tĕ′lô)	17a	20°37′S	41°13′W
Castilla, Peru (kȧs-tē′l-yä)	18	5°18′S	80°40′W
Castle, i., Bah. (kȧs′′l)	11	22°05′N	74°20′W
Castries, St. Luc. (kȧs-trē′)	9b	14°01′N	61°00′W
Castro, Braz. (käs′trô)	19	24°56′S	50°00′W
Castro, Chile (käs′tro)	20	42°27′S	73°48′W
Cat, i., Bah.	11	24°30′N	75°30′W
Catacamas, Hond. (kä-tä-kȧ′mäs)	8	14°52′N	85°55′W
Cataguases, Braz. (kä-tä-gwȧ′sĕs)	17a	21°23′S	42°42′W
Catalão, Braz. (kä-tä-louɴ′)	19	18°09′S	47°42′W
Catalina, i., Dom. Rep. (kä-tä-lē′nä)	11	18°20′N	69°00′W
Catamarca, Arg. (kä-rä-mȧ′r-kä)	20	28°29′S	65°45′W
Catamarca, prov., Arg. (kä-tä-mär′kä)	20	27°15′S	67°15′W
Catanduva, Braz. (kä-tȧn-do͞o′vä)	19	21°12′S	48°47′W
Catemaco, Mex. (kä-tā-mä′kō)	7	18°26′N	95°06′W
Catemaco, Lago, l., Mex. (lä′gô-kä-tā-mä′kō)	7	18°23′N	95°04′W
Catoche, Cabo, c., Mex. (kä-tô′chĕ)	4	21°30′N	87°15′W
Catorce, Mex. (kä-tôr′sȧ)	6	23°41′N	100°51′W
Catu, Braz. (kȧ-to͞o)	19	12°26′S	38°12′W

ng-sing; ŋ-baŋk; ɴ-nasalized n; nŏd; cŏmmit; ōld; ôbey; ôrder; oi-boil; fo͞od; ȯ-as oo in foot; ou-out; s-soft; sh-dish; th-thin; pūre; ůnite; ûrn; stŭd; circŭs; ü-as in French tu; ′-indeterminate vowel.

42 Cauca – Compostela

PRONOUNCING INDEX

PLACE (Pronunciation)	PAGE	LAT.	LONG.
Cauca, r., Col. (kou´kä)	18	7°30′N	75°26′W
Caucagua, Ven. (käȯ-kȧ´gwä)	19b	10°17′N	66°22′W
Cauquenes, Chile (kou-kā´nās)	20	35°54′S	72°14′W
Caura, r., Ven. (kou´rä)	18	6°48′N	64°40′W
Cauto, r., Cuba (kou´tō)	10	20°33′N	76°20′W
Cava, Braz. (kȧ´vä)	20b	22°41′S	43°26′W
Cavalcante, Braz. (kä-väl-kän´tā)	19	13°45′S	47°33′W
Caviana, Ilha, i., Braz. (kä-vyä´nä)	19	0°45′N	49°33′W
Caxambu, Braz. (kä-shȧ´m-bo͞o)	19	22°00′S	44°45′W
Caxias, Braz. (kä´shė-äzh)	19	4°48′S	43°16′W
Caxias do Sul, Braz. (kä´shė-äzh-do̊-so͞o´l)	20	29°13′S	51°03′W
Cayambe, Ec. (kȧ-īä´m-bĕ)	18	0°03′N	79°09′W
Cayenne, Fr. Gu. (kā-ĕn´)	19	4°56′N	52°18′W
Cayetano Rubio, Mex. (kä-yĕ-tȧ-no̊-ro͞o´byo̊)	6	20°37′N	100°21′W
Cayey, P.R.	5b	18°05′N	66°12′W
Cayman Brac, i., Cay. Is. (kī-män´ brȧk)	10	19°45′N	79°50′W
Cayman Islands, dep., N.A.	10	19°30′N	80°30′W
Cay Sal Bank, bk., (kē-săl)	10	23°55′N	80°20′W
Cazones, r., Mex. (kä-zô´nĕs)	7	20°37′N	97°28′W
Cazones, Ensenada de, b., Cuba (ĕn-sĕ-nä-dä-dĕ-kä-zō´nȧs)	10	22°05′N	81°30′W
Cazones, Golfo de, b., Cuba (go̊l-fô-dĕ-kä-zō´nȧs)	10	21°55′N	81°15′W
Ceará-Mirim, Braz. (sā-ä-rä´mē-rė´ɴ)	19	6°00′S	35°13′W
Cebaco, Isla, i., Pan. (ė´s-lä-så-bä´kō)	9	7°27′N	81°08′W
Cedral, Mex. (sā-dräl´)	6	23°47′N	100°42′W
Cedros, Hond. (sā´drōs)	8	14°36′N	87°07′W
Cedros, i., Mex.	4	28°10′N	115°10′W
Ceiba del Agua, Cuba (sā´bä-dĕl-ȧ´gwä)	11a	22°53′N	82°38′W
Celaya, Mex. (sā-lä´yä)	4	20°33′N	100°49′W
Celestún, Mex. (sĕ-lĕs-to͞o´n)	8a	20°57′N	90°18′W
Ceniza, Pico, mtn., Ven. (pē´ko̊-sĕ-nē´zä)	19b	10°24′N	67°26′W
Central, Cordillera, mts., Bol. (ko̊r-dēl-yĕ´rä-sĕn-trȧ´l)	18	19°18′S	65°29′W
Central, Cordillera, mts., Col.	18a	3°58′N	75°55′W
Central, Cordillera, mts., Dom. Rep.	11	19°05′N	71°30′W
Central America, reg., N.A. (ä-mĕr´ĭ-kȧ)	4	10°45′N	87°15′W
Cerralvo, Mex. (sĕr-räl´vō)	2	26°05′N	99°37′W
Cerralvo, i., Mex.	4	24°00′N	109°59′W
Cerrito, Col. (sĕr-rē´-to̊)	18a	3°41′N	76°17′W
Cerritos, Mex. (sĕr-rē´to̊s)	6	22°26′N	100°16′W
Cerro de Pasco, Peru (sĕr´rō dā päs´kō)	18	10°45′S	76°14′W
Cerro Gordo, Arroyo de, r., Mex. (är-rô-yŏ-dĕ-sĕ´r-ro̊-gôr-dŏ)	2	26°12′N	104°06′W
Certegui, Col. (sĕr-tĕ´gē)	18a	5°21′N	76°35′W
Chacabuco, Arg. (chä-kä-bo͞o´kō)	17c	34°37′S	60°27′W
Chacaltianguis, Mex. (chä-käl-tė-äŋ´gwės)	7	18°18′N	95°50′W
Chachapoyas, Peru (chä-chä-poi´yäs)	18	6°16′S	77°48′W
Chaco, prov., Arg. (chä´kō)	20	26°00′S	60°45′W
Chagres, r., Pan. (chä´grĕs)	9	9°18′N	79°22′W
Chalatenango, El Sal. (chäl-ä-tĕ-näŋ´gō)	8	14°04′N	88°54′W
Chalcatongo, Mex. (chäl-kä-tôŋ´gō)	7	17°04′N	97°41′W
Chalchihuites, Mex. (chäl-chė-wē´tås)	6	23°28′N	103°57′W
Chalchuapa, El Sal. (chäl-chwä´pä)	8	14°01′N	89°39′W
Chalco, Mex. (chäl-kō)	7a	19°15′N	98°54′W
Chaltel, Cerro (Monte Fitzroy), mtn., S.A. (sĕ´r-ro̊-chäl´tĕl)	20	48°10′S	73°18′W
Chama, Sierra de, mts., Guat. (sē-ĕ´r-rä-dĕ-chä-mä)	8	15°48′N	90°20′W
Chame, Punta, c., Pan. (pȯ´n-tä-chä´må)	9	8°41′N	79°27′W
Chamelecón, r., Hond. (chä-mĕ-lĕ-kô´n)	8	15°09′N	88°42′W
Champerico, Guat. (chăm-på-rē´kō)	8	14°18′N	91°55′W
Champotón, Mex. (chäm-pō-tōn´)	7	19°21′N	90°43′W
Champotón, r., Mex.	7	19°19′N	90°15′W
Chañaral, Chile (chän-yä-räl´)	20	26°20′S	70°46′W
Chances Peak, vol., Monts.	9b	16°43′N	62°10′W
Chapada, Serra da, mts., Braz. (sĕ´r-rä-dä-shä-pä´dä)	19	14°57′S	54°34′W
Chapadão, Serra do, mtn., Braz. (sĕ´r-rä-do̊-shä-pȧ-dou´ɴ)	17a	20°31′S	46°20′W
Chapala, Mex. (chä-pä´lä)	6	20°18′N	103°10′W
Chapala, Lago de, l., Mex. (lä´gô-dĕ-chä-pä´lä)	4	20°14′N	103°02′W
Chapalagana, r., Mex. (chä-pä-lä-gȧ´nä)	6	22°11′N	104°09′W
Chaparral, Col. (chä-pär-rȧ´l)	18	3°44′N	75°28′W
Chapultenango, Mex. (chä-pȯl-tē-näŋ´gō)	7	17°19′N	93°08′W
Charcas, Mex. (chär´käs)	6	23°09′N	101°09′W
Charco de Azul, Bahía, b., Pan.	9	8°14′N	82°45′W
Charlestown, St. K./N.	9b	17°10′N	62°32′W
Charlotte Amalie, V.I.U.S. (shär-lŏt´ĕ ä-mä´lĭ-ȧ)	5	18°21′N	64°54′W
Chascomús, Arg. (chäs-kō-mo͞os´)	20	35°32′S	58°01′W
Chavinda, Mex. (chä-vē´n-dä)	6	20°01′N	102°27′W
Chazumba, Mex. (chä-zȯm´bä)	7	18°11′N	97°41′W
Chepén, Peru (chĕ-pĕ´n)	18	7°17′S	79°24′W
Chepo, Pan. (chā´pō)	9	9°12′N	79°06′W
Chepo, r., Pan.	9	9°10′N	78°36′W
Cherán, Mex. (chā-rän´)	6	19°41′N	101°54′W
Cherokee Sound, Bah.	10	26°15′N	76°55′W
Chetumal, Bahía de, b., N.A. (bă-ē-ă dĕ chĕt-o͞o-mäl´)	4	18°07′N	88°05′W
Chiapa, Río de, r., Mex.	8	16°00′N	92°20′W
Chiapa de Corzo, Mex. (chė-ä´pä dā ko̊r´zō)	7	16°44′N	93°01′W
Chiapas, state, Mex. (chė-ä´päs)	4	17°10′N	93°00′W
Chiapas, Cordillera de, mts., Mex. (ko̊r-dēl-yĕ´rä-dĕ-chyȧ´räs)	7	15°55′N	93°15′W
Chiautla, Mex. (chyä-o͞ot´lȧ)	6	18°16′N	98°37′W
Chicbul, Mex. (chēk-bo͞o´l)	7	18°45′N	90°56′W
Chichancanab, Lago de, l., Mex. (lä´gô-dĕ-chē-chän-kä-nä´b)	8a	19°50′N	88°28′W
Chichén Itzá, hist., Mex.	8a	20°40′N	88°35′W
Chichimilá, Mex. (chē-chē-mė´lä)	8a	20°36′N	88°14′W
Chichiriviche, Ven. (chē-chē-rē-vē-chĕ)	19b	10°56′N	68°17′W
Chiclayo, Peru (chė-klä´yō)	18	6°46′S	79°50′W
Chico, r., Arg.	20	44°30′S	66°00′W
Chico, r., Arg.	20	49°15′S	69°30′W
Chicoloapan, Mex. (chė-kō-lwä´pän)	7a	19°24′N	98°54′W
Chiconautla, Mex.	7a	19°39′N	99°01′W
Chicontepec, Mex. (chė-kōn´tĕ-pĕk´)	6	20°58′N	98°08′W
Chicxulub, Mex. (chēk-so͞o-lo͞o´b)	8a	21°10′N	89°30′W
Chignanuapán, Mex. (chē´g-ŋä-nwä-pȧ´n)	6	19°49′N	98°02′W
Chihuahua, Mex. (chė-wä´wä)	4	28°37′N	106°06′W
Chihuahua, state, Mex.	4	29°00′N	107°30′W
Chilapa, Mex. (chė-lä´pä)	6	17°34′N	99°14′W
Chilchota, Mex. (chėl-chō´tä)	6	19°40′N	102°04′W
Chile, nation, S.A. (chē´lā)	20	35°00′S	72°00′W
Chilecito, Arg. (chė-lå-sē´tō)	20	29°06′S	67°25′W
Chilibre, Pan. (chē-lė´brĕ)	4a	9°09′N	79°37′W
Chillán, Chile (chēl-yän´)	20	36°44′S	72°06′W
Chiloé, Isla de, i., Chile	20	42°30′S	73°55′W
Chilpancingo de los Bravo, Mex.	4	17°32′N	99°30′W
Chimalpa, Mex. (chē-mäl´pä)	7a	19°26′N	99°22′W
Chimaltenango, Guat. (chė-mäl-tå-näŋ´gō)	8	14°39′N	90°48′W
Chimaltitán, Mex. (chėmäl-tē-tän´)	6	21°36′N	103°50′W
Chimborazo, mtn., Ec. (chėm-bo̊-rä´zō)	18	1°35′S	78°45′W
Chimbote, Peru (chėm-bō´tå)	18	9°02′S	78°33′W
China, Mex. (chē´nä)	2	25°43′N	99°13′W
Chinameca, El Sal. (Chĕ-nä-mā´kä)	8	13°31′N	88°18′W
Chinandega, Nic. (chē-nän-dā´gä)	8	12°38′N	87°08′W
Chincha Alta, Peru (chĭn´chä äl´tä)	18	13°24′S	76°04′W
Chinchas, Islas, is., Peru (ė´s-läs-chē´n-chäs)	18	11°27′S	79°05′W
Chinchorro, Banco, bk., Mex. (bä´n-kô-chēn-chô´r-rŏ)	8a	18°43′N	87°25′W
Chiquimula, Guat. (chē-kė-mo͞o´lä)	8	14°47′N	89°31′W
Chiquimulilla, Guat. (chē-kē-mo͞o-lė´l-yä)	8	14°08′N	90°23′W
Chiquinquira, Col. (chė-kēŋ´kė-rä´)	18	5°33′N	73°49′W
Chiriquí, Punta, c., Pan. (pȯ´n-tä-chē-rė-kē´)	9	9°13′N	81°39′W
Chiriquí Grande, Pan. (chē-rė-kē´ grän´dā)	9	8°57′N	82°08′W
Chirripó, Río, r., C.R.	9	9°50′N	83°20′W
Chivilcoy, Arg. (chē-vėl-koi´)	20	34°51′S	60°03′W
Chixoy, r., Guat. (chē-ᴋoi´)	8	15°40′N	90°35′W
Choapa, r., Chile (chȯ-ȧ´pä)	17b	31°56′S	70°48′W
Choele Choel, Arg. (chȯ-ĕ´lĕ-chȯĕ´l)	20	39°14′S	65°46′W
Cholula, Mex. (chō-lo͞o´lä)	6	19°04′N	98°19′W
Choluteca, Hond. (chō-lo͞o-tā´kä)	8	13°18′N	87°12′W
Choluteco, r., Hond.	8	13°34′N	86°59′W
Chone, Ec. (chô´nĕ)	18	0°48′S	80°06′W
Chonos, Archipiélago de los, is., Chile	20	44°35′N	76°15′W
Chorrillos, Peru (chôr-rē´l-yōs)	18	12°17′S	76°55′W
Christiansted, V.I.U.S.	5b	17°45′N	64°44′W
Chubut, prov., Arg. (chȯ-bo͞ot´)	20	44°00′S	69°15′W
Chubut, r., Arg. (chȯ-bo͞ot´)	20	43°05′S	69°00′W
Chucunaque, r., Pan. (cho͞o-ko͞o-nä´kå)	9	8°36′N	77°48′W
Chulucanas, Peru	18	5°13′S	80°13′W
Chuquicamata, Chile (cho͞o-kė-kä-mä´tä)	20	22°08′S	68°57′W
Churumuco, Mex. (cho͞o-ro͞o-mo͞o´kō)	6	18°39′N	101°40′W
Chuviscar, r., Mex. (cho͞o-vės-kär´)	2	28°34′N	105°36′W
Ciego de Ávila, Cuba (syā´gō dā ä´vė-lä)	5	21°50′N	78°45′W
Ciego de Ávila, prov., Cuba	10	22°00′N	78°40′W
Ciénaga, Col. (syā´nä-gä)	18	11°01′N	74°15′W
Cienfuegos, Cuba (syĕn-fwā´gōs)	5	22°10′N	80°30′W
Cienfuegos, prov., Cuba	10	22°15′N	80°40′W
Cienfuegos, Bahía, b., Cuba (bä-ē´ä-syĕn-fwā´gōs)	10	22°00′N	80°35′W
Ciervo, Isla de la, i., Nic. (ė´s-lä-dē-lä-syĕ´r-vô)	9	11°56′N	83°20′W
Cihuatlán, Mex. (sē-wä-tlȧ´n)	6	19°13′N	104°36′W
Cihuatlán, r., Mex.	6	19°11′N	104°30′W
Cinco Balas, Cayos, is., Cuba (kä´yōs-thēŋ´kō bä´läs)	10	21°05′N	79°25′W
Cintalapa, Mex. (sėn-tä-lä´pä)	7	16°41′N	93°44′W
Ciri Grande, r., Pan. (sē´rē-grȧ´n´dĕ)	4a	8°55′N	80°04′W
Cisneros, Col. (sės-nĕ´ro̊s)	18a	6°33′N	75°05′W
Ciudad Altamirano, Mex. (syo͞o-dä´d-äl-tä-mē-rä´no̊)	6	18°24′N	100°38′W
Ciudad Bolívar, Ven. (syo͞o-dhädh´ bo̊-lē´vär)	18	8°07′N	63°41′W
Ciudad Camargo, Mex.	4	27°42′N	105°10′W
Ciudad Chetumal, Mex.	4	18°30′N	88°17′W
Ciudad Darío, Nic. (syo͞o-dhädh´dä´rė-ō)	8	12°44′N	86°08′W
Ciudad de la Habana, prov., Cuba	10	23°20′N	82°10′W
Ciudad del Carmen, Mex. (syo͞o-dä´d-dĕl-kȧ´r-mĕn)	4	18°39′N	91°49′W
Ciudad del Maíz, Mex. (syo͞o-dhädh´del mä-ēz´)	6	22°24′N	99°37′W
Ciudad Fernández, Mex. (syo͞o-dhädh´fĕr-nän´dĕz)	6	21°56′N	100°03′W
Ciudad García, Mex. (syo͞o-dhädh´gär-sē´ä)	4	22°39′N	103°02′W
Ciudad Guayana, Ven.	18	8°30′N	62°45′W
Ciudad Guzmán, Mex. (syo͞o-dhädh´gȯz-män)	4	19°40′N	103°29′W
Ciudad Hidalgo, Mex. (syo͞o-dä´d-ė-dä´l-go̊)	6	19°41′N	100°35′W
Ciudad Juárez, Mex. (syo͞o-dhädh hwä´rāz)	4	31°44′N	106°28′W
Ciudad Madero, Mex. (syo͞o-dä´d-mä-dĕ´ro̊)	7	22°16′N	97°52′W
Ciudad Mante, Mex. (syo͞o-dä´d-män´tĕ)	4	22°34′N	98°58′W
Ciudad Manuel Doblado, Mex. (syo͞o-dä´d-män-wāl´dō-blä´dō)	6	20°43′N	101°57′W
Ciudad Obregón, Mex. (syo͞o-dhädh-ô-brĕ-gô´n)	4	27°40′N	109°58′W
Ciudad Serdán, Mex. (syo͞o-dä´d-sĕr-dȧ´n)	7	18°58′N	97°26′W
Ciudad Victoria, Mex. (syo͞o-dhädh´vėk-tō´rė-ä)	4	23°43′N	99°09′W
Clarence Town, Bah.	11	23°05′N	75°00′W
Clarines, Ven. (klä-rē´nĕs)	19b	9°57′N	65°10′W
Cláudio, Braz. (klou´-dēō)	17a	20°26′S	44°44′W
Coacalco, Mex. (kō-ä-käl´kō)	7a	19°37′N	99°06′W
Coahuayana, Río de, r., Mex. (rė´ō-dĕ-kô-ä-wä-yȧ´nä)	6	19°00′N	103°33′W
Coahuayutla, Mex. (kō´ä-wī-yo͞ot´lȧ)	6	18°19′N	101°44′W
Coahuila, state, Mex. (kō-ä-wē´lȧ)	4	27°30′N	103°00′W
Coalcomán, Río de, r., Mex. (rė´ō-dĕ-kō-äl-kō-män´)	6	18°45′N	103°15′W
Coalcomán, Sierra de, mts., Mex.	6	18°30′N	102°45′W
Coalcomán de Matamoros, Mex.	6	18°46′N	103°10′W
Coamo, P.R. (kō-ä´mō)	5b	18°05′N	66°21′W
Coari, Braz. (kō-är´ė)	18	4°06′S	63°10′W
Coatepec, Mex. (kō-ä-tā-pĕk)	6	19°23′N	98°44′W
Coatepec, Mex.	7	19°26′N	96°56′W
Coatepec, Mex.	7a	19°08′N	99°25′W
Coatepeque, El Sal.	8	13°56′N	89°30′W
Coatepeque, Guat. (kō-ä-tå-pā´kå)	8	14°40′N	91°52′W
Coatetelco, Mex. (kō-ä-tå-tĕl´kō)	6	18°43′N	99°17′W
Coatlinchán, Mex. (kô-ä-tlē´n-chä´n)	7a	19°26′N	98°52′W
Coatzacoalcos, Mex.	4	18°09′N	94°26′W
Coatzacoalcos, r., Mex.	7	17°40′N	94°41′W
Coba, hist., Mex. (kô´bä)	8a	20°23′N	87°23′W
Cobán, Guat. (kō-bän´)	4	15°28′N	90°19′W
Cobija, Bol. (kô-bē´hä)	18	11°12′S	68°49′W
Cobre, r., Jam. (kō´brā)	10	18°05′N	77°00′W
Cochabamba, Bol.	18	17°24′S	66°09′W
Cochinos, Bahía, b., Cuba (bä-ē´ä-ko̊-chė´nōs)	10	22°05′N	81°10′W
Cochinos Banks, bk.	10	22°20′N	76°15′W
Coco, r., N.A.	5	14°55′N	83°45′W
Coco, Cayo, i., Cuba (kä´-yō-ko̊´ko̊)	10	22°30′S	78°30′W
Coco, Isla del, i., C.R. (ė´s-lä-dĕl-ko̊-ko̊)	4	5°33′N	87°02′W
Coco Solito, Pan. (kô-kô-sŏ-lē´tŏ)	4a	9°21′N	79°53′W
Cocula, Mex. (kō-ko͞o´lä)	6	20°23′N	103°47′W
Cocula, r., Mex.	6	18°17′N	99°45′W
Codajás, Braz. (kō-dä-häzh´)	18	3°44′S	62°09′W
Codera, Cabo, c., Ven. (kȧ´bo̊-kō-dĕ´rä)	19b	10°35′N	66°06′W
Codrington, Antig. (kŏd´rĭng-tŭn)	9b	17°39′N	61°49′W
Coelho da Rocha, Braz.	20b	22°47′S	43°23′W
Coig, r., Arg. (ko̊´ėk)	20	51°15′N	71°00′W
Coipasa, Salar de, pl., Bol. (sä-lä´r-dĕ-koi-pä´-sä)	18	19°12′S	69°13′W
Coixtlahuaca, Mex. (kō-ēks´tlä-wä´kä)	7	17°42′N	97°17′W
Cojedes, dept., Ven. (ko̊-kĕ´dĕs)	19b	9°50′N	68°21′W
Cojimar, Cuba (kō-hė-mär´)	11a	23°10′N	82°19′W
Cojutepeque, El Sal. (ko̊-hȯ-tē-pā´kå)	8	13°45′N	88°50′W
Colatina, Braz. (kô-lä-tė´nä)	19	19°33′S	40°42′W
Colchagua, prov., Chile (ko̊l-chȧ´gwä)	17b	34°42′S	71°24′W
Colhué Huapi, l., Arg. (ko̊l-wā´ȯȧ´pė)	20	45°30′S	68°45′W
Colima, Mex. (kōlē´mä)	4	19°13′N	103°45′W
Colima, state, Mex.	6	19°10′N	104°00′W
Colima, Nevado de, mtn., Mex. (nĕ-vä´do̊-dĕ-ko̊-lē´mä)	4	19°30′N	103°38′W
Colombia, Col. (ko̊-lo̊m´bė-ä)	18a	3°23′N	74°48′W
Colombia, nation, S.A.	18	3°30′N	72°30′W
Colón, Arg. (kō-lōn´)	17c	33°55′S	61°08′W
Colón, Cuba (ko̊-lo̊´n)	10	22°45′N	80°55′W
Colón, Mex. (kō-lōn´)	6	20°46′N	100°02′W
Colón, Pan. (kŏ-lô´n)	5	9°22′N	79°54′W
Colón, Archipiélago de (Galapagos Islands), is., Ec.	18	0°10′S	87°45′W
Colón, Montañas de, mts., Hond. (môn-tä´n-yäs-dĕ-kŏ-lô´n)	9	14°58′N	84°39′W
Colonia, Ur. (kō-lō´nė-ä)	20	34°27′S	57°50′W
Colonia, dept., Ur.	17c	34°08′S	57°50′W
Colonia Suiza, Ur. (ko̊-lô´nėä-sȯē´zä)	17c	34°17′S	57°15′W
Coloradas, Lomas, Arg. (lo̊´mäs-kō-lō-rä´däs)	20	43°30′S	68°00′W
Colorado, r., Arg.	20	38°30′S	66°00′W
Colorados, Archipiélago de los, is., Cuba	10	22°25′N	84°25′W
Colotepec, r., Mex. (kô-lo̊´tĕ-pĕk)	7	15°56′N	96°57′W
Colotlán, Mex. (ko̊-lo̊-tlän´)	6	22°06′N	103°14′W
Colotlán, r., Mex.	6	22°09′N	103°17′W
Colquechaca, Bol. (ko̊l-kā-chä´kä)	18	18°47′S	66°02′W
Columbus Bank, bk., (ko̊-lŭm´bŭs)	11	22°05′N	75°30′W
Columbus Point, c., Bah.	11	24°10′N	75°15′W
Comalá, Mex. (kō-mä-lä´)	6	19°22′N	103°47′W
Comalapa, Guat. (kō-mä-lä´-pä)	8	14°43′N	90°56′W
Comalcalco, Mex. (kō-mäl-käl´kō)	7	18°16′N	93°13′W
Comayagua, Hond. (kō-mä-yä´gwä)	4	14°24′N	87°36′W
Comete, Cape, c., T./C. Is. (ko̊-mā´tå)	11	21°45′N	71°25′W
Comitán, Mex. (ko̊-mē-tän´)	4	16°16′N	92°09′W
Comodoro Rivadavia, Arg.	20	45°47′S	67°31′W
Comonfort, Mex. (ko̊-mōn-fô´rt)	6	20°43′N	100°47′W
Companario, Cerro, mtn., S.A. (sĕ´r-ro̊-kôm-pä-nä´ryo̊)	17b	35°54′S	70°23′W
Compostela, Mex. (kōm-po̊-stā´lä)	6	21°14′N	104°54′W

ăt; finăl; rāte; senåte; ärm; ȧsk; sofȧ; fâre; ch-choose; dh-as th in other; bē; ėvent; bĕt; recĕnt; cratẽr; g-gō; gh-guttural g; bĭt; ĭ-short neutral; rīde; ᴋ-guttural k as ch in German ich;

PRONOUNCING INDEX

Concepción – El Pilar 43

PLACE (Pronunciation)	PAGE	LAT.	LONG.
Concepción, Bol. (kôn-sĕp′syōn′)	19	15°47′S	61°08′W
Concepción, Chile	20	36°51′S	72°59′W
Concepción, Pan.	9	8°31′N	82°38′W
Concepción, Para.	20	23°29′S	57°18′W
Concepción, vol., Nic.	8	11°36′N	85°43′W
Concepción, r., Mex.	4	30°25′N	112°20′W
Concepción del Mar, Guat. (kôn-sĕp-syōn′dĕl mär′)	8	14°07′N	91°23′W
Concepción del Oro, Mex. (kôn-sĕp-syōn′ dĕl ō′rō)	4	24°39′N	101°24′W
Concepción del Uruguay, Arg. (kôn-sĕp-syô′n-dĕl-ōō-rōō-gwī′)	20	32°31′S	58°10′W
Conception, i., Bah.	11	23°50′N	75°05′W
Conchos, r., Mex.	4	29°30′N	105°00′W
Concordia, Arg. (kŏn-kôr′dĭ-á)	20	31°18′S	57°59′W
Concordia, Col.	18a	6°04′N	75°54′W
Concordia, Mex. (kôn-kô′r-dyä)	6	23°17′N	106°06′W
Condega, Nic. (kŏn-dĕ′gä)	8	13°20′N	86°27′W
Condeúba, Braz. (kōn-dā-ōō′bä)	19	14°47′S	41°44′W
Conselheiro Lafaiete, Braz.	19	20°40′S	43°46′W
Consolación del Sur, Cuba (kōn-sô-lä-syōn′)	10	22°30′N	83°55′W
Constitución, Chile (kōn′stĭ-tōō-syōn′)	20	35°24′S	72°25′W
Contagem, Braz. (kōn-tá′zhĕm)	17a	19°54′S	44°05′W
Contepec, Mex. (kōn-tĕ-pĕk′)	6	20°04′N	100°07′W
Contreras, Mex. (kôn-trĕ′räs)	7a	19°18′N	99°14′W
Copacabana, Braz. (kô′pä-kà-bá′nä)	20b	22°57′S	43°11′W
Copalita, r., Mex. (kô-pä-lē′tä)	7	15°55′N	96°06′W
Copán, hist., Hond. (kō-pän′)	8	14°50′N	89°10′W
Copiapó, Chile (kō-pyä-pō′)	20	27°16′S	70°28′W
Coquimbo, Chile (kô-kēm′bō)	20	29°58′S	71°31′W
Coquimbo, prov., Chile	17b	31°50′S	71°05′W
Coracora, Peru (kō′rä-kō′rä)	18	15°12′S	73°42′W
Corcovado, mtn., Braz. (kôr-kô-vä′dô)	20b	22°57′S	43°13′W
Corcovado, Golfo, b., Chile (kôr-kô-vä′dhō)	20	43°40′S	75°00′W
Cordeiro, Braz. (kôr-dā′rō)	17a	22°03′S	42°22′W
Córdoba, Arg. (kôr′dô-vä)	20	30°20′S	64°03′W
Córdoba, Mex. (kô′r-dô-bä)	4	18°53′N	96°54′W
Córdoba, prov., Arg. (kôr′dô-vä)	20	32°00′S	64°00′W
Córdoba, Sierra de, mts., Arg.	20	31°15′S	64°30′W
Corinto, Braz. (kô-rē′n-tō)	19	18°20′S	44°16′W
Corinto, Col.	18a	3°09′N	76°12′W
Corinto, Nic. (kōr-ĭn′to)	8	12°30′N	87°12′W
Cornwall, Bah.	10	25°55′N	77°15′W
Coro, Ven. (kō′rô)	18	11°22′N	69°43′W
Corocoro, Bol. (kō-rô-kō′rô)	18	17°15′S	68°21′W
Coronada, Bahía de, b., C.R. (bä-ē′ä-dĕ-kô-rô-nä′dō)	9	8°47′N	84°04′W
Coronel, Chile (kō-rô-nĕl′)	20	37°00′S	73°10′W
Coronel Brandsen, Arg. (kŏ-rô-nĕl-brá′nd-sĕn)	17c	35°09′S	58°15′W
Coronel Dorrego, Arg. (kŏ-rô-nĕl-dôr-rĕ′gô)	20	38°43′S	61°16′W
Coronel Oviedo, Para. (kŏ-rŏ-nĕl-ô-vêĕ′dô)	20	25°28′S	56°22′W
Coronel Pringles, Arg. (kŏ-rô-nĕl-prēn′glĕs)	20	37°54′S	61°22′W
Coronel Suárez, Arg. (kŏ-rô-nĕl-swä′rās)	20	37°27′S	61°49′W
Corozal, Belize (cŏr-ðth-äl′)	8a	18°25′N	88°23′W
Corral, Chile (kô-räl′)	20	39°57′S	73°15′W
Corralillo, Cuba (kô-rä-lē-yō)	10	23°00′N	80°40′W
Correntina, Braz. (kô-rĕn-tē-nȧ)	19	13°18′S	44°33′W
Corrientes, Arg. (kō-ryĕn′tās)	20	27°25′S	58°39′W
Corrientes, prov., Arg.	20	28°45′S	58°00′W
Corrientes, Cabo, c., Col. (ká′bô-kō-ryĕn′tās)	18	5°34′N	77°35′W
Corrientes, Cabo, c., Cuba (ká′bô-kôr-rē-ĕn′tĕs)	10	21°50′N	84°25′W
Corrientes, Cabo, c., Mex.	4	20°25′N	105°41′W
Cortazar, Mex. (kôr-tä-zär)	6	20°30′N	100°57′W
Cortés, Ensendada de, b., Cuba (ĕn-sĕ-nä-dä-dĕ-kôr-tās′)	10	22°05′N	83°45′W
Coruripe, Braz. (kō-rô-rē′pĭ)	19	10°09′S	36°13′W
Cosamaloápan, Mex. (kô-sá-mä-lwä′pän)	7	18°21′N	95°48′W
Coscomatepec, Mex. (kôs′kōmä-tĕ-pĕk′)	7	19°04′N	97°03′W
Cosigüina, vol., Nic.	8	12°59′N	87°35′W
Cosoleacaque, Mex. (kō sō lā-ä-kä′kē)	7	18°01′N	94°38′W
Costa Rica, nation, N.A. (kŏs′tȧ rē′kȧ)	5	10°30′N	84°30′W
Cotabambas, Peru (kô-tä-bäm′bäs)	18	13°49′S	72°17′W
Cotaxtla, Mex. (kō-täs′tlä)	7	18°49′N	96°22′W
Cotaxtla, r., Mex.	7	18°54′N	96°21′W
Coteaux, Haiti	11	18°15′N	74°05′W
Cotija de la Paz, Mex. (kô-tē′-ĸä-dĕ-lä-pá′z)	6	19°46′N	102°43′W
Cotopaxi, mtn., Ec. (kō-tô-păk′sê)	18	0°40′S	78°26′W
Cotorro, Cuba (kô-tôr-rō)	11a	23°03′N	82°17′W
Coulto, Serra do, mts., Braz. (sĕ′r-rä-dô-kô-ô′tô)	20b	22°33′S	43°27′W
Courantyne, r., S.A. (kôr′ăntĭn)	19	4°28′N	57°42′W
Coxim, Braz. (kō-shēɴ′)	19	18°32′S	54°43′W
Coxquihui, Mex. (kōz-kē-wē′)	7	20°10′N	97°34′W
Coyaima, Col. (kô-yáē′mä)	18a	3°48′N	75°11′W
Coyame, Mex. (kō-yä′mā)	2	29°26′N	105°05′W
Coyoacán, Mex. (kô-yô-á-kän′)	6	19°21′N	99°10′W
Coyuca de Benítez, Mex. (kō-yōō′kä dā bā-nē′tāz)	6	17°04′N	100°06′W
Coyuca de Catalán, Mex. (kō-yōō′kä dā kä-tä-län′)	6	18°19′N	100°41′W
Coyutla, Mex. (kō-yōō′tlä)	7	20°13′N	97°40′W
Cozoyoapán, Mex. (kô-zō-yô-ä-pá′n)	6	16°45′N	98°17′W
Cozumel, Mex. (kō-zōō-mĕ′l)	8a	20°31′N	86°55′W
Cozumel, Isla de, i., Mex. (ê′s-lä-dĕ-kō-zōō-mĕ′l)	4	20°26′N	87°10′W
Crateús, Braz. (krä-tå-ōōzh′)	19	5°09′S	40°35′W
Crato, Braz. (krä′tô)	19	7°19′S	39°13′W
Cristina, Braz. (krês-tē′-nä)	17a	22°13′S	45°15′W
Cristóbal Colón, Pico, mtn., Col. (pē′kô-krēs-tô′bäl-kō-lôn′)	18	11°00′N	74°00′W
Crooked, I., Bah.	11	22°45′N	74°10′W
Crooked Island Passage, strt., Bah.	11	22°40′N	74°50′W
Crown Mountain, mtn., V.I.U.S.	5c	18°22′N	64°58′W
Cruces, Cuba (krōō′sås)	10	22°20′N	80°20′W
Cruces, Arroyo de, r., Mex. (är-rŏ′yô-dĕ-krōō′sĕs)	2	26°17′N	104°32′W
Cruillas, Mex. (krōō-ēl′yäs)	2	24°45′N	98°31′W
Cruz, Cabo, c., Cuba (ká′-bô-krōōz)	5	19°50′N	77°45′W
Cruz, Cayo, i., Cuba (kä′yō-krōōz)	10	22°15′N	77°50′W
Cruz Alta, Braz. (krōōz äl′tä)	20	28°41′S	54°02′W
Cruz del Eje, Arg. (krōō′s-dĕl-ĕ-kĕ)	20	30°46′S	64°45′W
Cruzeiro, Braz. (krōō-zā′rô)	17a	22°36′S	44°57′W
Cruzeiro do Sul, Braz. (krōō-zā′rô dô sōōl)	18	7°34′S	72°40′W
Cúa, Ven. (kōō′ä)	19b	10°10′N	66°54′W
Cuajimalpa, Mex. (kwä-hê-mäl′pä)	7a	19°21′N	99°18′W
Cuale, Sierra del, mts., Mex. (sē-ĕ′r-rä-dĕl-kwá′lĕ)	6	20°20′N	104°58′W
Cuarto, r., Arg.	20	33°00′S	63°25′W
Cuatro Caminos, Cuba (kwä′trô-kä-mē′nōs)	11a	23°01′N	82°13′W
Cuatro Ciénegas, Mex. (kwä′trō syā′nå-gäs)	2	26°59′N	102°03′W
Cuauhtemoc, Mex. (kwä-ōō-tĕ-mŏk′)	7	15°43′N	91°57′W
Cuautepec, Mex. (kwä-ōō-tĕ-pĕk)	6	16°41′N	99°04′W
Cuautepec, Mex.	6	20°01′N	98°19′W
Cuautitlán, Mex. (kwä-ōō-tĕt-län′)	7a	19°40′N	99°12′W
Cuautla, Mex. (kwä-ōō′tlá)	6	18°47′N	98°57′W
Cuba, nation, N.A. (kū′bȧ)	5	22°00′N	79°00′W
Cubagua, Isla, i., Ven. (ê′s-lä-kōō-bä′gwä)	19b	10°48′N	64°10′W
Cuchillo Parado, Mex. (kōō-chē′lyô pä-rä′dō)	2	29°26′N	104°52′W
Cuchumatanes, Sierra de los, mts., Guat.	8	15°35′N	91°10′W
Cúcuta, Col. (kōō′kōō-tä)	18	7°56′N	72°30′W
Cuenca, Ec. (kwen′kä)	18	2°52′S	78°54′W
Cuencame, Mex. (kwĕn-kä-mā′)	2	24°52′N	103°42′W
Cuerámaro, Mex. (kwā-rä′mä-rô)	6	20°39′N	101°44′W
Cuernavaca, Mex. (kwĕr-nä-vä′kä)	4	18°55′N	99°15′W
Cuetzalá del Progreso, Mex. (kwĕt-zä-lä dĕl prō-grā′sō)	6	18°07′N	99°51′W
Cuetzalan del Progreso, Mex.	7	20°02′N	97°33′W
Cuicatlán, Mex. (kwē-kä-tlän′)	7	17°46′N	96°57′W
Cuilapa, Guat. (kô-ē-lä′pä)	8	14°16′N	90°20′W
Cuitzeo, Mex. (kwēt′zā-ō)	6	19°57′N	101°11′W
Cuitzeo, Laguna de, l., Mex. (lä-ô′nä-dĕ-kwēt′zā-ō)	6	19°58′N	101°05′W
Cul de Sac, pl., Haiti (kōō′l dĕ sä′k)	11	18°35′N	72°05′W
Culebra, i., P.R. (kōō-lā′brä)	5b	18°19′N	65°32′W
Culiacán, Mex. (kōō-lyä-kä′n)	4	24°45′N	107°30′W
Cumaná, Ven.	18	10°28′N	64°10′W
Cundinamarca, dept., Col. (kōōn-dê-nä-mä′r-kå)	18a	4°57′N	74°27′W
Cunduacán, Mex. (kôn-dōō á kän′)	7	18°04′N	93°23′W
Cunha, Braz. (kōō′nyá)	17a	23°05′S	44°56′W
Cupula, Pico, mtn., Mex. (pē′kō-kōō′pōō-lä)	4	24°45′N	111°10′W
Cuquío, Mex. (kōō-kē′ō)	6	20°55′N	103°03′W
Curaçao, i., Neth. Ant. (kōō-rä-sä′ō)	18	12°12′N	68°58′W
Curacautín, Chile (kä-rä-käōō-tē′n)	20	38°25′S	71°53′W
Curaumilla, Punta, c., Chile (kōō-rou-mē′lyä)	17b	33°05′S	71°44′W
Curepto, Chile (kōō-rĕp-tô)	17b	35°06′S	72°02′W
Curitiba, Braz. (kōō-rē-tē′bá)	19	25°20′S	49°15′W
Curly Cut Cays, is., Bah.	10	23°40′N	77°40′W
Currais Novos, Braz. (kōōr-rä′ēs nŏ-vōs)	19	6°02′S	36°39′W
Current, i., Bah. (kŭ-rĕnt)	10	25°20′N	76°50′W
Curupira, Serra, mts., S.A. (sĕr′rȧ kōō-rōō-pē′rȧ)	18	1°00′N	65°30′W
Cururupu, Braz. (kōō-rô-rô-pōō′)	19	1°40′S	44°56′W
Curvelo, Braz. (kôr-vĕl′ô)	19	18°47′S	44°14′W
Cusco, Peru	18	13°36′S	71°52′W
Cutzamalá, r., Mex. (kōō-tzä-mä-lä′)	6	18°57′N	100°41′W
Cutzamalá de Pinzón, Mex. (kōō-tzä-mä-lä′dĕ-pēn-zô′n)	6	18°28′N	100°36′W
Cuyotenango, Guat. (kōō-yô-tĕ-näŋ′gô)	8	14°30′N	91°35′W
Cuyuni, r., S.A. (kōō-yōō′nē)	19	6°40′N	60°44′W
Cuyutlán, Mex. (kōō-yōō-tlän′)	6	18°54′N	104°04′W

D

PLACE (Pronunciation)	PAGE	LAT.	LONG.
Dabeiba, Col. (dá-bā′bä)	18a	7°01′N	76°16′W
Dajabón, Dom. Rep. (dä-ĸä-bô′n)	11	19°35′N	71°40′W
Damas Cays, is., Bah. (dä′mäs)	10	23°50′N	79°50′W
Dame Marie, Cap, c., Haiti (dām márē′)	11	18°35′N	74°50′W
Darby, i., Bah.	10	23°50′N	76°20′W
Darien, Col. (dā-rĭ-ĕn′)	18a	3°56′N	76°30′W
Darién, Cordillera de, mts., Nic.	8	13°00′N	85°42′W
Darien, Serranía del, mts.	9	8°13′N	77°28′W
Darwin, Cordillera, mts., Chile (kôr-dēl-yĕ′rä-där′wên)	20	54°40′S	69°30′W
David, Pan. (dá-vēdh′)	5	8°27′N	82°27′W
Deán Funes, Arg. (dĕ-á′n-fōō-nĕs)	20	30°26′S	64°12′W
Dedo do Deus, mtn., Braz. (dĕ-dô-dô-dĕ′ōōs)	20b	22°30′S	43°02′W
Degollado, Mex. (dā-gô-lyä′dō)	6	20°27′N	102°11′W
Delgada, Punta, c., Arg. (pōō′n-tä-dĕl gä′dä)	20	43°46′S	63°46′W
Denham, Mount, mtn., Jam.	5	18°20′N	77°30′W
Descalvado, Braz. (dĕs-käl-vá-dô)	17a	21°55′S	47°37′W
Deseado, r., Arg. (dā-sā-ä′dhō)	20	46°50′S	67°45′W
Desirade Island, i., Guad. (dā-zē-rás′)	9b	16°21′N	60°51′W
Desolación, i., Chile (dĕ-sô-lä-syô′n)	20	53°05′S	74°00′W
Devil's Island *see* Diable, Île du, i., Fr. Gu.	19	5°15′N	52°40′W
Diable, Île du (Devil's Island), i., Fr. Gu.	19	5°15′N	52°40′W
Diablo Heights, Pan. (dyá′blŏ)	4a	8°58′N	79°34′W
Diablotins, Morne, mtn., Dom.	9b	15°31′N	61°24′W
Diamantina, Braz.	19	18°14′S	43°32′W
Diamantino, Braz. (dê-á-män-tē′no)	19	14°22′S	56°23′W
Diana Bank, bk., (dī′ăn′á)	11	22°30′N	74°45′W
Diego de Ocampo, Pico, mtn., Dom. Rep. (pê′-kô-dyĕ′gô-dĕ-ŏ-kä′m-pô)	11	19°40′N	70°45′W
Diego Ramírez, Islas, is., Chile (dê ā′gō rä-mē′rāz)	20	56°15′S	70°15′W
Dios, Cayo de, i., Cuba (kä′yō-dĕ-dē-ōs′)	10	22°05′N	83°05′W
Diquis, r., C.R. (dê-kēs′)	9	8°59′N	83°24′W
Diriamba, Nic. (dēr-yäm′bä)	8	11°52′N	86°15′W
Distrito Federal, state, Braz. (dēs-trē′tô-fĕ-dĕ-rä′l)	19	15°49′S	47°39′W
Distrito Federal, state, Mex.	6	19°14′N	99°08′W
Divinópolis, Braz. (dē-vē-nô′pō-lês)	19	20°10′S	44°53′W
Doce, r., Braz. (dō′så)	19	19°01′S	42°14′W
Doce, Canal Número, can., Arg.	17c	36°47′S	59°00′W
Doce Leguas, Cayos de las, is., Cuba	10	20°55′N	79°05′W
Doctor Arroyo, Mex. (dōk-tōr′ är-rō′yô)	6	23°41′N	100°10′W
Dollar Harbor, b., Bah.	10	25°30′N	79°15′W
Dolores, Arg. (dô-lō′rĕs)	20	36°20′S	57°42′W
Dolores, Col.	18a	3°33′N	74°54′W
Dolores, Ur.	17c	33°32′S	58°15′W
Dolores Hidalgo, Mex. (dô-lô′rĕs-ē-däl′gō)	6	21°09′N	100°56′W
Domeyko, Cordillera, mts., Chile (kôr-dēl-yĕ′rä-dô-mā′kô)	18	20°50′S	69°02′W
Dominica, nation, N.A. (dŏ-mĭ-nē′kȧ)	5	15°30′N	60°45′W
Dominica Channel, strt., N.A.	9b	15°00′N	61°30′W
Dominican Republic, nation, N.A. (dô-mĭn′ĭ-kăn)	5	19°00′N	70°45′W
Dom Silvério, Braz. (doɴ-sêl-vĕ′ryō)	17a	20°09′S	42°57′W
Don Martín, Presa de, res., Mex. (prĕ′sä-dĕ-dôn-mär-tē′n)	2	27°35′N	100°38′W
Dos Bahías, Cabo, c., Arg. (ká′bô-dôs-bä-ē′äs)	20	44°55′S	65°35′W
Dos Caminos, Ven. (dôs-kä-mê′nōs)	19b	9°38′N	67°17′W
Dourada, Serra, mts., Braz. (sĕ′r-rä-dôōō-rá′dă)	19	15°11′S	49°57′W
Drake Passage, strt., (drāk păs′ĭj)	15	57°00′S	65°00′W
Dr. Ir. W. J. van Blommestein Meer, res., Sur.	19	4°45′N	55°05′W
Duarte, Pico, mtn., Dom. Rep. (dŭŭ′ärtēh pĕcô)	5	19°00′N	71°00′W
Duas Barras, Braz. (dōō′äs-bá′r-räs)	17a	22°03′S	42°30′W
Duda, r., Col. (dōō′dä)	18a	3°25′N	74°23′W
Dulce, Golfo, b., C.R. (gōl′fô dōōl′sā)	5	8°25′N	83°13′W
Duque de Caxias, Braz. (dōō′kĕ-dĕ-ká′shyás)	17a	22°46′S	43°18′W
Durango, Mex. (dōō-rä′n-gô)	4	24°02′N	104°42′W
Durango, state, Mex.	4	25°00′N	106°00′W
Durazno, Ur. (dōō-räz′nō)	20	33°21′S	56°31′W
Durazno, dept., Ur.	17c	33°00′S	56°35′W
Dzibalchén, Mex. (zē-bäl-chĕ′n)	8a	19°25′N	89°39′W
Dzidzantún, Mex. (zēd-zän-tōō′n)	8a	21°18′N	89°00′W
Dzilam González, Mex. (ze-la′m-gōn-zä′lĕz)	8a	21°21′N	88°53′W
Dzitás, Mex. (zē-tá′s)	8a	20°47′N	88°32′W

E

PLACE (Pronunciation)	PAGE	LAT.	LONG.
East, Mount, mtn., Pan.	4a	9°09′N	79°46′W
East Caicos, i., T./C. Is. (kī′kōs)	11	21°40′N	71°35′W
Echandi, Cerro, mtn., N.A. (sĕ′r-rô-ĕ-chä′nd)	9	9°05′N	82°51′W
Ecuador, nation, S.A. (ĕk′wȧ-dôr)	18	0°00′N	78°30′W
Ejutla de Crespo, Mex. (å-hòt′lä dā krās′pö)	7	16°34′N	96°44′W
El Banco, Col. (ĕl băn′cô)	18	8°58′N	74°01′W
Elbow Cay, i., Bah.	10	26°25′N	76°55′W
El Cajón, Col. (ĕl-kä-kô′n)	18a	4°50′N	76°35′W
El Cambur, Ven. (käm-bōōr′)	19b	10°24′N	68°06′W
El Carmen, Chile (ká′r-mĕn)	17b	34°14′S	71°23′W
El Carmen, Col. (ká′r-mĕn)	18	9°54′N	75°12′W
El Cuyo, Mex.	8a	21°30′N	87°42′W
El Ebano, Mex. (ā-bä′nō)	6	22°13′N	98°26′W
El Espino, Nic. (ĕl-ĕs-pē′nŏ)	8	13°26′N	86°48′W
Eleuthera, i., Bah. (ê-lū′thĕr-ȧ)	5	25°05′N	76°10′W
Eleuthera Point, c., Bah.	11	24°35′N	76°05′W
El Grullo, Mex. (grōōl-yô)	6	19°46′N	104°10′W
El Guapo, Ven. (gwá′pô)	19b	10°07′N	66°00′W
El Hatillo, Ven. (ä-tē′l-yô)	19b	10°08′N	65°13′W
El Oro, Mex. (ô-rô)	6	19°49′N	100°04′W
El Pao, Ven. (ĕl pá′ō)	18	8°08′N	62°37′W
El Paraíso, Hond. (pä-rä-ē′sō)	8	13°55′N	86°35′W
El Pilar, Ven. (pē lä′r)	19b	9°56′N	64°48′W

ng-sing; ŋ-baŋk; ɴ-nasalized n; nŏd; cŏmmit; ōld; ôbey; ôrder; oi-boil; fōōd; ȯ-as oo in foot; ou-out; s-soft; sh-dish; th-thin; pūre; ůnite; ûrn; stŭd; circŭs; ü-as in French tu; ′-indeterminate vowel.

44 El Porvenir – Guacanayabo PRONOUNCING INDEX

PLACE (Pronunciation)	PAGE	LAT.	LONG.
El Porvenir, Pan. (pôr-vä-nēr′)	9	9°34′N	78°55′W
El Real, Pan. (rā-äl)	9	8°07′N	77°43′W
El Salto, Mex. (säl′tō)	6	23°48′N	105°22′W
El Salvador, nation, N.A.	4	14°00′N	89°30′W
El Sauce, Nic. (ĕl-sá′ō-sĕ)	8	13°00′N	86°40′W
El Tigre, Ven. (tē′grĕ)	18	8°49′N	64°15′W
El Triunfo, El Sal.	8	13°17′N	88°32′W
El Triunfo, Hond. (ĕl-trē-ōō′n-fō)	8	13°06′N	87°00′W
El Viejo, Nic. (ĕl-vyĕ′kō)	8	12°10′N	87°10′W
El Viejo, vol., Nic.	8	12°44′N	87°03′W
Emiliano Zapata, Mex. (ĕ-mē-lyä′nô-zä-pá′tà)	7	17°45′N	91°46′W
Encantada, Cerro de la, mtn., Mex. (sĕ′r-rô-dĕ-lä-ĕn-kän-tä′dä)	4	31°58′N	115°15′W
Encarnación, Para. (ĕn-kär-nä-syōn′)	20	27°26′S	55°52′W
Encarnación de Díaz, Mex. (ĕn-kär-nä-syōn dā dē′áz)	6	21°34′N	102°15′W
Encontrados, Ven. (ĕn-kōn-trá′dōs)	18	9°01′N	72°10′W
Enfer, Pointe d′, c., Mart.	9b	14°21′N	60°48′W
Engaño, Cabo, c., Dom. Rep. (ká′-bô- ĕn-gä-nô)	5	18°40′N	68°30′W
Enriquillo, Dom. Rep. (ĕn-rē-kē′l-yō)	11	17°55′N	71°15′W
Enriquillo, Lago, l., Dom. Rep. (lä′gô-ĕn-rē-kē′l-yô)	11	18°35′N	71°35′W
Enseñada, Arg.	17c	34°50′S	57°55′W
Ensenada, Mex. (ĕn-sĕ-nä′dä)	4	32°00′N	116°30′W
Entre Rios, prov., Arg.	20	31°30′S	59°00′W
Envigado, Col. (ĕn-vē-gá′dō)	18a	6°10′N	75°34′W
Erechim, Braz. (ĕ-rĕ-shē′n)	20	27°43′S	52°11′W
Escalón, Mex.	2	26°45′N	104°20′W
Escondido, r., Nic.	9	12°04′N	84°09′W
Escondido, Río, r., Mex. (rē′ō-ĕs-kōn-dē′dô)	2	28°30′N	100°45′W
Escudo de Veraguas, i., Pan. (ĕs-kōō′dä dā vä-rä′gwäs)	9	9°07′N	81°25′W
Escuinapa, Mex. (ĕs-kwē-nä′pä)	4	22°49′N	105°44′W
Escuintla, Guat. (ĕs-kwēn′tlä)	8	14°16′N	90°47′W
Ese, Cayos de, i., Col.	9	12°24′N	81°07′W
Esmeraldas, Ec. (ĕs-mā-räl′däs)	18	0°58′N	79°45′W
Esparta, C.R. (ĕs-pär′tä)	9	9°59′N	84°40′W
Esperanza, Cuba (ĕs-pĕ-rä′n-zä)	10	22°30′N	80°10′W
Espinal, Col. (ĕs-pē-näl′)	18	4°10′N	74°53′W
Espinhaço, Serra do, mts., Braz. (sĕ′r-rä-dô-ĕs-pē-ná-sô)	19	16°00′S	44°00′W
Espinillo, Punta, c., Ur. (pōō′n-tä-ĕs-pē-nē′l-yô)	17c	34°49′S	56°27′W
Espírito Santo, Braz. (ĕs-pē′rē-tô-sän′tō)	19	20°27′S	40°18′W
Espírito Santo, state, Braz.	19	19°57′S	40°58′W
Espíritu Santo, Bahía del, b., Mex.	8a	19°25′N	87°28′W
Espita, Mex. (ĕs-pē′tä)	8a	20°57′N	88°22′W
Esquel, Arg. (ĕs-kĕ′l)	20	42°47′S	71°22′W
Essequibo, r., Guy. (ĕs-ä-kē′bō)	19	4°26′N	58°17′W
Estância, Braz. (ĕs-tän′sĭ-ä)	19	11°17′S	37°18′W
Estrêla, r., Braz. (ĕs-trĕ′là)	20b	22°39′S	43°16′W
Estrondo, Serra do, mts., Braz. (sĕr′rá dô ĕs-trôn′-dō)	19	9°52′S	48°56′W
Etzatlán, Mex. (ĕt-zä-tlän′)	6	20°44′N	104°04′W
Extórrax, r., Mex. (ĕx-tó′ráx)	6	21°04′N	99°39′W
Extrema, Braz. (ĕsh-trĕ′mä)	17a	22°52′S	46°19′W
Exuma Sound, strt., Bah. (ĕk-sōō′mä)	10	24°20′N	76°20′W
Ezeiza, Arg. (ĕ-zā′zä)	20a	34°52′S	58°31′W

F

PLACE (Pronunciation)	PAGE	LAT.	LONG.
Fagnano, l., S.A. (fäk-nä′nō)	20	54°35′S	68°20′W
Fajardo, P.R.	5b	18°20′N	65°40′W
Falcón, dept., Ven. (fäl-kô′n)	19b	11°00′N	68°28′W
Falcon Reservoir, res., N.A. (fôk′n)	2	26°47′N	99°03′W
Falkland Islands, dep., S.A. (fôk′lănd)	20	50°45′S	61°00′W
Falmouth, Jam.	10	18°30′N	77°40′W
Famatina, Sierra de, mts., Arg.	20	29°00′S	67°50′W
Farallón, Punta, c., Mex. (pô′n-tä-fä-rä-lōn)	6	19°21′N	105°03′W
Faro, Braz. (fä′rō)	19	2°05′S	56°32′W
Fartura, Serra da, mts., Braz. (sĕ′r-rä-dá-fär-tōō′rä)	20	26°40′S	53°15′W
Federal, Distrito, dept., Ven. (dês-trē′tô-fĕ-dĕ-rä′l)	19b	10°34′N	66°55′W
Feia, Logoa, l., Braz. (lô-gôä-fĕ′yä)	17a	21°54′S	41°15′W
Feira de Santana, Braz. (fĕ′ê-rä dä sänt-än′ä)	19	12°16′S	38°46′W
Felipe Carrillo Puerto, Mex.	8a	19°36′N	88°04′W
Fernando de Noronha, Arquipélago, is., Braz.	19	3°51′S	32°25′W
Ferreñafe, Peru (fĕr-rĕn-yá′fĕ)	18	6°38′S	79°48′W
Filadelfia, C.R. (fīl-á-dĕl′fī-á)	8	10°26′N	85°37′W
Finlandia, Col. (fēn-lä′n-dēä)	18a	4°38′N	75°39′W
Fish Cay, i., Bah.	11	22°30′N	74°20′W
Fitzroy, Monte (Cerro Chaltel), mtn., S.A.	20	48°10′S	73°18′W
Flamingo Cay, i., Bah. (flá-mĭn′gō)	11	22°50′N	75°50′W
Flamingo Point, c., V.I.U.S.	5c	18°19′N	65°00′W
Florencia, Col. (flō-rĕn′sē-á)	18	1°31′N	75°13′W
Florencio Sánchez, Ur. (flō-rĕn-sēō-sá′n-chĕz)	17c	33°52′S	57°24′W
Florencio Varela, Arg. (flō-rĕn′sē-o vä-rä′lä)	20a	34°50′S	58°16′W
Flores, Braz. (flō′rĕzh)	19	7°57′S	37°48′W
Flores, Guat.	8a	16°53′N	89°54′W
Flores, dept., Ur.	17c	33°33′S	57°00′W
Flores, r., Arg.	17c	36°13′S	60°28′W
Floriano, Braz. (flō-rà-ä′nò)	19	6°17′S	42°58′W
Florianópolis, Braz. (flō-rē-ä-nō′pô-lēs)	20	27°30′S	48°30′W
Florida, Col. (flō-rē′dä)	18a	3°20′N	76°12′W
Florida, Cuba	10	22°10′N	79°50′W
Florida, Ur. (flô-rē-dhä)	20	34°06′S	56°14′W
Florida, dept., Ur. (flô-rē′dhä)	17c	33°48′S	56°15′W
Florida, Straits of, strt., N.A.	5	24°10′N	81°00′W
Florido, Río, r., Mex. (flô-rē′dō)	2	27°21′N	104°48′W
Fomento, Cuba (fō-mĕ′n-tō)	10	21°35′N	78°20′W
Fómeque, Col. (fō′mĕ-kĕ)	18a	4°29′N	73°52′W
Fonseca, Golfo de, b., N.A. (gôl-fô-dĕ-fōn-sā′kä)	4	13°09′N	87°55′W
Fonte Boa, Braz. (fōn′tá bō′á)	18	2°32′S	66°05′W
Fontera, Punta, c., Mex. (pōō′n-tä-fōn-tĕ′rä)	7	18°36′N	92°43′W
Fontibón, Col. (fōn-tē-bôn′)	18a	4°42′N	74°09′W
Formiga, Braz. (fōr-mē′gá)	19	20°27′S	45°25′W
Formigas Bank, bk., (fōr-mē′gäs)	11	18°30′N	75°40′W
Formosa, Arg. (fōr-mō′sä)	20	27°25′S	58°12′W
Formosa, Braz.	19	15°32′S	47°10′W
Formosa, prov., Arg.	20	24°30′S	60°45′W
Formosa, Serra, mts., Braz. (sĕ′r-rä)	19	12°59′S	55°11′W
Fortaleza, Braz. (fôr′tä-lā′zá)	19	3°35′S	38°31′W
Fort-de-France, Mart. (dē fräns)	5	14°37′N	61°06′W
Fortune, i., Bah.	11	22°35′N	74°20′W
Fragoso, Cayo, i., Cuba (kä′yō-frä-gō′sô)	10	22°45′N	79°30′W
Franca, Braz. (frä′n-kä)	19	20°28′S	47°20′W
Francés, Cabo, c., Cuba (kä′bô-frän-sĕ′s)	10	21°55′N	84°05′W
Francés, Punta, c., Cuba (pōō′n-tä-frän-sĕ′s)	10	21°45′N	83°10′W
Francés Viejo, Cabo, c., Dom. Rep. (ká′bô-frän′sås vyā′hō)	11	19°40′N	69°35′W
Francisco Sales, Braz. (frän-sē′s-kô-sá′lĕs)	17a	21°42′S	44°26′W
Fray Bentos, Ur. (frī bĕn′tōs)	20	33°10′S	58°19′W
Fraziers Hog Cay, i., Bah.	10	25°25′N	77°55′W
Fredonia, Col. (frē-dō′nyá)	18a	5°55′N	75°40′W
Freeport, Bah.	10	26°30′N	78°45′W
Freirina, Chile (frå-ĭ-rē′nä)	20	28°35′S	71°26′W
French Guiana, dep., S.A. (gē-ä′nä)	19	4°20′N	53°00′W
Fresnillo, Mex. (frás-nēl′yō)	4	23°10′N	102°52′W
Fresno, Col. (frĕs′nō)	18a	5°10′N	75°01′W
Frias, Arg. (frē-äs)	20	28°43′S	65°03′W
Frio, Cabo, c., Braz. (kä′bō-frē′ō)	19	22°58′S	42°08′W
Frontera, Mex. (frōn-tā′rä)	7	18°34′N	92°38′W
Fuego, vol., Guat. (fwā′gō)	8	14°29′N	90°52′W
Fuente, Mex. (fwĕ′n-tĕ′)	2	28°39′N	100°34′W
Fuerte, Río del, r., Mex. (rē′ō-dĕl-fōō-ĕ′r-tĕ)	4	26°15′N	108°50′W
Fuerte Olimpo, Para. (fwĕr′tā ō-lēm-pō)	20	21°10′S	57°49′W
Fundación, Col. (fōōn-dä-syō′n)	18	10°43′N	74°13′W
Furbero, Mex. (fōōr-bĕ′rō)	7	20°21′N	97°32′W
Furnas, Represa de, res., Braz.	19	21°00′S	46°00′W

G

PLACE (Pronunciation)	PAGE	LAT.	LONG.
Gaillard Cut, reg., Pan. (gä-ēl-yä′rd)	4a	9°03′N	79°42′W
Galapagos Islands *see* Colón, Archipiélago de, is., Ec.	18	0°10′S	87°45′W
Galeana, Mex. (gä-lā-ä′nä)	2	24°50′N	100°04′W
Galera, Cerro, mtn., Pan. (sĕ′r-rô-gä-lĕ′rä)	4a	8°55′N	79°38′W
Galeras, vol., Col. (gä-lĕ′räs)	18	0°57′N	77°27′W
Galina Point, c., Jam. (gä-lē′nä)	10	18°25′N	76°50′W
Gallinas, Punta de, c., Col. (gä-lyē′näs)	18	12°10′N	72°10′W
Garanhuns, Braz. (gä-rän-yônsh′)	19	8°49′S	36°28′W
García, Mex. (gär-sē′ä)	2	25°50′N	100°37′W
García de la Cadena, Mex.	6	21°14′N	103°26′W
Garín, Arg. (gä-rē′n)	20a	34°25′S	58°44′W
Garzón, Col. (gär-thōn′)	18	2°13′N	75°44′W
Gasper Hernández, Dom. Rep. (gäs-pär′ ĕr-nän′dāth)	11	19°40′N	70°15′W
Gastre, Arg. (gäs-trĕ)	20	42°12′S	68°50′W
Gatun, Pan. (gä-tōōn′)	9	9°16′N	79°25′W
Gatun, r., Pan.	4a	9°21′N	79°40′W
Gatún, Lago, l., Pan.	9	9°13′N	79°24′W
Gatun Locks, trans., Pan.	4a	9°16′N	79°57′W
General Alvear, Arg. (gĕ-nĕ-räl′äl-vē-á′r)	17c	36°04′S	60°02′W
General Arenales, Arg. (ä-rē-nä′lĕs)	17c	34°19′S	61°16′W
General Belgrano, Arg. (bĕl-grä′nô)	17c	35°45′S	58°32′W
General Cepeda, Mex. (sĕ-pĕ′dä)	2	25°24′N	101°29′W
General Conesa, Arg. (kô-nĕ′sä)	17c	36°30′S	57°19′W
General Guido, Arg. (gē′dô)	17c	36°41′S	57°48′W
General Lavalle, Arg. (lá-vá′l-yĕ)	17c	36°25′S	56°55′W
General Madariaga, Arg. (män-dá-rēä′gä)	20	36°59′S	57°14′W
General Paz, Arg. (pá′z)	17c	35°30′S	58°20′W
General Pedro Antonio Santos, Mex.	6	21°37′N	98°58′W
General Pico, Arg. (pē′kô)	20	36°46′S	63°44′W
General Roca, Arg. (rô-kä)	20	39°01′S	67°31′W
General San Martín, Arg. (sän-mär-tē′n)	20a	34°35′S	58°32′W
General Sarmiento (San Miguel), Arg.	20a	34°33′S	58°43′W
General Viamonte, Arg. (vēā′môn-tĕ)	17c	35°01′S	60°59′W
General Zuazua, Mex. (zwä′zwä)	2	25°54′N	100°07′W
Genovesa, i., Ec. (ĕ′s-lä-gĕ-nō-vē-sä)	18	0°08′N	90°15′W
George Town, Bah.	11	23°30′N	75°50′W
George Town, Cay. Is.	10	19°20′N	81°20′W
Georgetown, Guy. (jôrj′toun)	19	7°45′N	58°04′W
Geral, Serra, mts., Braz. (sĕr′rá zhâ-räl′)	20	28°30′S	51°00′W
Geral de Goiás, Serra, mts., Braz. (zhâ-räl′-dĕ-gô-yá′s)	19	14°22′S	45°40′W
Gibara, Cuba (hē-bä′rä)	10	21°05′N	76°10′W
Goascorán, Hond. (gō-äs′kō-rän′)	8	13°37′N	87°43′W
Gogorrón, Mex. (gō-gô-rōn′)	6	21°51′N	100°54′W
Goiânia, Braz. (gô-yá′nyä)	19	16°41′S	48°57′W
Goiás, Braz. (gô-yá′s)	19	15°57′S	50°10′W
Goiás, state, Braz.	19	16°00′S	48°00′W
Gold Hill, mtn., Pan.	4a	9°03′N	79°08′W
Golfito, C.R. (gôl-fē′tō)	9	8°40′N	83°12′W
Gómez Farías, Mex. (gō′māz fä-rē′äs)	2	24°59′N	101°02′W
Gómez Palacio, Mex. (pä-lä′syō)	4	25°35′N	103°30′W
Gonaïves, Haiti (gō-nà-ēv′)	5	19°25′N	72°45′W
Gonaïves, Golfe des, b., Haiti (gō-nà-ēv′)	11	19°20′N	73°20′W
Gonâve, Île de la, i., Haiti (gô-náv′)	5	18°50′N	73°30′W
Gonzales, Mex. (gôn-zá′lĕs)	6	22°47′N	98°26′W
González Catán, Arg. (gōn-zá′lĕz-kä-tá′n)	20a	34°47′S	58°39′W
Gorda, Punta, c., Cuba (pōō′n-tä-gôr-dä)	10	22°25′N	82°10′W
Gorda Cay, i., Bah. (gôr′dä)	10	26°05′N	77°30′W
Gospa, r., Ven. (gôs-pä)	19b	9°43′N	64°23′W
Gotera, El Sal. (gō-tā′rä)	8	13°41′N	88°06′W
Governador, Ilha do, i., Braz. (gô-vĕr-nä-dō-′r-ē-lá′dō)	20b	22°48′S	43°13′W
Governador Portela, Braz. (pōr-tĕ′lä)	20b	22°28′S	43°30′W
Governador Valadares, Braz. (vä-lä-dä′rĕs)	19	18°47′S	41°45′W
Governor′s Harbour, Bah.	10	25°15′N	76°15′W
Goya, Arg. (gō′yä)	20	29°06′S	59°12′W
Gracias, Hond. (grä′sē-äs)	8	14°35′N	88°37′W
Grama, Serra de, mtn., Braz. (sĕ′r-rä-dĕ-grá′má)	17a	20°42′S	42°28′W
Granada, Nic. (grä-nä′dhä)	4	11°55′N	85°58′W
Gran Bajo, reg., Arg. (grän′bà′kō)	20	47°35′S	68°45′W
Gran Chaco, reg., S.A. (grän′chá′kō)	20	25°30′S	62°15′W
Grand Bahama, i., Bah.	5	26°35′N	78°30′W
Grand Bourg, Guad. (grän bōōr′)	9b	15°54′N	61°20′W
Grand Caicos, i., T./C. Is. (grănd kä-ē′kōs)	11	21°45′N	71°50′W
Grand Cayman, i., Cay. Is. (kā′măn)	5	19°15′N	81°15′W
Grande, r., Arg.	17b	35°25′S	70°14′W
Grande, r., Bol.	18	16°49′S	63°19′W
Grande, r., Braz.	19	19°48′S	49°54′W
Grande, r., Mex.	7	17°37′N	96°41′W
Grande, r., Nic. (grän′dĕ)	9	13°01′N	84°21′W
Grande, r., Ur.	17c	33°19′S	57°15′W
Grande, Arroyo, r., Mex. (är-rō′yō-grä′n-dĕ)	6	23°30′N	98°45′W
Grande, Bahía, b., Arg. (bä-ē′ä-grán′dĕ)	20	50°45′S	68°00′W
Grande, Boca, mth., Ven. (bō′kä-grä′n-dĕ)	19	8°46′N	60°17′W
Grande, Cuchilla, mts., Ur. (kōō-chē′l-yä)	20	33°00′S	55°15′W
Grande, Ilha, i., Braz. (grän′dĕ)	17a	23°11′S	44°14′W
Grande, Salinas, l., Arg. (sä-lē′näs)	20	29°45′S	65°00′W
Grande, Salto, wtfl., Braz. (säl-tô)	19	16°18′S	39°38′W
Grande Cayemite, Île, i., Haiti	11	18°45′N	73°45′W
Grande de Otoro, r., Hond. (grä′dą dā ô-tô′rô)	8	14°42′N	88°21′W
Grande de Santiago, Río, r., Mex. (rēô-grä′n-dĕ-dĕ-sän-tyá′gô)	4	20°30′N	104°00′W
Grande Rivière du Nord, Haiti (rē-vyâr′ dü nôr′)	11	19°35′N	72°10′W
Grande Terre, i., Guad.	9b	16°28′N	61°13′W
Grande Vigie, Pointe de la, c., Guad. (gränd vē-gē′)	9b	16°32′N	61°25′W
Grand Turk, T./C. Is. (tûrk)	11	21°30′N	71°10′W
Grand Turk, i., T./C. Is.	11	21°30′N	71°10′W
Granito, Braz. (grä-nē′tō)	19	7°39′S	39°34′W
Granma, prov., Cuba	10	20°10′N	76°50′W
Gran Pajonal, reg., Peru (grä′n-pä-kō-näl′)	18	11°14′S	71°45′W
Gran Piedra, mtn., Cuba (grän-pyĕ′drä)	11	20°00′N	75°40′W
Grão Mogol, Braz. (groun′ mô-gôl′)	19	16°34′S	42°35′W
Grass Cay, i., V.I.U.S.	5c	18°22′N	64°50′W
Gravois, Pointe à, c., Haiti (grá-vwä′)	11	18°00′N	74°20′W
Great Abaco, i., Bah. (ä′bä-kō)	5	26°30′N	77°05′W
Great Bahama Bank, bk., (bá-hä′má)	10	25°00′N	78°50′W
Great Corn Island, i., Nic.	9	12°10′N	82°54′W
Greater Antilles, is., N.A.	5	20°30′N	79°15′W
Great Exuma, i., Bah. (ĕk-sōō′mä)	10	23°35′N	76°00′W
Great Guana Cay, i., Bah. (gwä′nä)	10	24°00′N	76°20′W
Great Harbor Cay, i., Bah. (kē)	10	25°45′N	77°50′W
Great Inagua, i., Bah. (ē-nä′gwä)	5	21°00′N	73°15′W
Great Isaac, i., Bah. (ī′zák)	10	26°05′N	79°05′W
Great Pedro Bluff, c., Jam.	10	17°50′N	78°05′W
Great Ragged, i., Bah.	11	22°10′N	75°45′W
Great Stirrup Cay, i., Bah. (stĭr-ŭp)	10	25°50′N	77°55′W
Green Cay, i., Bah.	10	24°05′N	77°10′W
Grenada, nation, N.A.	5	12°02′N	61°15′W
Grenadines, The, is., N.A. (grĕn′á-dēnz)	9b	12°37′N	61°35′W
Grenville, Gren.	9b	12°07′N	61°38′W
Grijalva, r., Mex. (grē-häl′vä)	7	17°25′N	93°23′W
Gruñidora, Mex. (grōō-nyē-dô′rō)	6	24°10′N	101°49′W
Guabito, Pan. (gwä-bē′tô)	9	9°30′N	82°33′W
Guacanayabo, Golfo de, b., Cuba (gôl-fô-dĕ-gwä-kä-nä-yä′bō)	10	20°30′N	77°40′W

ăt; finăl; rāte; senåte; ärm; åsk; sofá; fâre; ch-choose; dh-as th in other; bē; ēvent; bĕt; recĕnt; cratẽr; g-gō; gh-guttural g; bĭt; ĭ-short neutral; rīde; ĸ-guttural k as ch in German ich;

PRONOUNCING INDEX

Guacara – Iztaccíhuatl 45

PLACE (Pronunciation)	PAGE	LAT.	LONG.
Guacara, Ven. (gwá'kä-rä)	19b	10°16'N	67°48'W
Guadalajara, Mex. (gwä-dhä-lä-hä'rä)	4	20°41'N	103°21'W
Guadalcázar, Mex. (gwä-dhäl-kä'zär)	6	22°38'N	100°24'W
Guadalupe, Mex.	2	31°23'N	106°06'W
Guadalupe, i., Mex.	4	29°00'N	118°45'W
Guadeloupe, dep., N.A. (gwä-dē-lo͞op)	5	16°40'N	61°10'W
Guadeloupe Passage, strt., N.A.	9b	16°26'N	62°00'W
Guadiana, Bahía de, b., Cuba (bä-ē'ä-dē-gwä-dhē-ä'nä)	10	22°10'N	84°35'W
Guaira, Braz. (gwä-ē-rä)	19	24°03'S	54°02'W
Guaire, r., Ven. (gwī'rĕ)	19b	10°25'N	66°43'W
Guajaba, Cayo, i., Cuba (kä'yō-gwä-hä'bä)	10	21°50'N	77°35'W
Guajará Mirim, Braz. (gwä-zhä-rä'mē-rēN')	18	10°58'S	65°12'W
Guajira, Península de, pen., S.A.	18	12°35'N	73°00'W
Gualán, Guat. (gwä-län')	8	15°08'N	89°21'W
Gualeguay, Arg. (gwä-lĕ-gwä'y)	20	33°10'S	59°20'W
Gualeguay, r., Arg.	20	32°49'S	59°05'W
Gualicho, Salina, l., Arg. (sä-lē'nä-gwä-lē'chō)	20	40°20'S	65°15'W
Guamo, Col. (gwá'mō)	18a	4°02'N	74°58'W
Guanabacoa, Cuba (gwä-nä-bä-kō'ä)	5	23°08'N	82°19'W
Guanabara, Baía de, b., Braz.	17a	22°44'S	43°09'W
Guanacaste, Cordillera, mts., C.R.	8	10°54'N	85°27'W
Guanacevi, Mex. (gwä-nä-sĕ-vē')	4	25°30'N	105°45'W
Guanahacabibes, Península de, pen., Cuba	10	21°55'N	84°35'W
Guanajay, Cuba (gwä-nä-hī')	10	22°55'N	82°40'W
Guanajuato, Mex. (gwä-nä-hwä'tō)	4	21°01'N	101°16'W
Guanajuato, state, Mex.	4	21°00'N	101°00'W
Guanape, Ven. (gwä-nä'pĕ)	19b	9°55'N	65°32'W
Guanape, r., Ven.	19b	9°52'N	65°20'W
Guanare, Ven. (gwä-nä'rā)	18	8°57'N	69°47'W
Guanduçu, r., Braz. (gwä'n-do͞o'so͞o)	20b	22°50'S	43°40'W
Guane, Cuba (gwä'nā)	10	22°10'N	84°05'W
Guanta, Ven. (gwän'tä)	19b	10°15'N	64°35'W
Guantánamo, Cuba (gwän-tä'nä-mō)	11	20°10'N	75°10'W
Guantánamo, prov., Cuba	11	20°10'N	75°05'W
Guantánamo, Bahía de, b., Cuba	11	19°35'N	75°35'W
Guapiles, C.R. (gwä-pē-lĕs)	9	10°05'N	83°54'W
Guapimirim, Braz. (gwä-pē-mē-rē'N)	20b	22°31'S	42°59'W
Guaporé, r., S.A. (gwä-pō-rā')	18	12°11'S	63°47'W
Guaqui, Bol. (guä'kē)	18	16°42'S	68°47'W
Guarabira, Braz. (gwä-rä-bē'rá)	19	6°49'S	35°27'W
Guaranda, Ec. (gwä-rän'dä)	18	1°39'S	78°57'W
Guarapari, Braz. (gwä-rä-pä'rē)	19	20°34'S	40°31'W
Guarapiranga, Represa do, res., Braz.	17a	23°45'S	46°44'W
Guarapuava, Braz. (gwä-rä-pwä'vá)	20	25°29'S	51°26'W
Guaribe, r., Ven. (gwä-rē'bĕ)	19b	9°48'N	65°17'W
Guárico, dept., Ven.	19b	9°42'N	67°25'W
Guarulhos, Braz. (gwá-rō'l-yôs)	17a	23°28'S	46°30'W
Guarus, Braz. (gwá'ro͞os)	17a	21°44'S	41°19'W
Guasca, Col. (gwäs'kä)	18a	4°52'N	73°52'W
Guasipati, Ven. (gwä-sē-pä'tē)	19	7°26'N	61°57'W
Guatemala, Guat. (guä-tä-mä'lä)	4	14°37'N	90°32'W
Guatemala, nation, N.A.	4	15°45'N	91°45'W
Guatire, Ven. (gwä-tē'rĕ)	19b	10°28'N	66°34'W
Guaviare, r., Col.	18	3°35'N	69°28'W
Guayabal, Cuba (gwä-yä-bä'l)	10	20°40'N	77°40'W
Guayalejo, r., Mex. (gwä-yä-lĕ'hō)	6	23°24'N	99°09'W
Guayama, P.R. (gwä-yä'mä)	5b	18°00'N	66°08'W
Guayamouc, r., Haiti	11	19°05'N	72°00'W
Guayaquil, Ec. (gwī-ä-kēl')	18	2°16'S	79°53'W
Guayaquil, Golfo de, b., Ec. (gôl-fô-dĕ)	18	3°03'S	82°12'W
Guaymas, Mex. (gwá'y-mäs)	4	27°49'N	110°58'W
Guayubin, Dom. Rep. (gwä-yo͞o-bē'n)	11	19°40'N	71°25'W
Guazacapán, Guat. (gwä-zä-kä-pän')	8	14°04'N	90°26'W
Güere, r., Ven. (gwĕ'rĕ)	19b	9°39'N	65°00'W
Guerrero, Mex. (gĕr-rā'rō)	2	26°47'N	99°20'W
Guerrero, Mex.	2	28°20'N	100°24'W
Guerrero, state, Mex.	4	17°45'N	100°15'W
Guia de Pacobaíba, Braz. (gwē'ä-dĕ-pä'kō-bī'bä)	20b	22°42'S	43°10'W
Guiana Highlands, mts., S.A.	15	3°20'N	60°00'W
Guichicovi, Mex. (gwē-chē-kō'vē)	7	16°58'N	95°10'W
Güigüe, Ven. (gwē'gwē)	19b	10°05'N	67°48'W
Güija, Lago, l., N.A. (gē'hä)	8	14°16'N	89°21'W
Güira de Melena, Cuba (gwē'rä dā mā-lā'nä)	10	22°45'N	82°30'W
Güiria, Ven. (gwē-rē'ä)	18	10°43'N	62°16'W
Guisisil, vol., Nic. (gē-sē-sēl')	8	12°40'N	86°11'W
Gurgueia, r., Braz.	19	8°12'S	43°49'W
Guri, Embalse, res., Ven.	18	7°30'N	63°00'W
Gurupi, Serra do, mts., Braz. (sĕ'r-rä-dō-go͞o-ro͞o-pē')	19	5°32'S	47°02'W
Gustavo A. Madero, Mex. (go͞os-tä'vô-á-mä-dĕ'rô)	6	19°29'N	99°07'W
Gutiérrez Zamora, Mex. (go͞o-tī-âr'rāz zä-mō'rä)	7	20°27'N	97°17'W
Guyana, nation, S.A. (gŭy'ănä)	19	7°45'N	59°00'W

H

PLACE (Pronunciation)	PAGE	LAT.	LONG.
Habana (La Habana), prov., Cuba (hä-vä'nä)	10	22°45'N	82°25'W
Haiti, nation, N.A. (hā'tĭ)	5	19°00'N	72°15'W
Hanábana, r., Cuba (hä-nä-bä'nä)	10	22°30'N	80°55'W
Hanover, i., Chile	20	51°00'S	74°45'W
Hans Lollick, i., V.I.U.S. (häns'lŏl'ĭk)	5c	18°24'N	64°55'W
Havana, Cuba	5	23°08'N	82°23'W
Hawks Nest Point, c., Bah.	11	24°05'N	75°30'W
Hecelchakán, Mex. (ā-sĕl-chä-kän')	7	20°10'N	90°09'W
Heredia, C.R. (ā-rā'dhē-ä)	9	10°04'N	84°06'W
Hermosillo, Mex. (ĕr-mô-sē'l-yō)	4	29°00'N	110°57'W
Herrero, Punta, Mex. (po͞o'n-tä-ĕr-rĕ'rô)	8a	19°18'N	87°24'W
Hidalgo, Mex. (ē-dhäl'gō)	6	24°14'N	99°25'W
Hidalgo, Mex.	2	27°49'N	99°53'W
Hidalgo, state, Mex.	4	20°45'N	99°30'W
Hidalgo del Parral, Mex. (ē-dä'l-gō-dĕl-pär-rá'l)	4	26°55'N	105°40'W
Hidalgo Yalalag, Mex. (ē-dhäl'gō-yä-lä-läg)	7	17°12'N	96°11'W
Highborne Cay, i., Bah. (hī'bôrn kē)	10	24°45'N	76°50'W
Higuero, Punta, c., P.R.	5b	18°21'N	67°11'W
Higuerote, Ven. (ē-gē-rô'tĕ)	19b	10°29'N	66°06'W
Higüey, Dom. Rep. (ē-gwĕ'y)	11	18°40'N	68°45'W
Hillaby, Mount, mtn., Barb. (hĭl'á-bĭ)	9b	13°15'N	59°35'W
Hinche, Haiti (hēn'chā) (änsh)	11	19°10'N	72°05'W
Hispaniola, i., N.A. (hĭ'spăn-ĭ-ō-lá)	5	17°30'N	73°15'W
Hoctún, Mex. (ôk-to͞o'n)	8a	20°52'N	89°10'W
Hog Cay, i., Bah.	11	23°35'N	75°30'W
Hogsty Reef, rf., Bah.	11	21°45'N	73°50'W
Holbox, Mex. (ôl-bō'x)	8a	21°33'N	87°19'W
Holbox, Isla, i., Mex. (ē's-lä-ôl-bō'x)	8a	21°40'N	87°21'W
Holguín, Cuba (ôl-gēn')	5	20°55'N	76°15'W
Holguín, prov., Cuba	11	20°40'N	76°15'W
Honda, Col. (hōn'dá)	18	5°13'N	74°45'W
Honda, Bahía, b., Cuba (bä-ē'ä-ô'n-dä)	10	23°10'N	83°20'W
Hondo, Río, r., N.A. (hon-dō')	8a	18°16'N	88°32'W
Honduras, nation, N.A. (hŏn-do͞o'rás)	4	14°30'N	88°00'W
Honduras, Gulf of, b., N.A.	4	16°30'N	87°30'W
Hool, Mex. (o͞o'l)	8a	19°32'N	90°22'W
Hopelchén, Mex. (o-pĕl-chĕ'n)	8a	19°47'N	89°51'W
Horconcitos, Pan. (ŏr-kôn-sē'-tôs)	9	8°18'N	82°11'W
Horn, Cape *see* Hornos, Cabo de, c., Chile	20	56°00'S	67°00'W
Hornos, Cabo de (Cape Horn), c., Chile	20	56°00'S	67°00'W
Horqueta, Para. (ôr-kĕ'tä)	20	23°20'S	57°00'W
Hoste, i., Chile (ôs'tā)	20	55°20'S	70°45'W
Hostotipaquillo, Mex. (ôs-tō'tĭ-pä-kēl'yō)	6	21°09'N	104°05'W
Hoto Mayor, Dom. Rep. (ô-tô-mä-yō'r)	11	18°45'N	69°10'W
Hotte, Massif de la, mts., Haiti	11	18°25'N	74°00'W
Huajicori, Mex. (wä-jē-kō'rē)	6	22°41'N	105°24'W
Huajuapan de León, Mex. (wäj-wä'pän dā lā-ón')	7	17°46'N	97°45'W
Huallaga, r., Peru (wäl-yä'gä)	18	8°12'S	76°34'W
Huamachuco, Peru (wä-mä-cho͞o'kō)	18	7°52'S	78°11'W
Huamantla, Mex. (wä-män'tlá)	7	19°18'N	97°54'W
Huamuxtitlán, Mex. (wä mo͞os tē tlän')	6	17°49'N	98°38'W
Huancavelica, Peru (wän'kä-vä-lē'kä)	18	12°47'S	75°02'W
Huancayo, Peru (wän-kä'yō)	18	12°09'S	75°04'W
Huanchaca, Bol. (wän-chä'kä)	18	20°09'S	66°40'W
Huánuco, Peru (wä-no͞o'kô)	18	9°50'S	76°17'W
Huánuni, Bol. (wä-no͞o'nē)	18	18°11'S	66°43'W
Huaquechula, Mex. (wä-kē-cho͞o'lä)	6	18°44'N	98°37'W
Huaral, Peru (wä-rä'l)	18	11°28'S	77°11'W
Huarás, Peru (öä'rá's)	18	9°32'S	77°29'W
Huascarán, Nevados, mts., Peru (wäs-kä-rän')	18	9°05'S	77°50'W
Huasco, Chile (wäs'kō)	20	28°32'S	71°16'W
Huatla de Jiménez, Mex. (wá'tlä-dĕ-kē-mĕ'nĕz)	7	18°08'N	96°49'W
Huatlatlauch, Mex. (wä'tlä-tlä-o͞o'ch)	6	18°40'N	98°04'W
Huatusco, Mex. (wä-to͞os'kô)	7	19°09'N	96°57'W
Huauchinango, Mex. (wä-o͞o-chē-näŋ'gô)	6	20°09'N	98°03'W
Huaunta, Nic. (wä-ô'n-tä)	9	13°30'N	83°32'W
Huaunta, Laguna, l., Nic. (lä-gô'nä-wä-ô'n-tä)	9	13°35'N	83°46'W
Huautla, Mex. (wä-o͞o'tlä)	6	21°04'N	98°13'W
Huaynamota, Río de, r., Mex. (rē'ō-dĕ-wäy-nä-mō'tä)	6	22°10'N	104°36'W
Huazolotitlán, Mex. (wäzô-lô-tlē-tlän')	7	16°18'N	97°55'W
Huehuetenango, Guat. (wā-wå-tå-näŋ'gô)	8	15°19'N	91°26'W
Huejotzingo, Mex. (wā-hô-tzĭŋ'gō)	6	19°09'N	98°24'W
Huejúcar, Mex. (wā-ho͞o'kär)	6	22°26'N	103°12'W
Huejuquilla el Alto, Mex. (wā-ho͞o-kēl'yä ĕl äl'tō)	6	22°42'N	103°54'W
Huejutla, Mex. (wā-ho͞o'tlä)	6	21°08'N	98°26'W
Huetamo de Núñez, Mex.	6	18°34'N	100°53'W
Hueycatenango, Mex. (wĕy-kä-tē-nä'n-gô)	6	17°31'N	99°10'W
Hueytlalpan, Mex. (wā'ī-tlăl'pän)	7	20°03'N	97°41'W
Huichapán, Mex. (wē-chä-pän')	6	20°22'N	99°39'W
Huila, dept., Col. (wē'lä)	18a	3°10'N	75°20'W
Huila, Nevado de, mtn., Col. (nĕ-vä-dô-dĕ-wē'lä)	18a	2°59'N	76°01'W
Huimanguillo, Mex. (wē-män-gēl'yō)	7	17°50'N	93°16'W
Huitzilac, Mex. (oē't-zē-lä'k)	7a	19°01'N	99°16'W
Huitzitzilingo, Mex. (wē-tzē-tzē-lē'n-go)	6	21°11'N	98°42'W
Huitzuco, Mex. (wē-tzo͞o'kō)	6	18°16'N	99°20'W
Huixquilucán, Mex. (oē'x-kē-lo͞o-kä'n)	7a	19°21'N	99°22'W
Humacao, P.R. (o͞o-mä-kä'ō)	5b	18°09'N	65°49'W
Humuya, r., Hond. (o͞o-mo͞o'yä)	8	14°38'N	87°36'W
Hurlingham, Arg. (o͞o'r-lēn-gäm)	20a	34°36'S	58°38'W
Hurricane Flats, bk., (hŭ-rĭ-kăn flăts)	10	23°35'N	78°30'W

I

PLACE (Pronunciation)	PAGE	LAT.	LONG.
Ibagué, Col.	18	4°27'N	75°14'W
Ibarra, Ec. (ė-bär'rä)	18	0°19'N	78°08'W
Ibiapaba, Serra da, mts., Braz. (sĕ'r-rä-dä-ē-byä-pá'bä)	19	3°30'S	40°55'W
Ica, Peru (ē'kä)	18	14°09'S	75°42'W
Içá (Putumayo), r., S.A.	18	3°00'S	69°00'W
Içana, Braz. (ē-sä'nä)	18	0°15'N	67°19'W
Iguaçu, r., Braz. (ē-gwä-so͞o')	20b	22°42'S	43°19'W
Iguala, Mex. (ė-gwä'lä)	6	18°18'N	99°34'W
Iguassú, r., S.A. (ē-gwä-so͞o')	20	25°45'S	52°30'W
Iguassú Falls, wtfl., S.A.	19	25°40'S	54°16'W
Iguatama, Braz. (ē-gwä-tá'mä)	17a	20°13'S	45°40'W
Iguatu, Braz. (ė-gwä-to͞o')	19	6°22'S	39°17'W
Ilhabela, Braz. (ē'lä-bĕ'lä)	17a	23°47'S	45°21'W
Ilha Grande, Baía de, b., Braz. (ēl'yá grän'dĕ)	17a	23°17'S	44°25'W
Ilhéus, Braz. (ē-lĕ'o͞os)	19	14°52'S	39°00'W
Illampu, Nevado, mtn., Bol. (nĕ-vá'dô-ēl-yäm-po͞o')	18	15°50'S	68°15'W
Illapel, Chile (ē-zhä-pĕ'l)	20	31°37'S	71°10'W
Illimani, Nevado, mtn., Bol. (nĕ-vá'dô-ēl-yė-mä'nė)	18	16°50'S	67°38'W
Ilo, Peru	18	17°46'S	71°13'W
Ilobasco, El Sal. (ė-lô-bäs'kô)	8	13°57'N	88°46'W
Ilopango, Lago, l., El Sal. (ē-lô-päŋ'gō)	8	13°48'N	88°50'W
Impameri, Braz.	19	17°44'S	48°03'W
Indio, r., Pan. (ē'n-dyô)	4a	9°13'N	79°28'W
Inferror, Laguna, l., Mex. (lä-gô'nä-ėn-fĕr-rôr)	7	16°18'N	94°40'W
Infiernillo, Presa de, res., Mex.	6	18°50'N	101°50'W
Ingles, Cayos, is., Cuba (kä-yōs-ē'n-glĕ's)	10	21°55'N	82°35'W
Inhambupe, Braz. (ėn-yäm-bo͞o'pä)	19	11°47'S	38°13'W
Inhomirim, Braz. (ē-nô-mė-rē'N)	20b	22°34'S	43°11'W
Inírida, r., Col. (ē nė rē'dä)	18	2°25'N	70°38'W
Inner Brass, i., V.I.U.S. (bräs)	5c	18°23'N	64°58'W
Ipiales, Col. (ē-pē-ä'lås)	18	0°48'N	77°45'W
Ipu, Braz. (ē-po͞o)	19	4°11'S	40°45'W
Iquique, Chile (ē-kē'kĕ)	18	20°16'S	70°07'W
Iquitos, Peru (ė-kē'tôs)	18	3°39'S	73°18'W
Irapuato, Mex. (ē-rä-pwä'tō)	6	20°41'N	101°24'W
Irazú, vol., C.R. (ē-rä-zo͞o')	9	9°58'N	83°54'W
Iriona, Hond. (ē-rē-ō'nä)	8	15°53'N	85°12'W
Irois, Cap des, c., Haiti	11	18°25'N	74°50'W
Isaacs, Mount, mtn., Pan. (ē-sä-á'ks)	4a	9°22'N	79°31'W
Isabela, i., Ec. (ē-sä-bā'lä)	18	0°47'S	91°35'W
Isabela, Cabo, c., Dom. Rep. (ká'bô-ē-sä-bĕ'lä)	11	20°00'N	71°00'W
Isabella, Cordillera, mts., Nic. (kôr-dēl-yĕ'rä-ē-sä-bĕlä)	8	13°20'N	85°37'W
Isla Mujeres, Mex. (ē's lä mo͞o hĕ'rĕs)	8a	21°26'N	86°53'W
Istmina, Col. (ēst-mē'nä)	18a	5°10'N	76°40'W
Itabaiana, Braz. (ē-tä-bä-yá-nä)	19	10°42'S	37°17'W
Itabapoana, Braz. (ē-tä'-bä-pôä'nä)	17a	21°19'S	40°58'W
Itabapoana, r., Braz.	17a	21°11'S	41°18'W
Itabirito, Braz. (ē-tä-bė-rē'tô)	17a	20°15'S	43°46'W
Itabuna, Braz. (ē-tä-bo͞o'nä)	19	14°47'S	39°17'W
Itacoara, Braz. (ē-tä-kô'ä-rä)	17a	21°41'S	42°04'W
Itacoatiara, Braz. (ē-tá-kwá-tyä'rá)	19	3°03'S	58°18'W
Itagüí, Col. (ē-tä'gwė)	18a	6°11'N	75°36'W
Itagui, r., Braz.	20b	22°53'S	43°43'W
Itaipava, Braz. (ē-tī-pá'-vä)	20b	22°23'S	43°09'W
Itaipu, Braz. (ē-tī'po͞o)	20b	22°58'S	43°02'W
Itaituba, Braz. (ē-tä'ī-to͞o'bá)	19	4°12'S	56°00'W
Itajaí, Braz. (ē-tä-zhī')	20	26°52'S	48°39'W
Itambi, Braz. (ē-tä'm-bė)	20b	22°44'S	42°57'W
Itapecerica, Braz. (ē-tä-pĕ-sĕ-rē'kä)	17a	20°29'S	45°08'W
Itapecuru-Mirim, Braz. (ē-tä-pĕ'ko͞o-ro͞o-mė-rēN')	19	3°17'S	44°15'W
Itaperuna, Braz. (ē-tä'pä-ro͞o'nä)	19	21°12'S	41°53'W
Itapetininga, Braz. (ē-tä-pĕ-tė-nē'N-gä)	19	23°37'S	48°03'W
Itapira, Braz. (ē-tá-pē'rá)	19	20°42'S	51°19'W
Itapira, Braz.	17a	22°27'S	46°47'W
Itatiaia, Pico da, mtn., Braz. (pē'-kô-dá-ē-tä-tyá'ēä)	19	22°18'S	44°41'W
Itatiba, Braz. (ē-tä-tē'bä)	17a	23°01'S	46°48'W
Itaúna, Braz. (ē-tä-o͞o'nä)	17a	20°05'S	44°35'W
Itu, Braz. (ē-to͞o')	17a	23°16'S	47°16'W
Ituango, Col. (ē-twäŋ'gō)	18	7°07'N	75°44'W
Ituiutaba, Braz. (ē-to͞o-ēo͞o-tä'bä)	19	18°56'S	49°17'W
Itumirim, Braz. (ē-to͞o-mė-rē'N)	17a	21°20'S	44°51'W
Itundujía Santa Cruz, Mex. (ė-to͞on-do͞o-hē'ä sä'n-tä kro͞o'z)	7	16°50'N	97°43'W
Iturbide, Mex. (ē'to͞or-bē'dhā)	8a	19°38'N	89°31'W
Ituzaingo, Arg. (ē-to͞o-zä-ē'n-gô)	20a	34°40'S	58°40'W
Iúna, Braz. (ē-o͞o'-nä)	17a	20°22'S	41°32'W
Ixcateopán, Mex. (ēs-kä-tä-ō-pän')	6	18°29'N	99°49'W
Ixhautlán, Mex. (ēs-wät-län')	6	20°41'N	98°01'W
Ixhuatán, Mex. (ēs-hwä-tän')	7	16°19'N	94°30'W
Ixmiquilpan, Mex. (ēs-mė-kēl'pän)	6	20°30'N	99°12'W
Ixtacalco, Mex. (ēs-tá-käl'kō)	7a	19°23'N	99°07'W
Ixtaltepec (Asunción), Mex. (ēs-täl-tē-pĕk')	7	16°33'N	95°04'W
Ixtapalapa, Mex. (ēs'tä-pä-lä'pä)	7a	19°21'N	99°06'W
Ixtapaluca, Mex. (ēs'tä-pä-lo͞o'kä)	7a	19°18'N	98°53'W
Ixtepec, Mex. (ėks-tĕ'pĕk)	7	16°37'N	95°09'W
Ixtlahuaca, Mex. (ēs-tlä-wä'kä)	6	19°34'N	99°46'W
Ixtlán de Juárez, Mex. (ēs-tlän' dā hwä'râz)	7	17°20'N	96°29'W
Ixtlán del Río, Mex. (ēs-tlän'dĕl rē'ō)	6	21°05'N	104°22'W
Izabal, Guat. (ē'zä-bäl')	8	15°23'N	89°10'W
Izabal, Lago, l., Guat.	8	15°30'N	89°04'W
Izalco, El Sal. (ē-zäl'kō)	8	13°50'N	89°40'W
Izamal, Mex. (ē-zä-mä'l)	8a	20°55'N	89°00'W
Iztaccíhuatl, mtn., Mex.	6	19°10'N	98°38'W

46 Jaboatão – Lobería

PRONOUNCING INDEX

J

PLACE (Pronunciation)	PAGE	LAT.	LONG.
Jaboatão, Braz. (zhä-bô-å-touɴ)	19	8°14'S	35°08'W
Jacala, Mex. (hä-kȧ'lä)	6	21°01'N	99°11'W
Jacaltenango, Guat. (hä-kȧl-tĕ-nȧŋ'gō)	8	15°39'N	91°41'W
Jacarézinho, Braz. (zhä-kä-rĕ'zê-nyô)	19	23°13'S	49°58'W
Jacmel, Haiti (zhȧk-mĕl')	11	18°15'N	72°30'W
Jaco, l., Mex. (hä'kō)	2	27°51'N	103°50'W
Jacobina, Braz. (zhä-kô-bē'nȧ)	19	11°13'S	40°30'W
Jacutinga, Braz. (zhȧ-ko͞o-tēn'gä)	17a	22°17'S	46°36'W
Jaén, Peru (ĸä-ĕ'n)	18	5°38'S	78°49'W
Jagüey Grande, Cuba (hä'gwā grän'dā)	10	22°35'N	81°05'W
Jaibo, r., Cuba (hä-ē'bō)	11	20°10'N	75°20'W
Jalacingo, Mex. (hä-lä-sĭŋ'gō)	7	19°47'N	97°16'W
Jalapa, Guat. (hä-lä'pȧ)	8	14°38'N	89°58'W
Jalapa de Díaz, Mex.	7	18°06'N	96°33'W
Jalapa del Marqués, Mex. (dĕl mär-kās')	7	16°30'N	95°29'W
Jalisco, Mex. (hä-lēs'kō)	6	21°27'N	104°54'W
Jalisco, state, Mex.	4	20°07'N	104°45'W
Jalostotitlán, Mex. (hä-lōs-tē-tlȧn')	6	21°09'N	102°30'W
Jalpa, Mex. (häl'pā)	7	18°12'N	93°06'W
Jalpa, Mex. (häl'pä)	6	21°40'N	103°04'W
Jalpan, Mex. (häl'pän)	6	21°13'N	99°31'W
Jaltepec, Mex. (häl-tå-pĕk')	7	17°20'N	95°15'W
Jaltipán, Mex. (häl-tå-pän')	7	17°59'N	94°42'W
Jaltocán, Mex. (häl-tô-kän')	6	21°08'N	98°32'W
Jamaica, nation, N.A.	5	17°45'N	78°00'W
Jamaica Cay, i., Bah.	11	22°45'N	75°55'W
Jamay, Mex. (hä-mī')	6	20°16'N	102°43'W
James Point, c., Bah.	10	25°20'N	76°30'W
James Ross, i., Ant.	15	64°20'S	58°20'W
Jamiltepec, Mex. (hä-mēl-tå-pĕk)	7	16°16'N	97°54'W
Januária, Braz. (zhä-nwä'rê-ä)	19	15°31'S	44°17'W
Japeri, Braz. (zhȧ-pĕ'rê)	20b	22°38'S	43°40'W
Japurá (Caquetá), r., S.A.	18	2°00'S	68°00'W
Jarabacoa, Dom. Rep. (ĸä-rä-bä-kô'ä)	11	19°05'N	70°40'W
Jaral del Progreso, Mex. (hä-räl dĕl prô-grā'sō)	6	20°21'N	101°05'W
Jardines, Banco, bk., Cuba (bä'n-kō-här-dē'nås)	10	21°45'N	81°40'W
Jari, r., Braz. (zhä-rē)	19	0°28'N	53°00'W
Jatibonico, Cuba (hä-tê-bô-nē'kô)	10	22°00'N	79°15'W
Jauja, Peru (ĸä-ò'ĸ)	18	11°43'S	75°32'W
Jaumave, Mex. (hou-mä'vå)	6	23°23'N	99°24'W
Javari, r., S.A. (ĸȧ-vä-rē)	18	4°25'S	72°07'W
Jequitinhonha, r., Braz. (zhĕ-kē-têŋ-ŏ'n-yä)	19	16°47'S	41°19'W
Jérémie, Haiti (zhā-rå-mē')	11	18°40'N	74°10'W
Jeremoabo, Braz. (zhĕ-rā-mō-ȧ'bō)	19	10°03'S	38°13'W
Jerez, Punta, c., Mex. (po͞o'n-tä-ĸĕ-rāz')	7	23°04'N	97°44'W
Jesús Carranza, Mex. (hē-so͞o's-kär-rȧ'n-zä)	7	17°26'N	95°01'W
Jicarón, Isla, i., Pan. (ĸē-kä-rōn')	9	7°14'N	81°41'W
Jiguani, Cuba (ĸē-gwä-nē')	10	20°20'N	76°30'W
Jigüey, Bahía, b., Cuba (bä-ĕ'ä-ĸē'gwā)	10	22°15'N	78°10'W
Jilotepeque, Guat. (ĸē-lô-tĕ-pĕ'kĕ)	8	14°39'N	89°36'W
Jiménez, Mex. (ĸê-mā'nåz)	6	24°12'N	98°29'W
Jiménez, Mex.	2	29°03'N	100°42'W
Jiménez, Mex.	2	27°09'N	104°55'W
Jiménez del Téul, Mex. (tĕ-o͞o'l)	6	21°28'N	103°51'W
Jinotega, Nic. (ĸē-nô-tā'gä)	8	13°07'N	86°00'W
Jinotepe, Nic. (ĸē-nô-tā'på)	8	11°52'N	86°12'W
Jipijapa, Ec. (ĸē-pê-hä'pä)	18	1°36'S	80°52'W
Jiquilisco, El Sal. (ĸē-kē-lē's-kô)	8	13°18'N	88°32'W
Jiquilpan de Juárez, Mex. (ĸē-kēl'pän dā hwä'rāz)	6	20°00'N	102°43'W
Jiquipilco, Mex. (hē-kē-pē'l-kô)	7a	19°32'N	99°37'W
Jitotol, Mex. (ĸē-tô-tōl')	7	17°03'N	92°54'W
João Pessoa, Braz.	19	7°09'S	34°45'W
João Ribeiro, Braz. (zhô-uɴ-rē-bā'rō)	17a	20°42'S	44°03'W
Jobabo, r., Cuba (hō-bä'bä)	10	20°50'N	77°15'W
Jocotepec, Mex. (jô-kō-tå-pĕk')	6	20°17'N	103°26'W
Joinville, Braz. (zhwăɴ-vēl')	20	26°18'S	48°47'W
Joinville, i., Ant.	15	63°00'S	53°30'W
Jojutla, Mex. (hō-ho͞o'tlȧ)	6	18°39'N	99°11'W
Jola, Mex. (ĸô'lä)	6	21°08'N	104°26'W
Jomulco, Mex. (hô-mo͞ol'kô)	6	21°08'N	104°24'W
Jonacatepec, Mex.	6	18°39'N	98°46'W
Jonuta, Mex. (hô-no͞o'tä)	7	18°07'N	92°09'W
Jorullo, Volcán de, vol., Mex. (vôl-kȧ'n-dĕ-hô-ro͞ol'yō)	6	18°54'N	101°38'W
José C. Paz, Arg.	20a	34°32'S	58°44'W
Joulter's Cays, is., Bah. (jōl'tẽrz)	10	25°20'N	78°10'W
Jovellanos, Cuba (hō-vĕl-yä'nōs)	10	22°50'N	81°10'W
Juan Aldama, Mex. (kòȧ'n-äl-dȧ'mä)	6	24°16'N	103°21'W
Juan Díaz, r., Pan. (ĸo͞oȧ'n-dê'äz)	4a	9°05'N	79°30'W
Juan Fernández, Islas de, is., Chile	15	33°30'S	79°00'W
Juan L. Lacaze, Ur. (ho͞oȧ'n-ĕ'lĕ-lä-kä'zĕ)	17c	34°25'S	57°28'W
Juan Luis, Cayos de, is., Cuba (ka-yōs-dĕ-hwän lo͞o-ēs')	10	22°15'N	82°00'W
Juárez, Arg. (ho͞oȧ'rĕz)	20	37°42'S	59°46'W
Juázeiro, Braz. (zho͞oȧ'zā'rô)	19	9°27'S	40°28'W
Juazeiro do Norte, Braz. (zho͞oȧ'zā'rô-dô-nôr-tĕ)	19	7°16'S	38°57'W
Júcaro, Cuba (ho͞o'kȧ-rô)	10	21°40'N	78°50'W
Juchipila, Mex. (ho͞o-chê-pē'lȧ)	6	21°26'N	103°09'W
Juchitán, Mex. (ho͞o-chê-tän')	4	16°15'N	95°00'W
Juchitlán, Mex. (ho͞o-chē-tlän)	6	20°05'N	104°07'W
Jucuapa, El Sal. (ĸo͞o-kwä'pä)	8	13°30'N	88°24'W
Juigalpa, Nic. (hwê-gäl'pä)	8	12°02'N	85°24'W
Juiz de Fora, Braz. (zhò-ēzh' dā fō'rä)	19	21°47'S	43°20'W
Jujuy, Arg. (ho͞o-hwē')	20	24°14'S	65°15'W
Jujuy, prov., Arg. (ho͞o-hwē')	20	23°00'S	65°45'W
Juliaca, Peru (ho͞o-lê-ä'kä)	18	15°26'S	70°12'W
Jumento Cays, is., Bah. (ho͞o-mĕn'tō)	11	23°05'N	75°40'W
Jundiaí, Braz.	19	23°11'S	46°52'W
Junín, Arg. (ho͞o-nē'n)	20	34°35'S	60°56'W
Junín, Col.	18a	4°47'N	73°39'W
Juruá, r., S.A.	18	5°30'S	67°30'W
Juruena, r., Braz. (zho͞o-ro͞oĕ'nä)	19	12°22'S	58°34'W
Jutiapa, Guat. (ho͞o-tê-ä'pä)	8	14°16'N	89°55'W
Juticalpa, Hond. (ho͞o-tê-käl'pä)	4	14°35'N	86°17'W
Juventino Rosas, Mex.	6	20°38'N	101°02'W
Juventud, Isla de la, i., Cuba	5	21°40'N	82°45'W
Juxtlahuaca, Mex. (ho͞os-tlȧ-hwä'kä)	6	17°20'N	98°02'W

K

PLACE (Pronunciation)	PAGE	LAT.	LONG.
Kaieteur Fall, wtfl., Guy. (kī-ĕ-to͞or')	19	4°48'N	59°24'W
Kámuk, Cerro, mtn., C.R. (sĕ'r-rô-kä-mo͞o'k)	9	9°18'N	83°02'W
Kanasín, Mex. (kä-nä-sē'n)	8a	20°54'N	89°31'W
Kantunilkin, Mex. (kän-to͞o-nēl-kē'n)	8a	21°07'N	87°30'W
Kingston, Jam.	5	18°00'N	76°45'W
Kingstown, St. Vin. (kĭngz'toun)	5	13°10'N	61°14'W

L

PLACE (Pronunciation)	PAGE	LAT.	LONG.
La Asunción, Ven. (lä ä-so͞on-syōn')	18	11°02'N	63°57'W
La Banda, Arg. (lä bän'dä)	20	27°48'S	64°12'W
La Barca, Mex. (lä bär'kȧ)	6	20°17'N	102°33'W
Laberinto de las Doce Leguas, is., Cuba	10	20°40'N	78°35'W
Laboulaye, Arg. (lä-bô'o͞o-lä-yĕ)	20	34°01'S	63°10'W
Lábrea, Braz. (lä-brā'ä)	18	7°28'S	64°39'W
La Calera, Chile (lä-kä-lĕ-rä)	17b	32°47'S	71°11'W
La Calera, Col.	18a	4°43'N	73°58'W
Lacantum, r., Mex. (lä-kän-to͞o'm)	7	16°13'N	90°52'W
La Catedral, Cerro, mtn., Mex. (sĕ'r-rô-lä-kä-tĕ-drȧ'l)	7a	19°32'N	99°31'W
La Ceiba, Hond. (lȧ sēbä)	4	15°45'N	86°52'W
La Ceja, Col. (lä-sĕ-ĸä)	18a	6°02'N	75°25'W
La Chorrera, Pan. (lȧchôr-rā'rä)	9	8°54'N	79°47'W
La Cruz, Col. (l*ȧ* kro͞oz')	18	1°37'N	77°00'W
La Cruz, C.R. (lä-kro͞o'z)	8	11°05'N	85°37'W
La Cuesta, C.R. (lä-kwĕ's-tä)	9	8°32'N	82°51'W
La Dorada, Col. (lä dô-rä'dȧ)	18	5°28'N	74°42'W
La Esperanza, Hond. (lä ĕs-på-rän'zä)	8	14°20'N	88°21'W
La Gaiba, Braz. (lä-gī'bä)	19	17°54'S	57°32'W
Lagarto, r., Pan. (lä-gä'r-tô)	4a	9°08'N	80°05'W
Lagartos, l., Mex. (lä-gä'r-tôs)	8a	21°32'N	88°15'W
Lagoa da Prata, Braz. (lȧ-gô'ä-dȧ-prä'tȧ)	17a	20°04'S	45°33'W
Lagoa Dourada, Braz. (lȧ-gô'ä-dò-rä'dä)	17a	20°55'S	44°03'W
Lagos de Moreno, Mex. (lä'gōs dā mô-rā'nō)	4	21°21'N	101°55'W
La Grita, Ven. (lä grē'tȧ)	18	8°02'N	71°59'W
La Guaira, Ven. (lä gwä'ê-rä)	18	10°36'N	66°54'W
Laguna, Braz. (lä-go͞o'nä)	20	28°19'S	48°42'W
Laguna, Cayos, is., Cuba (kä'yōs-lä-gò'nă)	10	22°15'N	82°45'W
Lagunillas, Bol. (lä-go͞o-nēl'yäs)	18	19°42'S	63°38'W
Lagunillas, Mex. (lä-go͞o-nê'l-yäs)	6	21°34'N	99°41'W
La Habana *see* Havana, Cuba	5	23°08'N	82°23'W
Laja, Río de la, r., Mex. (rê'ō-dĕ-lä-lȧ'ĸä)	6	21°17'N	100°57'W
Lajas, Cuba (lä'häs)	10	22°25'N	80°20'W
Lajeado, Braz. (lä-zhĕȧ'dô)	20	29°24'S	51°46'W
Lajes, Braz. (lȧ'zhĕs)	20	27°47'S	50°17'W
Lajinha, Braz. (lä-zhē'nyä)	17a	20°08'S	41°36'W
La Libertad, El Sal.	8	13°29'N	89°20'W
La Libertad, Guat. (lä lē-bĕr-tädh')	8	15°31'N	91°44'W
La Libertad, Guat.	8a	16°46'N	90°12'W
La Ligua, Chile (lä lē'gwä)	17b	32°21'S	71°13'W
La Luz, Mex. (lä lo͞oz')	6	21°04'N	101°19'W
Lamas, Peru (lä'mȧs)	18	6°24'S	76°41'W
Lambari, Braz. (läm-bȧ'rē)	17a	21°58'S	45°22'W
Lambayeque, Peru (läm-bä-yā'kå)	18	6°41'S	79°58'W
La Mesa, Col.	18a	4°38'N	74°27'W
La Mora, Chile (lä-mō'rä)	17b	32°28'S	70°56'W
Lampa, r., Chile (lȧ'm-pä)	17b	33°15'S	70°55'W
Lampazos, Mex. (läm-pä'zōs)	4	27°03'N	100°30'W
Lanús, Arg. (lä-no͞os')	20a	34°42'S	58°24'W
La Oroya, Peru (lä-ô-rŏ'yă)	18	11°30'S	76°00'W
La Palma, Pan. (lä-päl'mä)	9	8°25'N	78°07'W
La Pampa, prov., Arg.	20	37°25'S	67°00'W
Lapa Rio Negro, Braz. (lä-pä-rē'ō-nĕ'grô)	20	26°12'S	49°56'W
La Paz, Arg. (lä päz')	20	30°48'S	59°47'W
La Paz, Bol.	18	16°31'S	68°03'W
La Paz, Hond.	8	14°15'N	87°40'W
La Paz, Mex. (lä-pȧ'z)	6	23°39'N	100°44'W
La Paz, Mex.	4	24°00'N	110°15'W
La Piedad Cabadas, Mex. (lä pyå-dhädh' kä-bä'dhäs)	6	20°20'N	102°04'W
La Plata, Arg. (lä plä'tä)	20	34°54'S	57°57'W
La Quiaca, Arg. (lä kê-ä'kä)	20	22°15'S	65°44'W
Largo, Cayo, Cuba (kä'yō-lär'gō)	10	21°40'N	81°30'W
La Rioja, Arg. (lä rê-ōhä)	20	29°18'S	67°42'W
La Rioja, prov., Arg. (lä-rê-ô'kä)	20	28°45'S	68°00'W
La Romona, Dom. Rep. (lä-rä-mô'nä)	11	18°25'N	69°00'W
La Sabana, Ven. (lä-sä-bȧ'nä)	19b	10°38'N	66°24'W
La Sabina, Cuba (lä-sä-bē'nä)	11a	22°51'N	82°05'W
Lascahobas, Haiti (läs-kä-ō'bȧs)	11	19°00'N	71°55'W
Las Cruces, Mex. (läs-kro͞o'sĕs)	7	16°37'N	93°54'W
La Selle, Massif de, mtn., Haiti (lä'sĕl')	11	18°25'N	72°05'W
La Serena, Chile (lä-sĕ-rĕ'nä)	20	29°55'S	71°24'W
Las Flores, Arg. (läs flo'rês)	20	36°01'S	59°07'W
Las Juntas, C.R. (läs-ĸo͞o'n-täs)	8	10°15'N	85°00'W
Las Palmas, Pan.	9	8°08'N	81°30'W
Las Piedras, Ur. (läs-pyĕ'drȧs)	17c	34°42'S	56°08'W
Las Pilas, vol., Nic. (läs-pē'läs)	8	12°32'N	86°43'W
Las Rosas, Mex. (läs rō thäs)	7	16°24'N	92°23'W
Las Tablas, Pan. (läs tä'bläs)	9	7°48'N	80°16'W
Las Tres Vírgenes, Volcán, vol., Mex. (vē'r-hĕ-nĕs)	4	26°00'N	111°45'W
Las Tunas, prov., Cuba	10	21°05'N	77°00'W
Las Vacas, Mex. (läs-vȧ'käs)	7	16°24'N	95°48'W
Las Vegas, Chile (läs-vĕ'gäs)	17b	32°50'S	70°59'W
Las Vegas, Ven. (läs-vĕ'gäs)	19b	10°26'N	64°08'W
Las Vigas, Mex.	7	19°38'N	97°03'W
Las Vizcachas, Meseta de, plat., Arg.	20	49°35'S	71°00'W
Latacunga, Ec. (lä-tä-kòŋ'gä)	18	1°02'S	78°33'W
La Tortuga, Isla, i., Ven. (ê's-lä-lä-tôr-to͞o'gä)	18	10°55'N	65°18'W
La Unión, Chile (lä-o͞o-nyô'n)	20	40°15'S	73°04'W
La Unión, El Sal.	8	13°18'N	87°51'W
La Unión, Mex. (lä o͞on-nyōn')	6	17°59'N	101°48'W
Lautaro, Chile (lou-tä'rô)	20	38°40'S	72°24'W
La Vega, Dom. Rep. (lä-vĕ'gä)	11	19°15'N	70°35'W
La Victoria, Ven. (lä vêk-tō'rê-ä)	18	10°14'N	67°20'W
Lavras, Braz. (lä'vräzh)	17a	21°15'S	44°59'W
Le Borgne, Haiti (lĕ bōrn'y')	11	19°50'N	72°30'W
Leche, Laguna de, l., Cuba (lä-gò'nä-dĕ-lĕ'chĕ)	10	22°10'N	78°30'W
Lee Stocking, i., Bah.	10	23°45'N	76°05'W
Leeward Islands, is., N.A. (lē'wẽrd)	5	17°00'N	62°15'W
Le François, Mart.	9b	14°37'N	60°55'W
Le Maire, Estrecho de, strt., Arg. (ĕs-trĕ'chô-dĕ-lĕ-mī'rĕ)	20	55°15'S	65°30'W
Le Marin, Mart.	9b	14°28'N	60°55'W
Le Môle, Haiti (lĕ mōl')	11	19°50'N	73°20'W
Le Moule, Guad. (lĕ mo͞ol')	9b	16°19'N	61°22'W
Lempa, r., N.A. (lĕm'pä)	8	13°20'N	88°46'W
Lençóes Paulista, Braz. (lĕɴ-sôɴs' pou-lēs'tȧ)	20	22°30'S	48°45'W
Lençóis, Braz. (lĕn-sóis)	19	12°38'S	41°28'W
Léogane, Haiti (lā-ô-gan')	11	18°30'N	72°35'W
León, Mex. (lå-ōn')	4	21°08'N	101°41'W
León, Nic. (lĕ-ô'n)	4	12°28'N	86°53'W
Leopoldina, Braz. (lā-ô-pôl-dē'nä)	17a	21°32'S	42°38'W
Lerdo, Mex. (lĕr'dō)	4	25°31'N	103°30'W
Lerma, Mex. (lĕr'mä)	7	19°49'N	90°34'W
Lerma, Mex.	7a	19°17'N	99°30'W
Lerma, r., Mex.	6	20°14'N	101°50'W
Les Cayes, Haiti	11	18°15'N	73°45'W
Les Saintes Islands, is., Guad. (lā-săɴt')	9b	15°50'N	61°40'W
Lesser Antilles, is.	5	12°15'N	65°00'W
Leticia, Col. (lĕ-tê'syȧ)	18	4°04'S	69°57'W
Libano, Col. (lē'bȧ-nô)	18a	4°55'N	75°05'W
Liberia, C.R.	8	10°38'N	85°28'W
Libertad, Arg.	20a	34°42'S	58°42'W
Libertad de Orituco, Ven. (lē-bĕr-tä'd-dĕ-ŏ-rê-to͞o'kô)	19b	9°32'N	66°24'W
Libón, r., N.A.	11	19°30'N	71°45'W
Libres, Mex. (lē'brās)	7	19°26'N	97°41'W
Licancábur, Cerro, mtn., S.A. (sĕ'r-rô-lē-kȧn-kȧ'bo͞or)	20	22°45'S	67°45'W
Licantén, Chile (lē-kän-tĕ'n)	17b	34°58'S	72°00'W
Lima, Peru (lē'mä)	18	12°06'S	76°55'W
Lima Duarte, Braz. (dwä'r-tĕ)	17a	21°52'S	43°47'W
Limay, r., Arg. (lē-mä'ê)	20	39°50'S	69°15'W
Limeira, Braz. (lē-mā'rä)	17a	22°34'S	47°24'W
Limón, C.R. (lê-mōn')	5	10°01'N	83°02'W
Limón, Hond. (lē-mô'n)	8	15°53'N	85°34'W
Limón, r., Dom. Rep.	11	18°20'N	71°40'W
Limón, Bahía, b., Pan.	4a	9°21'N	79°58'W
Linares, Chile (lê-nä'rās)	20	35°51'S	71°35'W
Linares, Mex.	4	24°53'N	99°34'W
Linares, prov., Chile	17b	35°53'S	71°30'W
Lincoln, Arg. (lĭŋ'k*ŭ*n)	20	34°51'S	61°29'W
Lins, Braz. (lê'ɴs)	19	21°42'S	49°41'W
Little Abaco, i., Bah. (ä'bä-kō)	10	26°55'N	77°45'W
Little Bahama Bank, bk., (b*ȧ*-hā'm*ȧ*)	10	26°55'N	78°40'W
Little Cayman, i., Cay. Is. (kā'm*ȧ*n)	10	19°40'N	80°05'W
Little Corn Island, i., Nic.	9	12°19'N	82°50'W
Little Exuma, i., Bah. (ĕk-so͞o'mä)	11	23°25'N	75°40'W
Little Hans Lollick, i., V.I.U.S. (häns lôl'lĭk)	5c	18°25'N	64°54'W
Little Inagua, i., Bah. (ê-nä'gwä)	11	21°30'N	73°00'W
Little Isaac, i., Bah. (ī'z*ȧ*k)	10	25°55'N	79°00'W
Little San Salvador, i., Bah. (săn săl'vȧ-dôr)	11	24°35'N	75°55'W
Livingston, Guat.	8	15°50'N	88°45'W
Livramento, Braz. (lē-vrȧ-mĕ'n-tô)	20	30°46'S	55°21'W
Llanos, reg., S.A. (lyȧ'nŏs)	18	4°00'N	71°15'W
Llera, Mex. (lyā'rä)	6	23°16'N	99°03'W
Llullaillaco, Volcán, vol., S.A. (lyo͞o-lyī-lyä'kō)	20	24°50'S	68°30'W
Lobería, Arg. (lô-bĕ'rē'ä)	20	38°13'S	58°48'W

ăt; fin*ă*l; rāte; senåte; ärm; ȧsk; sof*ȧ*; fâre; ch-choose; dh-as th in other; bē; ēvent; bĕt; rec*ĕ*nt; cratẽr; g-gō; gh-guttural g; bĭt; *ĭ*-short neutral; rīde; ĸ-guttural k as ch in German ich;

PRONOUNCING INDEX Lobos – Misantla 47

PLACE (Pronunciation)	PAGE	LAT.	LONG.
Lobos, Arg. (lō´bȯs)	17c	35°10'S	59°08'W
Lobos, Cayo, i., Bah. (lō´bȯs)	10	22°25'N	77°40'W
Lobos, Isla de, i., Mex. (ē´s-lä-dĕ-lô´bōs)	7	21°24'N	97°11'W
Lobos de Tierra, i., Peru (lô´bō-dĕ-tyĕ´r-rä)	18	6°29'S	80°55'W
Loja, Ec. (lō´hä)	18	3°49'S	79°13'W
Lomas de Zamora, Arg. (lō´mäs dā zä mō´rä)	17c	34°46'S	58°24'W
Londrina, Braz. (lôn-drē´nä)	19	21°53'S	51°17'W
Lone Star, Nic.	9	13°58'N	84°25'W
Long, i., Bah.	5	23°25'N	75°10'W
Lontue, r., Chile (lōn-tōē´)	17b	35°20'S	70°45'W
Lorena, Braz. (lȯ-rā´nȧ)	17a	22°45'S	45°07'W
Loreto, Braz. (lȯ-rā´tō)	19	7°09'S	45°10'W
Lorica, Col. (lȯ-rē´kä)	18	9°14'N	75°54'W
Los Andes, Chile (ȧn´dĕs)	17b	32°44'S	70°36'W
Los Angeles, Chile (äŋ´hȧ-lās)	20	37°27'S	72°15'W
Los Bronces, Chile (lȯs brō´n-sĕs)	17b	33°09'S	70°18'W
Los Estados, Isla de, i., Arg. (ē´s-lä dĕ lȯs ĕs-dȯs)	20	54°45'S	64°25'W
Los Herreras, Mex. (ĕr-rā-räs)	2	25°55'N	99°23'W
Los Indios, Cayos de, is., Cuba (kä´vōs dĕ lȯs ē´n-dvȯ´s)	10	21°50'N	83°10'W
Los Llanos, Dom. Rep. (lȯs ē-lä´nōs)	11	18°35'N	69°30'W
Los Palacios, Cuba	10	22°35'N	83°15'W
Los Reyes, Mex.	4	19°35'N	102°29'W
Los Reyes, Mex.	7a	19°21'N	98°58'W
Los Santos, Pan. (săn´tȯs)	9	7°57'N	80°24'W
Los Teques, Ven. (tĕ´kĕs)	18	10°22'N	67°04'W
Los Vilos, Chile (vē´lȯs)	20	31°56'S	71°29'W
Lota, Chile (lŏ´tä)	20	37°11'S	73°14'W
Loxicha, Mex.	7	16°03'N	96°46'W
Lucea, Jam.	10	18°25'N	78°10'W
Lucrecia, Cabo, c., Cuba	11	21°05'N	75°30'W
Luis Moya, Mex. (lo͞oē´s-mô-yä)	6	22°26'N	102°14'W
Luján, Arg. (lo͞o´hän´)	17c	34°36'S	59°07'W
Luján, r., Arg.	17c	34°33'S	58°59'W
Luminárias, Braz. (lo͞o-mē-nȧ´ryäs)	17a	21°32'S	44°53'W
Luque, Para. (loo´kā)	20	25°18'S	57°17'W
Luziânia, Braz. (lo͞o-zyȧ´nēȧ)	19	16°17'S	47°44'W

M

PLACE (Pronunciation)	PAGE	LAT.	LONG.
Macau, Braz. (mä-kȧ´ō)	19	5°12'S	36°34'W
Macaya, Pico de, mtn., Haiti	11	18°25'N	74°00'W
Maceió, Braz.	19	9°40'S	35°43'W
Machado, Braz. (mä-shȧ-dô)	17a	21°42'S	45°55'W
Machala, Ec. (mä-chȧ´lä)	18	3°18'S	78°54'W
Machu Picchu, Peru (mȧ´chō-pē´k-chō)	18	13°07'S	72°34'W
Macuelizo, Hond. (mä-kwĕ-lē´zȏ)	8	15°22'N	88°22'W
Madeira, r., S.A.	18	6°48'S	62°43'W
Madera, vol., Nic.	8	11°27'N	85°30'W
Madre, Laguna, l., Mex. (lă-go͞o´nă mă´drȧ)	3	25°08'N	97°41'W
Madre, Sierra, mts., N.A. (sē-ĕ´r-rä-mȧ´drĕ)	7	15°55'N	92°40'W
Madre de Dios, r., S.A. (mä´drȧ dā dē-ōs´)	18	12°07'S	68°02'W
Madre de Dios, Archipiélago, is., Chile (mȧ´drȧ dā dē-ōs´)	20	50°40'S	76°30'W
Madre del Sur, Sierra, mts., Mex. (sē-ĕ´r-rä-mä´drȧ dĕlso͞or´)	4	17°35'N	100°35'W
Madre Occidental, Sierra, mts., Mex.	4	29°30'N	107°30'W
Madre Oriental, Sierra, mts., Mex.	4	25°30'N	100°45'W
Madureira, Serra do, mtn., Braz. (sĕ´r-rä-dȯ-mä-do͞o-rā´rȧ)	20b	22°49'S	43°30'W
Maestra, Sierra, mts., Cuba (sē-ĕ´r-rȧ-mä-ās´trä)	5	20°05'N	77°05'W
Mafra, Braz. (mä´frä)	20	26°21'N	49°59'W
Magallanes, Estrecho de (Strait of Magellan), strt., S.A.	20	52°30'S	68°45'W
Magdalena, Arg. (mäg-dä-lā´nä)	17c	35°05'S	57°32'W
Magdalena, Bol.	18	13°17'S	63°57'W
Magdalena, i., Chile	20	44°45'S	73°15'W
Magdalena, r., Col.	18	7°45'N	74°04'W
Magdalena, Bahía, b., Mex. (bä-ē´ä-mäg-dä-lā´nä)	4	24°30'N	114°00'W
Magellan, Strait of *see* Magallanes, Estrecho de, strt., S.A.	20	52°30'S	68°45'W
Magiscatzín, Mex. (mä-kēs-kät-zēn´)	6	22°48'N	98°42'W
Maigualida, Sierra, mts., Ven. (sē-ĕ´r-rȧ-mī-gwä´lē-dĕ)	18	6°30'N	65°50'W
Maipo, S.A.	20	34°08'S	69°51'W
Maipo, r., Chile (mī´pȯ)	17b	33°45'S	71°08'W
Maiquetía, Ven. (mī-kĕ-tē´ä)	18	10°37'N	66°56'W
Mala, Punta, c., Pan. (pō´n-tä-mä´lä)	9	7°32'N	79°44'W
Málaga, Col. (mȧ´lä-gȧ)	18	6°41'N	72°46'W
Maldonado, Ur. (mäl-dŏ-nȧ´dȏ)	20	34°54'S	54°57'W
Maldonado, Punta, c., Mex. (po͞o´n-tä)	6	16°18'N	98°34'W
Malinalco, Mex. (mä-lē-näl´kō)	6	18°54'N	99°31'W
Malinaltepec, Mex. (mä-lē-näl-tȧ-pĕk´)	6	17°01'N	98°41'W
Malpelo, Isla de, i., Col. (mäl-pā´lō)	18	3°55'N	81°30'W
Maltrata, Mex. (mäl-trä´tä)	7	18°48'N	97°16'W
Mamantel, Mex. (mä-mȧn-tĕl´)	7	18°36'N	91°06'W
Mamoré, r., S.A.	18	13°00'S	65°20'W
Managua, Cuba (mä-nä´gwä)	11a	22°58'N	82°17'W
Managua, Nic.	4	12°10'N	86°16'W
Managua, Lago de, l., Nic. (lȧ´gô-dĕ)	8	12°28'N	86°10'W
Manaus, Braz. (mä-nä´o͞ozh)	19	3°01'S	60°00'W
Mandinga, Pan. (män-dĭŋ´gä)	9	9°32'N	79°04'W
Mangabeiras, Chapada das, pl., Braz.	19	8°05'S	47°32'W
Mangaratiba, Braz. (män-gä-rä-tē´bȧ)	17a	22°56'S	44°03'W
Mangles, Islas de, Cuba (ē´s-läs-dĕ-mäŋ´glās) (măn´g´lz)	10	22°05'N	82°50'W
Mangueira, Lagoa da, l., Braz.	20	33°15'S	52°45'W
Manhuaçu, Braz. (män-ōȧ´so͞o)	17a	20°17'S	42°01'W
Manhumirim, Braz. (män-o͞o-mē-rē´ɴ)	17a	22°30'S	41°57'W
Manicuare, Ven. (mä-nē-kwä´rĕ)	19b	10°35'N	64°10'W
Manitqueira, Serra da, mts., Braz.	17a	22°40'S	45°12'W
Manizales, Col. (mä-nē-zä´lȧs)	18	5°05'N	75°31'W
Man of War Bay, b., Bah.	11	21°05'N	74°05'W
Man of War Channel, strt., Bah.	10	22°45'N	76°10'W
Manseriche, Pongo de, reg., Peru (pŏ´n-gȯ-dĕ-män-sĕ-rē´chĕ)	18	4°15'S	77°45'W
Manta, Ec. (män´tä)	18	1°03'S	80°16'W
Mantua, Cuba (män-to͞o´ȧ)	10	22°20'N	84°15'W
Manzanares, Col. (män-sä-nä´rĕs)	18a	5°15'N	75°09'W
Manzanillo, Cuba (män´zä-nēl´yō)	5	20°20'N	77°05'W
Manzanillo, Mex.	4	19°02'N	104°21'W
Manzanillo, Bahía de, b., Mex. (bä-ē´ä-dĕ-män-zä-nē´l-yō)	6	19°00'N	104°38'W
Manzanillo, Bahía de, b., N.A.	11	19°55'N	71°50'W
Manzanillo, Punta, c., Pan.	9	9°40'N	79°33'W
Mao, Dom. Rep.	11	19°35'N	71°10'W
Mapastepec, Mex. (ma-päs-tȧ-pĕk´)	7	15°24'N	92°52'W
Mapimí, Mex. (mä-pē-mē´)	2	25°50'N	103°50'W
Mapimí, Bolsón de, des., Mex. (bôl-sŏ´n-dĕ-mä-pē´mē)	2	27°27'N	103°20'W
Mar, Serra do, mts., Braz. (sĕr´rȧ dȯ mär´)	20	26°30'S	49°15'W
Maracaibo, Ven. (mä-rä-kī´bō)	18	10°38'N	71°45'W
Maracaibo, Lago de, l., Ven. (lä´gô-dĕ-mä-rä-kī´bō)	18	9°55'N	72°13'W
Maracay, Ven. (mä-rä-käy´)	18	10°15'N	67°35'W
Marajó, Ilha de, i., Braz.	19	1°00'S	49°30'W
Maranguape, Braz. (mä-räŋ-gwä´pĕ)	19	3°48'S	38°38'W
Maranhão, state, Braz. (mä-rän-youɴ)	19	5°15'S	45°52'W
Marañón, r., Peru (mä-rä-nyōn´)	18	4°26'S	75°08'W
Marapanim, Braz. (ma-ra-pa-nē´ɴ)	19	0°45'S	47°42'W
Maravatio, Mex. (mä-rȧ-vä´tē-ō)	6	19°54'N	100°25'W
Marcala, Hond. (mär-kä-lä)	8	14°08'N	88°01'W
Marchena, i., Ec. (ē´s-lä-mär-chĕ´nä)	18	0°29'N	90°31'W
Mar Chiquíta, Laguna, l., Arg. (lä-go͞o´nä-mȧr-chē-kē´tä)	17c	34°25'S	61°10'W
Marcos Paz, Arg. (mär-kōs´ päz)	17c	34°49'S	58°51'W
Mar de Espanha, Braz. (mär-dĕ-ĕs-pȧ´nyȧ)	17a	21°53'S	43°00'W
Mar del Plata, Arg. (mär dĕl- plä´ta)	20	37°59'S	57°35'W
Margarita, Pan. (mär-go͞o-rē´tä)	4a	9°20'N	79°55'W
Margarita, Isla de, i., Ven. (mä-gȧ-rē´tȧ)	18	11°00'N	64°15'W
Mariana, Braz. (mä-ryȧ´nä)	17a	20°23'S	43°24'W
Marianao, Cuba (mä-rē-ä-nä´ō)	5	23°05'N	82°26'W
Mariano Acosta, Arg. (mä-rēä´nō ä-kōs´tä)	20a	34°20'S	58°10'W
Marias, Islas, is., Mex. (mä-rē´äs)	4	21°30'N	106°40'W
Mariato, Punta, c., Pan.	9	7°17'N	81°09'W
Marie Galante, i., Guad. (mȧ-rē´ gȧ-länt´)	9b	15°58'N	61°05'W
Marília, Braz. (mä-rē´lyȧ)	19	22°02'S	49°48'W
Mariposa, Chile (mä-rē-pȯ´sä)	17b	35°33'S	71°21'W
Mariquita, Col. (mä-rē-kē´tä)	18a	5°13'N	74°52'W
Mariscal Estigarribia, Para.	20	22°03'S	60°28'W
Marisco, Ponta do, c., Braz. (pô´n-tä-dȯ-mä-rē´s-kō)	20b	23°01'S	43°17'W
Marls, The, b., Bah. (märls)	10	26°30'N	77°15'W
Mar Muerto, l., Mex. (mär-mōĕ´r-tô)	7	16°13'N	94°22'W
Maroa, Ven. (mä-rō´ä)	18	2°43'N	67°37'W
Maro Jarapeto, mtn., Col. (mä-rȯ-hä-rä-pĕ´tȯ)	18a	6°29'N	76°39'W
Maroni, r., S.A. (mȧ-rō´nē)	19	3°02'N	53°54'W
Marquês de Valença, Braz. (mȧr-kĕ´s-dĕ-vä-lĕ´n-sȧ)	17a	22°16'S	43°42'W
Marsh Harbour, Bah.	10	26°30'N	77°00'W
Martinique, dep., N.A. (mȧr-tē-nēk´)	5	14°50'N	60°40'W
Masatepe, Nic. (mä-sä-tĕ´pĕ)	8	11°57'N	86°10'W
Masaya, Nic. (mä-sä´yä)	8	11°58'N	86°05'W
Mascota, Mex. (mäs-kō´tä)	6	20°33'N	104°45'W
Mascota, r., Mex.	6	20°33'N	104°52'W
Matagalpa, Nic. (mä-tä-gäl´pä)	4	12°52'N	85°57'W
Matamoros, Mex. (mä-tä-mō´rȯs)	2	25°32'N	103°13'W
Matamoros, Mex.	4	25°52'N	97°30'W
Matanzas, Cuba (mä-tän´zäs)	5	23°05'N	81°35'W
Matanzas, prov., Cuba	10	22°45'N	81°20'W
Matanzas, Bahía, b., Cuba (bä-ē´ä)	10	23°10'N	81°30'W
Matapalo, Cabo, c., C.R. (kȧ´bȯ-mä-tä-pä´lō)	9	8°22'N	83°25'W
Mataquito, r., Chile (mä-tä-kē´tȯ)	17b	35°08'S	71°35'W
Matehuala, Mex. (mä-tȧ-wä´lä)	4	23°38'N	100°39'W
Matias Barbosa, Braz. (mä-tē´äs-bȧr-bô-sä)	17a	21°53'S	43°19'W
Matillas, Laguna, l., Mex. (lä-gō´nä-mä-tē´l-yäs)	7	18°02'N	92°36'W
Matina, C.R. (mä-tē´nä)	9	10°06'N	83°20'W
Matlalcueyetl, Cerro, mtn., Mex. (sĕ´r-rä-mä-tläl-kwĕ´yĕtl)	6	19°13'N	98°02'W
Mato Grosso, Braz. (mät´ō grōs´ō)	19	15°04'S	59°58'W
Mato Grosso, state, Braz.	19	14°38'S	55°36'W
Mato Grosso, Chapada de, hills, Braz. (shä-pä´dä-dĕ)	19	13°39'S	55°42'W
Mato Grosso do Sul, state, Braz.	19	20°00'S	56°00'W
Matthew Town, Bah. (măth´ū toun)	11	21°00'N	73°40'W
Maturín, Ven. (mä-to͞o-rēn´)	18	9°48'N	63°16'W
Maués, Braz. (mȧ-wĕ´s)	19	3°34'S	57°30'W
Maule, r., Chile (mȧ´ō-lĕ)	17b	35°45'S	70°50'W
Mayaguana, i., Bah.	11	22°25'N	73°00'W
Mayaguana Passage, strt., Bah.	11	22°20'N	73°25'W
Mayagüez, P.R. (mä-yä-gwāz´)	5	18°12'N	67°10'W
Mayari, r., Cuba	11	20°25'N	75°35'W
Mayas, Montañas, mts., N.A. (mŏntăŋ´äs mā´äs)	8a	16°43'N	89°00'W
May Pen, Jam.	10	18°00'N	77°25'W
Mayran, Laguna de, l., Mex. (lä-ō´nä-dĕ-mī-rän´)	4	25°40'N	102°35'W
Mazagão, Braz. (mä-zȧ-gou´ɴ)	19	0°05'S	51°27'W
Mazapil, Mex. (mä-zä-pēl´)	2	24°40'N	101°30'W
Mazatenango, Guat. (mä-zä-tȧ-näŋ´gō)	4	14°30'N	91°30'W
Mazatla, Mex.	7a	19°30'N	99°24'W
Mazatlán, Mex.	4	23°14'N	106°27'W
Mazatlán (San Juan), Mex. (mä-zä-tlän´) (san hwan´)	7	17°05'N	95°26'W
Mecapalapa, Mex. (mă-kä-pä-lä´pä)	7	20°32'N	97°52'W
Medanosa, Punta, c., Arg. (po͞o´n-tä-mĕ-dä-nô´sä)	20	47°50'S	65°53'W
Medellín, Col. (mȧ-dhĕl-yēn´)	18	6°15'N	75°34'W
Medellín, Mex. (mĕ-dĕl-yē´n)	7	19°03'N	96°08'W
Mejillones, Chile (mȧ-κē-lyō´nȧs)	20	23°07'S	70°31'W
Melipilla, Chile (mȧ-lē-pē´lyä)	20	33°40'S	71°12'W
Melo, Ur. (mā´lō)	20	32°18'S	54°07'W
Memo, r., Ven. (mĕ´mō)	19b	9°32'N	66°30'W
Mendes, Braz. (mĕ´n-dĕs)	20b	22°32'S	43°44'W
Mendoza, Arg. (mĕn-dō´sä)	20	32°48'S	68°45'W
Mendoza, prov., Arg.	20	35°10'S	69°00'W
Meogui, Mex. (mȧ-ō´gē)	2	28°17'N	105°28'W
Mercedario, Cerro, mtn., Arg. (mĕr-sȧ-dhä´rē-ō)	20	31°58'S	70°07'W
Mercedes, Arg.	17c	34°41'S	59°26'W
Mercedes, Arg. (mĕr-sā´dhās)	20	29°04'S	58°01'W
Mercedes, Ur.	20	33°17'S	58°04'W
Mercedita, Chile (mĕr-sĕ-dē´tä)	17b	33°51'S	71°10'W
Mercês, Braz. (mĕ-sĕ´s)	17a	21°13'S	43°20'W
Merendón, Serranía de, mts., Hond.	8	15°01'N	89°05'W
Mérida, Mex.	4	20°58'N	89°37'W
Mérida, Ven.	18	8°30'N	71°15'W
Mérida, Cordillera de, mts., Ven. (mĕ´rē dhä)	18	8°30'N	70°45'W
Merlo, Arg. (mĕr-lȯ)	20a	34°40'S	58°44'W
Merume Mountains, mts., Guy. (mĕr-ü´mĕ)	19	5°45'N	60°15'W
Mesquita, Braz.	20b	22°48'S	43°26'W
Meta, dept., Col. (mĕ´tä)	18a	3°28'N	74°07'W
Meta, r., S.A.	18	4°33'N	72°09'W
Metán, Arg. (mĕ-tȧ´n)	20	25°32'S	64°51'W
Metapán, El Sal. (mȧ-täpän´)	8	14°21'N	89°26'W
Metepec, Mex. (mȧ-tĕ-pĕk´)	6	18°56'N	98°31'W
Metepec, Mex.	6	19°15'N	99°36'W
Metztitlán, Mex. (mĕtz-tēt-län)	6	20°36'N	98°45'W
Mexicalcingo, Mex. (mĕ-kē-käl-sēn´go)	7a	19°13'N	99°34'W
Mexicali, Mex. (mȧk-sē-kä´lē)	4	32°28'N	115°29'W
Mexicana, Altiplanicie, plat., Mex.	6	22°00'N	102°00'W
Mexico, nation, N.A.	4	23°45'N	104°00'W
Mexico, Gulf of, b., N.A.	4	25°15'N	93°45'W
Mexico City, Mex. (mĕk´sĭ-kō)	4	19°28'N	99°09'W
Mexticacán, Mex. (mĕs´tē-kä-kän´)	6	21°12'N	102°43'W
Mezquital, Mex. (mȧz-kē-täl´)	6	23°30'N	104°20'W
Mezquitic, Mex. (mȧz-kē-tēk´)	6	22°25'N	103°43'W
Mezquitic, r., Mex.	6	22°25'N	103°45'W
Miacatlán, Mex. (mē´ä-kä-tlän´)	6	18°42'N	99°17'W
Miahuatlán, Mex. (mē´ä-wä-tlän´)	7	16°20'N	96°30'W
Miches, Dom. Rep. (mē´chĕs)	11	19°00'N	69°05'W
Mico, Punta, c., Nic. (po͞o´n-tä-mē´kô)	9	11°38'N	83°24'W
Mier, Mex. (myâr)	2	26°26'N	99°08'W
Mier y Noriega, Mex. (myâr´ē nȯ-rē-ā´gä)	6	23°28'N	100°08'W
Miguel Auza, Mex.	6	24°17'N	103°27'W
Miguel Pereira, Braz.	20b	22°27'S	43°28'W
Milpa Alta, Mex. (mē´l-pä-ȧ´l-tä)	7a	19°11'N	99°01'W
Mimiapán, Mex. (mē-myä-pȧn´)	7a	19°26'N	99°28'W
Mimoso do Sul, Braz. (mē-mô´sō-dȯ-so͞o´l)	17a	21°03'S	41°21'W
Minas, Cuba (mē´näs)	10	21°30'N	77°35'W
Minas, Ur. (mē´näs)	20	34°18'S	55°12'W
Minas, Sierra de las, mts., Guat. (syĕr´rȧ dā läs mē´näs)	8	15°08'N	90°25'W
Minas de Oro, Hond. (mē´näs-dĕ-dĕ-ŏ-rô)	8	14°52'N	87°19'W
Minas Novas, Braz. (mē´näzh nō´väzh)	19	17°20'S	42°19'W
Minatitlán, Mex. (mē-nä-tē-tlän´)	4	17°59'N	94°33'W
Minatitlán, Mex.	6	19°21'N	104°02'W
Mineral del Chico, Mex. (mē-nȧ-räl´dĕl chē´kȯ)	6	20°13'N	98°46'W
Mineral del Monte, Mex. (mē-nȧ-räl dĕl mōn´tȧ)	6	20°18'N	98°39'W
Miquihuana, Mex. (mē-kē-wä´nä)	6	23°36'N	99°45'W
Miracema, Braz. (mē-rä-sĕ´mä)	17a	21°24'S	42°10'W
Miracema do Tocantins, Braz.	19	9°34'S	48°24'W
Mirador, Braz. (mē-rȧ-dōr´)	19	6°19'S	44°12'W
Miraflores, Col. (mē-rä-flō´rās)	18	5°10'N	73°13'W
Miraflores, Peru	18	16°19'S	71°20'W
Miraflores Locks, trans., Pan.	4a	9°00'N	79°35'W
Miragoâne, Haiti (mē-rȧ-gwän´)	11	18°25'N	73°05'W
Miranda, Col. (mē-rä´n-dä)	18a	3°14'N	76°11'W
Miranda, Ven.	19b	10°09'N	68°24'W
Miranda, dept., Ven.	19b	10°17'N	66°41'W
Mira Por Vos Islets, is., Bah. (mē´rä pȯr vōs)	11	22°05'N	74°30'W
Mira Por Vos Pass, strt., Bah.	11	22°10'N	74°35'W
Mirebalais, Haiti (mēr-bȧ-lĕ´)	11	18°50'N	72°05'W
Mirim, Lagoa, l., S.A. (mē-rēɴ´)	20	33°00'S	53°15'W
Misantla, Mex. (mē-sän´tlä)	7	19°55'N	96°49'W

ng-sing; ŋ-baŋk; ɴ-nasalized n; nŏd; cŏmmit; ōld; ȯbey; ôrder; oi-boil; fo͞od; ō-as oo in foot; ou-out; s-soft; sh-dish; th-thin; pūre; ůnite; ûrn; stŭd; circŭs; ü-as in French tu; ´-indeterminate vowel.

48 Misery – Órganos PRONOUNCING INDEX

PLACE (Pronunciation)	PAGE	LAT.	LONG.
Misery, Mount, mtn., St. K./N. (mĭz´rē-ĭ)	9b	17°28'N	62°47'W
Misiones, prov., Arg. (mē-syō´näs)	20	27°00'S	54°30'W
Miskito, Cayos, is., Nic.	9	14°34'N	82°30'W
Misteriosa, Lago, l., Mex. (mēs-tē-ryô´sä)	8a	18°05'N	90°15'W
Misti, Volcán, vol., Peru	18	16°04'S	71°20'W
Mita, Punta de, c., Mex. (pōō´n-tä-dě-mē´tä)	6	20°44'N	105°34'W
Mixico, Guat. (mēs´kô)	8	14°37'N	90°37'W
Mixquiahuala, Mex. (mēs-kē-wä´lä)	6	20°12'N	99°13'W
Mixteco, r., Mex. (mēs-tā´kō)	6	17°45'N	98°10'W
Moca, Dom. Rep. (mô´kä)	11	19°25'N	70°35'W
Mochitlán, Mex. (mō-chē-tlän´)	6	17°10'N	99°19'W
Mococa, Braz. (mô-kô´kà)	17a	21°29'S	46°58'W
Moctezuma, Mex. (môk´tā-zōō´mä)	6	22°44'N	101°06'W
Moengo, Sur.	19	5°43'N	54°19'W
Mogi das Cruzes, Braz. (mô-gê-däs-krōō´sěs)	19	23°33'S	46°10'W
Mogi-Guaçu, r., Braz. (mô-gê-gwá´sōō)	17a	22°06'S	47°12'W
Mogi-Mirim, Braz. (mô-gê-mē-rē´ɴ)	17a	22°26'S	46°57'W
Molina, Chile (mō-lē´nä)	17b	35°07'S	71°17'W
Mollendo, Peru (mô-lyěn´dō)	18	17°02'S	71°59'W
Momostenango, Guat. (mô-môs-tā-näŋ´gô)	8	15°02'N	91°25'W
Momotombo, Nic.	8	12°25'N	86°43'W
Mompos, Col. (mōm-pōs´)	18	9°05'N	74°30'W
Mona Passage, strt., N.A. (mō´nä)	5	18°00'N	68°10'W
Monção, Braz. (mon-souɴ´)	19	3°39'S	45°23'W
Monclova, Mex. (môn-klō´vä)	4	26°53'N	101°25'W
Monkey River, Belize (mŭŋ´kĭ)	8a	16°22'N	88°33'W
Montalbán, Ven. (mōnt-äl-bän)	19b	10°14'N	68°19'W
Monte, Arg. (mŏ´n-tě)	17c	35°25'S	58°49'W
Monteagudo, Bol. (mōn´tå-ä-gōō´dhō)	18	19°49'S	63°48'W
Monte Caseros, Arg. (mŏ´n-tě-kä-sě´rôs)	20	30°16'S	57°39'W
Montecillos, Cordillera de, mts., Hond.	8	14°19'N	87°52'W
Monte Cristi, Dom. Rep. (mŏ´n-tě-krě´s-tē)	11	19°50'N	71°40'W
Monte Escobedo, Mex. (mōn´tå ěs-kô-bā´dhō)	6	22°18'N	103°34'W
Montego Bay, Jam. (mŏn-tē´gō)	5	18°30'N	77°55'W
Montemorelos, Mex. (mōn´tå-mō-rā´lōs)	4	25°14'N	99°50'W
Montería, Col. (mōn-tā-rā´ä)	18	8°47'N	75°57'W
Monteros, Arg. (môn-tě´rôs)	20	27°14'S	65°29'W
Monterrey, Mex. (mōn-těr-rā´)	4	25°43'N	100°19'W
Montes Claros, Braz. (môn-těs-klä´rôs)	19	16°44'S	43°41'W
Montevideo, Ur. (mōn´tå-vê-dhā´ō)	20	34°50'S	56°10'W
Montijo, Bahía, b., Pan. (bä-ē´ä môn-tē´hō)	5	7°36'N	81°11'W
Montserrat, dep., N.A. (mŏnt-sě-răt´)	5	16°48'N	63°15'W
Moquegua, Peru (mô-kā´gwä)	18	17°15'S	70°54'W
Morales, Guat. (mô-rä´lěs)	8	15°29'N	88°46'W
Morant Point, c., Jam. (mô-rănt´)	10	17°55'N	76°10'W
Morawhanna, Guy. (mô-rá-hwä´ná)	19	8°12'N	59°33'W
Morelia, Mex. (mô-rā´lyä)	4	19°43'N	101°12'W
Morelos, Mex. (mô-rā´lōs)	6	22°46'N	102°36'W
Morelos, Mex.	7a	19°41'N	99°29'W
Morelos, Mex.	2	28°24'N	100°51'W
Morelos, r., Mex.	2	25°27'N	99°35'W
Moreno, Arg. (mô-rě´nō)	20a	34°39'S	58°47'W
Mores, i., Bah. (mōrz)	10	26°20'N	77°35'W
Morne Gimie, St. Luc. (môrn´ zhē-mē´)	9b	13°53'N	61°03'W
Moroleón, Mex. (mô-rō-lā-ōn´)	6	20°07'N	101°15'W
Morón, Arg. (mo-rŏ´n)	17c	34°39'S	58°37'W
Morón, Cuba (mô-rōn´)	10	22°05'N	78°35'W
Morón, Ven. (mô-rŏ´n)	19b	10°29'N	68°11'W
Morrinhos, Braz. (mô-rěn´yōzh)	19	17°45'S	48°56'W
Morro do Chapéu, Braz. (mōr-ŏ dô-shä-pě´ōō)	19	11°34'S	41°03'W
Morteros, Arg. (môr-tě´tôs)	20	30°47'S	62°00'W
Mortes, Rio das, r., Braz. (rēō-däs-mô´r-těs)	17a	21°04'S	44°29'W
Mosquitos, Costa de, cst., Nic. (kôs-tä-dě-mōs-kē´tō)	9	12°05'N	83°49'W
Mosquitos, Gulfo de los, b., Pan. (gōō´l-fô-dě-lôs-mōs-kē´tōs)	5	9°17'N	80°59'W
Motagua, r., N.A. (mô-tä´gwä)	8	15°29'N	88°39'W
Motul, Mex. (mō-tōō´l)	8a	21°07'N	89°14'W
Mouchoir Bank, bk., (mōō-shwär´)	11	21°35'N	70°40'W
Mouchoir Passage, strt., T./C. Is.	11	21°05'N	71°05'W
Moura, Braz. (mō´rá)	19	1°33'S	61°38'W
Moyahua, Mex. (mô-yä´wä)	6	21°16'N	103°10'W
Moyobamba, Peru (mō-yô-bäm´bä)	18	6°12'S	76°56'W
Moyuta, Guat. (mô-ē-ōō´tä)	8	14°01'N	90°05'W
Muleros, Mex. (mōō-lā´rōs)	6	23°44'N	104°00'W
Mullins River, Belize	8a	17°08'N	88°18'W
Muna, Mex. (mōō´nä)	8a	20°28'N	89°42'W
Mundonueva, Pico de, mtn., Col. (pē´kô-dě-mōō´n-dô-nwě´vä)	18a	4°18'N	74°12'W
Muneco, Cerro, mtn., Mex. (sě´r-rô-mōō-ně´kō)	7a	19°13'N	99°20'W
Muniz Freire, Braz.	17a	20°29'S	41°25'W
Muqui, Braz. (mōō-kôê)	17a	20°56'S	41°20'W
Muriaé, r., Braz.	17a	21°20'S	41°40'W
Musinga, Alto, mtn., Col. (ä´l-tô-mōō-sē´n-gä)	18a	6°40'N	76°13'W
Mustique, i., St. Vin. (müs-tēk´)	9b	12°53'N	61°03'W
Mutum, Braz. (mōō-tōō´m)	17a	19°48'S	41°24'W
Muzquiz, Mex. (mōōz´kēz)	2	27°53'N	101°31'W

N

PLACE (Pronunciation)	PAGE	LAT.	LONG.
Nacaome, Hond. (nä-kä-ō´må)	8	13°32'N	87°28'W
Nadadores, Mex. (nä-dä-dō´räs)	2	27°04'N	101°36'W
Nadir, V.I.U.S.	5c	18°19'N	64°53'W
Nagarote, Nic. (nä-gä-rô´tě)	8	12°17'N	86°35'W
Nagua, Dom. Rep. (ná´gwä)	11	19°20'N	69°40'W
Nahuel Huapi, l., Arg. (nä´wĕl wä´pê)	20	41°00'S	71°30'W
Nahuizalco, El Sal. (nä-wê-zäl´kō)	8	13°50'N	89°43'W
Naica, Mex. (nä-ē´kä)	2	27°53'N	105°30'W
Naiguata, Pico, mtn., Ven. (pē´kô)	19b	10°32'N	66°44'W
Najusa, r., Cuba (nä-hōō´sä)	10	20°55'N	77°55'W
Nanacamilpa, Mex. (nä-nä-kä-mē´l-pä)	7a	19°30'N	98°33'W
Naolinco, Mex. (nä-o-lēŋ´kō)	7	19°39'N	96°50'W
Napo, r., S.A. (nä´pō)	18	1°49'S	74°20'W
Nare, Col. (nä´rě)	18a	6°12'N	74°37'W
Nassau, Bah. (năs´ô)	5	25°05'N	77°20'W
Natagaima, Col. (nä-tä-gī´mä)	18a	3°38'N	75°07'W
Natal, Braz. (nä-täl´)	19	6°00'S	35°13'W
Natividade, Braz. (nä-tê-vê-dä´dě)	19	11°43'S	47°34'W
Naucalpan de Juárez, Mex.	7a	19°28'N	99°14'W
Nauchampatepetl, mtn., Mex. (näōō-chäm-pä-tě´pětl)	7	19°32'N	97°09'W
Nautla, Mex. (nä-ōōt´lä)	4	20°14'N	96°44'W
Nava, Mex. (nä´vä)	2	28°25'N	100°44'W
Navajas, Cuba (nä-vä-häs´)	10	22°40'N	81°20'W
Navarino, i., Chile (nä-vä-rē´nô)	20	55°30'S	68°15'W
Navarro, Arg. (nä-vá´r-rō)	17c	35°00'S	59°16'W
Navassa, i., N.A. (ná-văs´á)	11	18°25'N	75°15'W
Navidad, Chile (nä-vē-dä´d)	17b	33°57'S	71°51'W
Navidad Bank, bk., (nä-vê-dädh´)	11	20°05'N	69°00'W
Navidade do Carangola, Braz. (nä-vē-dä´dô-ká-rän-gô´la)	17a	21°04'S	41°58'W
Navojoa, Mex. (nä-vô-kô´ä)	4	27°00'N	109°40'W
Nayarit, state, Mex. (nä-yä-rēt´)	4	22°00'N	105°15'W
Nayarit, Sierra de, mts., Mex. (sē-ĕ´r-rä-dě)	6	23°20'N	105°07'W
Nazaré da Mata, Braz. (dä-mä-tä)	19	7°46'S	35°13'W
Nazas, Mex. (nä´zäs)	2	25°14'N	104°08'W
Nazas, r., Mex.	4	25°30'N	104°40'W
Necochea, Arg. (nā-kô-chā´ä)	20	38°30'S	58°45'W
Negro, r., Arg.	20	39°50'S	65°00'W
Negro, r., N.A.	8	13°01'N	87°10'W
Negro, r., S.A.	17c	33°17'S	58°18'W
Negro, r., S.A. (nā´grò)	18	0°18'S	63°21'W
Negro, Cerro, mtn., Pan. (sě´-rrô-nā´grô)	9	8°44'N	80°37'W
Neiba, Dom. Rep. (nå-ē´bä)	11	18°30'N	71°20'W
Neiba, Bahía de, b., Dom. Rep.	11	18°10'N	71°00'W
Neiba, Sierra de, mts., Dom. Rep. (sē-ĕr´rä-dě)	11	18°40'N	71°40'W
Neira, Col. (nā´rä)	18a	5°10'N	75°32'W
Neiva, Col. (nå-ē´vä)(nā´vä)	18	2°55'N	75°16'W
Nepomuceno, Braz. (ně-pô-mōō-sě´no)	17a	21°15'S	45°13'W
Netherlands Guiana *see* Suriname, nation, S.A.	19	4°00'N	56°00'W
Neuquén, Arg. (nĕ-ô-kän´)	20	38°52'S	68°12'W
Neuquén, prov., Arg.	20	39°40'S	70°45'W
Neuquén, r., Arg.	20	38°45'S	69°00'W
Nevado, Cerro el, mtn., Col. (sě´r-rô-ĕl-nĕ-vä´dô)	18a	4°02'N	74°08'W
Neveri, r., Ven. (nĕ-vĕ-rē)	19b	10°13'N	64°18'W
Neves, Braz.	20b	22°51'S	43°06'W
Nevis, i., St. K./N. (nē´vĭs)	5	17°05'N	62°38'W
Nevis Peak, mtn., St. K./N.	9b	17°11'N	62°33'W
New Amsterdam, Guy. (ăm´stěr-dăm)	19	6°14'N	57°30'W
New Providence, i., Bah. (prŏv´ĭ-děns)	10	25°00'N	77°25'W
Nexapa, r., Mex. (něks-ä´pä)	6	18°32'N	98°29'W
Nezahualcóyotl, Mex.	7a	19°27'N	99°03'W
Nicaragua, nation, N.A. (nĭk-á-rä´gwá)	4	12°45'N	86°15'W
Nicaragua, Lago de, l., Nic. (lä´gô dě)	4	11°45'N	85°28'W
Nicchehabin, Punta, c., Mex. (pōō´n-tä-něk-chě-ä-bē´n)	8a	19°50'N	87°20'W
Nicholas Channel, strt., N.A. (nĭk´ô-lás)	10	23°30'N	80°20'W
Nicolás Romero, Mex. (nē-kô-lá´s rô-mě´rô)	7a	19°38'N	99°20'W
Nicolls Town, Bah.	10	25°10'N	78°00'W
Nicoya, C.R. (nē-kō´yä)	8	10°08'N	85°27'W
Nicoya, Golfo de, b., C.R. (gôl-fô-dě)	8	10°03'N	85°04'W
Nicoya, Península de, pen., C.R.	8	10°05'N	86°00'W
Nieuw Nickerie, Sur. (nē-nē´kě-rē´)	19	5°51'N	57°00'W
Nieves, Mex. (nyā´vås)	6	24°00'N	102°57'W
Nilahue, r., Chile (nē-lá´wě)	17b	34°36'S	71°50'W
Nilópolis, Braz. (nē-lô´pō-lēs)	17a	22°48'S	43°25'W
Nioaque, Braz. (nēō-ä´-kě)	19	21°14'S	55°41'W
Nipe, Bahía de, b., Cuba (bä-ē´ä-dě-nē´pä)	11	20°50'N	75°30'W
Nipe, Sierra de, mts., Cuba (sē-ĕ´r-rá-dě)	11	20°20'N	75°50'W
Niquero, Cuba (nē-kā´rō)	10	20°00'N	77°35'W
Niterói, Braz. (nē-tě-rô´ĭ)	19	22°53'S	43°07'W
Nochistlán (Asunción), Mex. (nô-chēs-tlän´)	6	21°23'N	102°52'W
Nochixtlón, Mex. (ä-sòn-syōn´)	7	17°28'N	97°12'W
Nogales, Mex. (nō-gä´lěs)	7	18°49'N	97°09'W
Nogales, Mex.	4	31°15'N	111°00'W
Nombre de Dios, Mex. (nôm-brě-dě-dyô´s)	6	23°50'N	104°14'W
Nombre de Dios, Pan. (nŏ´m-brě)	9	9°34'N	79°28'W
Norte, Punta, c., Arg. (pōō´n-tä-nōr´tě)	17c	36°17'S	56°46'W
Norte, Serra do, mts., Braz. (sě´r-rä-dô-nôr´te)	19	12°04'S	59°08'W
North Bight, b., Bah. (bīt)	10	24°30'N	77°40'W
North Bimini, i., Bah. (bĭ´mĭ-nê)	10	25°45'N	79°20'W
North Caicos, i., T./C. Is. (kī´kôs)	11	21°55'N	72°00'W
North Cat Cay, i., Bah.	10	25°35'N	79°20'W
Northeast Point, c., Bah.	11	21°25'N	73°00'W
Northeast Point, c., Bah.	11	22°45'N	73°50'W
Northeast Providence Channel, strt., Bah. (prŏv´ĭ-děns)	10	25°45'N	77°00'W
North Elbow Cays, is., Bah.	10	23°55'N	80°30'W
North Gamboa, Pan. (gäm-bô´ä)	9	9°07'N	79°40'W
North Point, c., Barb.	9b	13°22'N	59°36'W
Northwest Providence Channel, strt., Bah. (prŏv´ĭ-děns)	10	26°15'N	78°45'W
Nova Cruz, Braz. (nō´vá-krōō´z)	19	6°22'S	35°20'W
Nova Friburgo, Braz. (frê-bōōr´gô)	19	22°18'S	42°31'W
Nova Iguaçu, Braz. (nō´vä-ē-gwä-sōō´)	19	22°45'S	43°27'W
Nova Lima, Braz. (lē´mä)	17a	19°59'S	43°51'W
Nova Resende, Braz.	17a	21°12'S	46°25'W
Nudo Coropuna, mtn., Peru (nōō´dô kō-rō-pōō´nä)	18	15°53'S	72°04'W
Nudo de Pasco, mtn., Peru (dě pás´kō)	18	10°34'S	76°12'W
Nueva Armenia, Hond. (nwā´vä är-mā´nē-á)	8	15°47'N	86°32'W
Nueva Esparta, dept., Ven. (nwě´vä ěs-pä´r-tä)	19b	10°50'N	64°35'W
Nueva Gerona, Cuba (kě-rō´nä)	10	21°55'N	82°45'W
Nueva Palmira, Ur. (päl-mē´rä)	17c	33°53'S	58°23'W
Nueva San Salvador, El Sal.	8	13°41'N	89°16'W
Nueve, Canal Número, can., Arg.	17c	36°22'S	58°19'W
Nueve de Julio, Arg. (nwā´vå dā hōō´lyô)	20	35°26'S	60°51'W
Nuevitas, Cuba (nwā-vē´täs)	5	21°35'N	77°15'W
Nuevitas, Bahía de, b., Cuba (bä-ē´ä dě nwā-vē´täs)	10	21°30'N	77°05'W
Nuevo Laredo, Mex. (lä-rā´dhō)	4	27°29'N	99°30'W
Nuevo León, state, Mex. (lå-ōn´)	4	26°00'N	100°00'W
Nuevo San Juan, Pan. (nwě´vô sän kōō-ä´n)	4a	9°14'N	79°43'W
Nurse Cay, i., Bah.	11	22°30'N	75°50'W

O

PLACE (Pronunciation)	PAGE	LAT.	LONG.
Oaxaca, Mex.	4	17°03'N	96°42'W
Oaxaca, state, Mex. (wä-hä´kä)	4	16°45'N	97°00'W
Oaxaca, Sierra de, mts., Mex. (sē-ĕ´r-rä dě)	7	16°15'N	97°25'W
Óbidos, Braz. (ō´bê-dòzh)	19	1°57'S	55°32'W
Ocampo, Mex. (ô-käm´pō)	6	22°49'N	99°23'W
Ocaña, Col. (ô-kän´yä)	18	8°15'N	73°37'W
Occidental, Cordillera, mts., Col.	18a	5°05'N	76°04'W
Occidental, Cordillera, mts., Peru	18	10°12'S	76°58'W
Ocean Bight, b., Bah.	11	21°15'N	73°15'W
Ocoa, Bahía de, b., Dom. Rep.	11	18°20'N	70°40'W
Ococingo, Mex. (ô-kô-sē´n-gô)	7	17°03'N	92°18'W
Ocom, Lago, l., Mex. (ô-kô´m)	8a	19°26'N	88°18'W
Ocós, Guat. (ô-kōs´)	8	14°31'N	92°12'W
Ocotal, Nic. (ō-kō-täl´)	8	13°36'N	86°31'W
Ocotepeque, Hond. (ô-kō-tā-pā´kå)	8	14°25'N	89°13'W
Ocotlán, Mex. (ō-kō-tlän´)	6	20°19'N	102°44'W
Ocotlán de Morelos, Mex. (dā mô-rā´lōs)	7	16°46'N	96°41'W
Ocozocoautla, Mex. (ô-kô´zô-kwä-ōō´tlä)	7	16°44'N	93°22'W
Ocumare del Tuy, Ven. (ō-kōō-mä´ra del twē´)	18	10°07'N	66°47'W
Oeiras, Braz. (wå-ē-räzh´)	19	7°05'S	42°01'W
O'Higgins, prov., Chile (ô-kē´gēns)	17b	34°17'S	70°52'W
Ojinaga, Mex. (ō-kē-nä´gä)	4	29°34'N	104°26'W
Ojitlán, Mex. (ōkē-tlän´)(sän-lōō´käs)	7	18°04'N	96°23'W
Ojo Caliente, Mex. (ōkō käl-yěn´tå)	6	21°50'N	100°43'W
Ojocaliente, Mex. (ô-kō-kä-lyě´n-tě)	6	22°39'N	102°15'W
Ojo del Toro, Pico, mtn., Cuba (pē´kô-ô-kō-děl-tô´rô)	10	19°55'N	77°25'W
Olanchito, Hond. (ō´län-chē´tô)	8	15°28'N	86°35'W
Olavarría, Arg. (ô-lä-vär-rē´ä)	20	36°49'N	60°15'W
Olazoago, Arg. (ô-läz-kôä´gô)	17c	35°14'S	60°37'W
Old Bahama Channel, strt., N.A. (bá-hā´má)	10	22°45'N	78°30'W
Old Bight, Bah.	11	24°15'N	75°20'W
Olinda, Braz. (ô-lē´n-dä)	19	8°00'S	34°58'W
Olinda, Braz.	20b	22°49'S	43°25'W
Oliveira, Braz. (ô-lē-vā´rä)	17a	20°42'S	44°49'W
Olivos, Arg. (ôlē´vōs)	20a	34°30'S	58°29'W
Ollagüe, Chile (ô-lyä´gā)	18	21°17'S	68°17'W
Omealca, Mex. (ōmå-äl´kō)	7	18°44'N	96°45'W
Ometepec, Mex. (ô-mā-tå-pěk´)	6	16°41'N	98°27'W
Omoa, Hond. (ô-mō´rä)	8	15°43'N	88°03'W
Omotepe, Isla de, i., Nic. (ē´s-lä-dě-ō-mō-tā´pä)	8	11°32'N	85°30'W
Onoto, Ven. (ô-nō´tô)	19b	9°38'N	65°03'W
Opalaca, Sierra de, mts., Hond. (sē-ĕ´r-rä-dě-ô-pä-lä´kä)	8	14°30'N	88°29'W
Opico, El Sal. (ô-pē´kō)	8	13°50'N	89°23'W
Orán, Arg. (ô-rá´n)	20	23°13'S	64°17'W
Orange, Cabo, c., Braz. (kå-bô-rá´n-zhě)	19	4°25'N	51°30'W
Orange Cay, i., Bah. (ŏr-ĕnj kē)	10	24°55'N	79°05'W
Orange Walk, Belize (wôl´k)	8a	18°09'N	88°32'W
Orchila, Isla, i., Ven.	18	11°47'N	66°34'W
Órganos, Sierra de los, mts., Cuba (sē-ĕ´r-rä-dě-lôs-ô´r-gä-nôs)	10	22°20'N	84°10'W

ăt; finăl; rāte; senåte; ärm; åsk; sofá; fâre; ch-choose; dh-as th in other; bē; ēvent; bět; recĕnt; cratẽr; g-gō; gh-guttural g; bĭt; ĭ-short neutral; rīde; ĸ-guttural k as ch in German ich;

PRONOUNCING INDEX

Orgãos – Pochutla 49

PLACE (Pronunciation)	PAGE	LAT.	LONG.
Orgãos, Serra das, mtn., Braz. (sĕ´r-rä-däs-ôr-gouɴ´s)	17a	22°30'S	43°01'W
Oriental, Cordillera, mts., Col. (kȯr-dēl-yĕ´rä)	18a	3°30'N	74°27'W
Oriental, Cordillera, mts., Dom. Rep. (kȯr-dēl-yĕ´rä-ô-ryĕ´n-täl)	11	18°55'N	69°40'W
Oriental, Cordillera, mts., S.A. (kȯr-dēl-yĕ´rä ō-rē-ĕn-täl´)	18	14°00'S	68°33'W
Orinoco, r., Ven. (ȯ-rĭ-nō´kȯ)	18	8°32'N	63°13'W
Orituco, r., Ven. (ȯ-rē-tōō´kō)	19b	9°37'N	66°25'W
Oriuco, r., Ven. (ȯ-rēōō´kō)	19b	9°36'N	66°25'W
Orizaba, Mex. (ȯ-rē-zä´bä)	4	18°52'N	97°05'E
Orizaba, Pico de, vol., Mex.	4	19°04'N	97°14'W
Oro, Río del, r., Mex. (rē´ō dĕl ō´rō)	6	18°04'N	100°59'W
Oro, Río del, r., Mex.	2	26°04'N	105°40'W
Orosi, vol., C.R. (ō-rō´sē)	8	11°00'N	85°30'W
Ortega, Col. (ôr-tĕ´gä)	18a	3°56'N	75°12'W
Oruro, Bol. (ȯ-rōō´rō)	18	17°57'S	66°59'W
Osa, Península de, pen., C.R. (ō´sä)	9	8°30'N	83°25'W
Osorno, Chile (ȯ-sō´r-nō)	20	40°42'S	73°13'W
Otavalo, Ec. (ōtä-vä'lō)	18	0°14'N	78°16'W
Otumba, Mex. (ȯ-tŭm´bä)	6	19°41'N	98°46'W
Otway, Seno, b., Chile (sĕ´nō-ô´t-wä´y)	20	53°00'S	73°00'W
Ouanaminthe, Haiti	11	19°35'N	71°45'W
Ouest, Point, c., Haiti	11	19°00'N	73°25'W
Ourinhos, Braz. (ôȯ-rē´nyôs)	19	23°04'S	49°45'W
Ouro Fino, Braz. (ōŭ-rô-fē´nō)	17a	22°18'S	46°21'W
Ouro Prêto, Braz. (ō´rȯ prā´tȯ)	20	20°24'S	43°30'W
Outer Brass, i., V.I.U.S. (bräs)	5c	18°24'N	64°58'W
Ovalle, Chile (ō-väl´yā)	20	30°43'S	71°16'W
Ovando, Bahía de, b., Cuba (bä-ē´ä-dĕ-ô-vä´n-dō)	11	20°10'N	74°05'W
Oviedo, Dom. Rep. (ȯ-vyĕ´dō)	11	17°50'N	71°25'W
Oxchuc, Mex. (ȯs-chōōk´)	7	16°47'N	92°24'W
Oxkutzcab, Mex. (ȯx-kōō´tz-käb)	8a	20°18'N	89°22'W
Oyapock, r., S.A. (ō-yȧ-pŏk´)	19	2°45'N	52°15'W
Ozama, r., Dom. Rep. (ȯ-zä´mä)	11	18°45'N	69°55'W
Ozuluama, Mex.	7	21°34'N	97°52'W
Ozumba, Mex.	7a	19°02'N	98°48'W

P

PLACE (Pronunciation)	PAGE	LAT.	LONG.
Pacaás Novos, Massiço de, mts., Braz.	18	11°03'S	64°02'W
Pacaraima, Serra, mts., S.A. (sĕr´rȧ pä-kä-rä-ē´mȧ)	18	3°45'N	62°30'W
Pacasmayo, Peru (pä-käs-mä´yō)	18	7°24'S	79°30'W
Pachuca, Mex. (pä-chōō´kä)	4	20°07'N	98°43'W
Padilla, Mex. (pä-dēl´yä)	6	24°00'N	98°45'W
Paine, Chile (pī´nĕ)	17b	33°49'S	70°44'W
Paita, Peru (pa-e´ta)	18	5°11'S	81°12'W
Pajapan, Mex. (pä-hä´pän)	7	18°16'N	94°41'W
Palencia, Guat. (pä-lĕn´sē-ȧ)	8	14°40'N	90°22'W
Palenque, Mex. (pä-lĕŋ´kȧ)	7	17°34'N	91°58'W
Palenque, Punta, c., Dom. Rep. (pōō´n-tä)	11	18°10'N	70°10'W
Palermo, Col. (pä lĕr´mô)	18a	2°53'N	75°26'W
Palín, Guat. (pä-lēn´)	8	14°42'N	90°42'W
Palizada, Mex. (pä-lē-zä´dä)	7	18°17'N	92°04'W
Palma, Braz. (päl´mä)	17a	21°23'S	42°18'W
Palmares, Braz. (päl-mȧ´rĕs)	19	8°46'S	35°28'W
Palmas, Braz. (päl´mäs)	20	26°20'S	51°56'W
Palmas, Braz.	19	10°08'S	48°18'W
Palma Soriano, Cuba (sȯ-ré-ä´nō)	10	20°15'N	76°00'W
Palmeira dos Índios, Braz. (pä-mā´rä-dȯs-ē´n-dyȯs)	19	9°26'S	36°33'W
Palmetto Point, c., Bah.	11	21°15'N	73°25'W
Palmira, Col. (päl-mē´rä)	18	3°33'N	76°17'W
Palmira, Cuba	10	22°15'N	80°25'W
Paloma, l., Mex. (pä-lō´mä)	2	26°53'N	104°02'W
Palomo, Cerro el, mtn., Chile (sĕ´r-rȯ-ĕl-pä-lô´mȯ)	17b	34°36'S	70°20'W
Pampa de Castillo, pl., Arg. (pä´m-pä-dĕ-käs-tē´l-yȯ)	20	45°30'S	67°30'W
Pampas, reg., Arg. (päm´päs)	20	37°00'S	64°30'W
Pamplona, Col. (päm-plō´nä)	18	7°19'N	72°41'W
Panamá, Pan.	5	8°58'N	79°32'W
Panamá, nation, N.A.	5	9°00'N	80°00'W
Panamá, Istmo de, isth., Pan.	5	9°00'N	80°00'W
Panama Canal, can., Pan.	4a	9°20'N	79°55'W
Pan de Guajaibón, mtn., Cuba (pän dā gwä-jȧ-bōn´)	10	22°50'N	83°20'W
Panimávida, Chile (pä-nē-mȧ´vē-dä)	17b	35°44'S	71°26'W
Pantepec, Mex. (pän-tȧ-pĕk´)	7	17°11'N	93°04'W
Pánuco, Mex. (pä´nōō-kȯ)	6	22°04'N	98°11'W
Pánuco, Mex. (pä´nōō-kȯ)	6	23°25'N	105°55'W
Pánuco, r., Mex.	4	21°59'N	98°20'W
Pánuco de Coronado, Mex. (pä´nōō-kȯ dā kō-rȯ-nä´dhō)	2	24°33'N	104°20'W
Panzós, Guat. (pä-zós´)	8	15°26'N	89°40'W
Pao, r., Ven. (pȧ´ō)	19b	9°52'N	67°57'W
Papagayo, r., Mex. (pä-pä-gä´yō)	6	16°52'N	99°41'W
Papagayo, Golfo del, b., C.R. (gȯl-fô-dĕl-pä-pä-gȧ´yō)	8	10°44'N	85°56'W
Papagayo, Laguna, l., Mex. (lä-ȯ-nä)	6	16°44'N	99°44'W
Papantla de Olarte, Mex. (pä-pän´tlä dā-ô-lä´r-tĕ)	4	20°30'N	97°15'W
Papatoapán, r., Mex. (pä-pä-tô-ā-pȧ´n)	7	18°00'N	96°22'W
Papinas, Arg. (pä-pē´näs)	17c	35°30'S	57°19'W
Papudo, Chile (pä-pōō´dȯ)	17b	32°30'S	71°25'W
Paquequer Pequeno, Braz. (pä-kĕ-kĕ´r-pĕ-kĕ´nȯ)	20b	22°19'S	43°02'W
Paracambi, Braz.	20b	22°36'S	43°43'W
Paracatu, Braz. (pä-rä-kä-tōō´)	19	17°17'S	46°43'W
Para de Minas, Braz. (pä-rä-dĕ-mē´näs)	19	19°52'S	44°37'W
Paradise, i., Bah.	10	25°05'N	77°20'W
Parados, Cerro de los, mtn., Col. (sĕ´r-rȯ-dĕ-lȯs-pä-rä´dōs)	18a	5°44'N	75°13'W
Paraguaçu, r., Braz. (pä-rä-gwä-zōō´)	19	12°25'S	39°46'W
Paraguay, nation, S.A. (păr´ȧ-gwā)	20	24°00'S	57°00'W
Paraguay, r., S.A. (pä-rä-gwä´y)	20	21°12'S	57°31'W
Paraíba, state, Braz. (pä-rä-ē´bä)	19	7°11'S	37°05'W
Paraíba, r., Braz.	17a	23°02'S	45°43'W
Paraíba do Sul, Braz. (dȯ-sōō´l)	17a	22°10'S	43°18'W
Paraibuna, Braz. (pä-räē-bōō´nä)	17a	23°23'S	45°38'W
Paraíso, C.R.	9	9°50'N	83°53'W
Paraíso, Mex.	7	18°24'N	93°11'W
Paraíso, Pan. (pä-rä-ē´sō)	4a	9°02'N	79°38'W
Paraisópolis, Braz. (pä-räē-sô´pō-lēs)	17a	22°35'S	45°45'W
Paraitinga, r., Braz. (pä-rä-ē-tē´n-gä)	17a	23°15'S	45°24'W
Paramaribo, Sur. (pȧ-rȧ-mȧ´rē-bō)	19	5°50'N	55°15'W
Paramillo, mtn., Col. (pä-rä-mē´l-yō)	18a	7°06'N	75°55'W
Paraná, Arg.	20	31°44'S	60°32'W
Paraná, r., S.A.	20	24°00'S	54°00'W
Paranaíba, Braz. (pä-rä-nȧ-ē´bȧ)	19	19°43'S	51°13'W
Paranaíba, r., Braz.	19	18°58'S	50°44'W
Paraná Ibicuy, r., Arg.	17c	33°27'S	59°26'W
Paranam, Sur.	19	5°39'N	55°13'W
Paránápanema, r., Braz. (pä-rä´nȧ´pä-nĕ-mä)	19	22°28'S	52°15'W
Paraopeda, r., Braz. (pä-rä-o-pĕ´dä)	17a	20°09'S	44°14'W
Parapara, Ven. (pä-rä-pä-rä)	19b	9°44'N	67°17'W
Parati, Braz. (pä-rätē)	17a	23°14'S	44°43'W
Pardo, r., Braz. (pär´dō)	19	15°25'S	39°40'W
Pardo, r., Braz.	17a	21°32'S	46°40'W
Parecis, Serra dos, mts., Braz. (sĕr´rȧ dōs pä-rȧ-sēzh´)	19	13°45'S	59°28'W
Paredón, Mex.	2	25°56'N	100°58'W
Paria, Golfo de, b., (gȯl-fȯ-dĕ-br-pä-rē-ä)	18	10°33'N	62°14'W
Paricutín, Volcán, vol., Mex.	6	19°27'N	102°14'W
Parida, Río de la, r., Mex. (rē´ō-dĕ-lä-pä-rē´dä)	2	26°23'N	104°40'W
Parima, Serra, mts., S.A. (sĕr´rȧ pä-rē´mȧ)	18	3°45'N	64°00'W
Pariñas, Punta, c., Peru (pōō´n-tä-pä-rē´n-yäs)	18	4°30'S	81°23'W
Parintins, Braz. (pä-rĭn-tĭɴzh´)	19	2°34'S	56°30'W
Parita, Golfo de, b., Pan. (gȯl-fô-dĕ-pä-rē´tä)	9	8°06'N	80°10'W
Parnaíba, Braz. (pär-nä-ē´bä)	19	3°00'S	41°42'W
Parnaiba, r., Braz.	19	3°57'S	42°30'W
Parral, Chile (pär-rä´l)	20	36°07'S	71°47'W
Parral, r., Mex.	2	27°25'N	105°08'W
Parras, Mex. (pär-räs´)	2	25°28'N	102°08'W
Parrita, C.R. (pär-rē´tä)	9	9°32'N	84°17'W
Pasión, Río de la, r., Guat. (rē´ō-dĕ-lä-pä-syōn´)	8a	16°31'N	90°11'W
Paso de los Libres, Arg. (pä-sȯ-dĕ-lȯs-lē´brĕs)	20	29°33'S	57°05'W
Paso de los Toros, Ur. (tŏ´rȯs)	17c	32°43'S	56°33'W
Passa Tempo, Braz. (pä´s-sä-tĕ´m-pô)	17a	20°40'S	44°29'W
Passo Fundo, Braz. (pä´sō fōn´dȯ)	20	28°16'S	52°13'W
Passos, Braz. (pä´s-sōs)	19	20°45'S	46°37'W
Pastaza, r., S.A. (päs-tä´zä)	18	3°05'S	76°18'W
Pasto, Col. (päs´tȯ)	18	1°15'N	77°19'W
Pastora, Mex. (päs-tȯ-rä)	6	22°08'N	100°04'W
Patagonia, reg., Arg. (păt-ȧ-gō´nĭ-ȧ)	20	46°45'S	69°30'W
Pati do Alferes, Braz. (pä-tē-dȯ-ȧl-fē´rĕs)	20b	22°25'S	43°25'W
Patos, Braz. (pä´tōzh)	19	7°03'S	37°14'W
Patos, Lagoa dos, l., Braz. (lä´gō-ȧ dozh pä´tōzh)	20	31°15'S	51°30'W
Patos de Minas, Braz. (dĕ-mē´näzh)	19	18°39'S	46°31'W
Patrocínio, Braz. (pä-trō-sē´nē-ȯ)	19	18°48'S	46°47'W
Patuca, r., Hond.	9	15°22'N	84°31'W
Patuca, Punta, c., Hond. (pōō´n-tä-pä-tōō´kä)	9	15°55'N	84°05'W
Pátzcuaro, Mex. (päts´kwä-rȯ)	6	19°30'N	101°36'W
Pátzcuaro, Lago de, l., Mex. (lä´gô-dĕ)	6	19°36'N	101°38'W
Patzicia, Guat. (pät-zē´syä)	8	14°36'N	90°57'W
Patzún, Guat. (pät-zōōn´)	8	14°40'N	91°00'W
Paulistano, Braz. (pȧ´ȯ-lēs-tȧ-nä)	19	8°13'S	41°06'W
Paulo Afonso, Salto, wtfl., Braz. (säl-tȯ-pou´lȯ äf-fôɴ´sȯ)	19	9°33'S	38°32'W
Pavarandocito, Col. (pä-vä-rän-dȯ-sē´tȯ)	18a	7°18'N	76°32'W
Pavuna, Braz. (pä-vōō´nȧ)	20b	22°48'S	43°21'W
Paysandú, Ur. (pī-sän-dōō´)	20	32°16'S	57°55'W
Peçanha, Braz. (pȧ-kän´yȧ)	19	18°37'S	42°26'W
Pedra Azul, Braz. (pā´drä-zōō´l)	19	16°03'S	41°13'W
Pedreiras, Braz. (pĕ-drā´räs)	19	4°30'S	44°31'W
Pedro Antonio Santos, Mex.	8a	18°55'N	88°13'W
Pedro Betancourt, Cuba (bā-täŋ-kōrt´)	10	22°40'N	81°15'W
Pedro de Valdivia, Chile (pĕ´drȯ-dĕ-väl-dē´vē-ä)	20	22°32'S	69°55'W
Pedro do Rio, Braz. (dȯ-rē´rô)	20b	22°20'S	43°09'W
Pedro II, Braz. (pā´drȯ sȧ-gȯn´dȯ)	19	4°20'S	41°27'W
Pedro Juan Caballero, Para. (hȯȧ´n-kä-bäl-yĕ´rō)	20	22°40'S	55°42'W
Pedro Miguel, Pan. (mē-gāl´)	4a	9°01'N	79°36'W
Pedro Miguel Locks, trans., Pan. (mē-gāl´)	4a	9°01'N	79°36'W
Pelée, Mont, mtn., Mart. (pē-lā´)	9b	14°49'N	61°10'W
Pelequén, Chile (pĕ-lĕ-kĕ´n)	17b	34°26'S	71°52'W
Pelican Harbor, b., Bah. (pĕl´ĭ-kăn)	10	26°20'N	76°45'W
Pena Nevada, Cerro, Mex.	6	23°47'N	99°52'W
Penas, Golfo de, b., Chile (gȯl fô dĕ pĕ´n-äs)	20	47°15'S	77°30'W
Penderisco, r., Col. (pĕn-dĕ-rē´s-kȯ)	18a	6°30'N	76°21'W
Penedo, Braz. (pȧ-nā´dȯ)	19	10°17'S	36°28'W
Penjamillo, Mex. (pĕn-hä-mēl´yō)	6	20°06'N	101°56'W
Pénjamo, Mex. (pān´hä-mȯ)	6	20°27'N	101°43'W
Pensilvania, Col. (pĕn-sēl-vȧ´nyä)	18a	5°31'N	75°05'W
Peotillos, Mex. (pȧ-ō-tel´yōs)	6	22°30'N	100°39'W
Pepe, Cabo, c., Cuba (kä´bȯ-pĕ´pĕ)	10	21°30'N	83°10'W
Perdões, Braz. (pĕr-dô´ēs)	17a	21°05'S	45°05'W
Pereira, Col. (pȧ-rā´rä)	18	4°49'N	75°42'W
Pergamino, Arg. (pĕr-gä-mē´nō)	20	33°53'S	60°36'W
Perija, Sierra de, mts., Col. (sē-ĕ´r-rȧ-dĕ-pĕ-rē´kä)	18	9°25'N	73°30'W
Perlas, Archipiélago de las, is., Pan.	9	8°29'N	79°15'W
Perlas, Laguna las, l., Nic. (lä-gȯ´nä-dĕ-läs)	9	12°34'N	83°19'W
Pernambuco *see* Recife, Braz.	19	8°09'S	34°59'W
Pernambuco, state, Braz. (pĕr-näm-bōō´kō)	19	8°08'S	38°54'W
Perote, Mex. (pĕ-rȯ´tĕ)	7	19°33'N	97°13'W
Perros, Bahía, b., Cuba (bä-ē´ä-pā´rōs)	10	22°25'N	78°35'W
Peru, nation, S.A.	18	10°00'S	75°00'W
Peru-Chile Trench, deep	15	25°00'S	71°30'W
Pescado, r., Ven. (pĕs-kä´dō)	19b	9°33'N	65°32'W
Pespire, Hond. (pȧs-pē´rȧ)	8	13°35'N	87°20'W
Pesquería, r., Mex. (pȧs-kȧ-rē´ȧ)	2	25°55'N	100°25'W
Petacalco, Bahía de, b., Mex. (bä-ē´ä-dĕ-pĕ-tä-kȧl´kȯ)	6	17°55'N	102°00'W
Petare, Ven. (pĕ-tä´rĕ)	19b	10°28'N	66°48'W
Petatlán, Mex. (pā-tä-tlän´)	6	17°31'N	101°17'W
Petén, Laguna de, l., Guat. (lä-gȯ´nä-dĕ-pȧ-tān´)	8a	17°05'N	89°54'W
Pétionville, Haiti	11	18°30'N	72°20'W
Petite Terre, i., Guad. (pĕ-tēt´târ´)	9b	16°12'N	61°00'W
Petit Goâve, Haiti (pē-tē´ gŏ-ȧv´)	11	18°25'N	72°50'W
Petlalcingo, Mex. (pĕ-tläl-sĕŋ´gȯ)	7	18°05'N	97°53'W
Peto, Mex. (pĕ´tȯ)	8a	20°07'N	88°49'W
Petorca, Chile (pā-tōr´kȧ)	17b	32°14'S	70°55'W
Petrolina, Braz. (pĕ-trō-lē´nȧ)	19	9°18'S	40°28'W
Petrópolis, Braz. (pȧ-trȯ-pȯ-lēzh´)	19	22°31'S	43°10'W
Picara Point, c., V.I.U.S. (pē-kä´rä)	5c	18°23'N	64°57'W
Pichilemu, Chile (pē-chē-lĕ´mōō)	17b	34°22'S	72°01'W
Pichucalco, Mex. (pē-chōō-käl´kȯ)	7	17°34'N	93°06'W
Picos, Braz. (pē´kōzh)	19	7°13'S	41°23'W
Piedade, Braz. (pyā-dä´dĕ)	17a	23°42'S	47°25'W
Piedras, Punta, c., Arg. (pōō´n-tä-pyĕ´dräs)	17c	35°25'S	57°10'W
Piedras Negras, Mex. (pyā´dräs nā´gräs)	4	28°41'N	100°33'W
Pijijiapan, Mex. (pēkē-kē-ä´pän)	7	15°40'N	93°12'W
Pilar, Arg. (pē´lär)	17c	34°27'S	58°55'W
Pilar, Para.	20	27°00'S	58°15'W
Pilar de Goiás, Braz. (dĕ-gô´yȧ´s)	19	14°47'S	49°33'W
Pilcomayo, r., S.A. (pēl-cō-mī´ȯ)	20	24°45'S	59°15'W
Pilón, r., Mex. (pĕ-lōn´)	6	24°13'N	99°03'W
Pimal, Cerra, mtn., Mex. (sĕ´r-rä-pē-mäl´)	6	22°58'N	104°19'W
Pinacate, Cerro, mtn., Mex. (sĕ´r-rȯ-pē-nä-kä´tĕ)	4	31°45'N	113°30'W
Pinar del Río, Cuba (pē-när´ dĕl rē´ȯ)	5	22°25'N	83°35'W
Pinar del Río, prov., Cuba	10	22°45'N	83°25'W
Pindamonhangaba, Braz. (pē´n-dä-mōnyȧ´n-gä-bä)	17a	22°56'S	45°26'W
Pinder Point, c., Bah.	10	26°35'N	78°35'W
Pinhal, Braz. (pē-nyȧ´l)	17a	22°11'S	46°43'W
Pinotepa Nacional, Mex. (pē-nȯ-tā´pä nä-syȯ-näl´)	6	16°21'N	98°04'W
Pinta, i., Ec.	18	0°41'N	90°47'W
Piracaia, Braz. (pē-rä-kȧ´yä)	17a	23°04'S	46°20'W
Piracicaba, Braz. (pē-rä-sē-kä´bä)	19	22°43'S	47°39'W
Piraíba, r., Braz. (pä-rä-ē´bȧ)	17a	21°38'S	41°29'W
Piranga, Braz. (pē-rä´n-gä)	17a	20°41'S	43°17'W
Pirapetinga, Braz. (pē-rä-pĕ-tē´n-gä)	17a	21°40'S	42°20'W
Pirapora, Braz. (pē-rȧ-pō´rȧ)	19	17°39'S	44°54'W
Pirassununga, Braz. (pē-rä-sōō-nōō´n-gä)	17a	22°00'S	47°24'W
Pirenópolis, Braz. (pē-rĕ-nô´pō-lēs)	19	15°56'S	48°49'W
Piritu, Laguna de, l., Ven. (lä-gȯ´nä-dĕ-pē-rē´tōō)	19b	10°00'N	64°57'W
Pisagua, Chile (pē-sä´gwä)	18	19°43'S	70°12'W
Pisco, Peru (pēs´kō)	18	13°43'S	76°07'W
Pisco, Bahía de, b., Peru	18	13°43'S	77°48'W
Pitalito, Col. (pē-tä-lē´tō)	18	1°45'N	75°09'W
Piũi, Braz. (pē-ōō´ē)	17a	20°27'S	45°57'W
Piura, Peru (pē-ōō´rä)	18	5°13'S	80°46'W
Placetas, Cuba (plä-thā´täs)	10	22°10'N	79°40'W
Plana or Flat Cays, is., Bah. (plä´nä)	11	22°35'N	73°35'W
Plata, Río de la, est., S.A. (dālä plä´tä)	20	34°35'S	58°15'W
Plateforme, Pointe, c., Haiti	11	19°35'N	73°50'W
Plato, Col. (plä´tō)	18	9°49'N	74°48'W
Platón Sánchez, Mex. (plä-tōn´ sän´chĕz)	6	21°14'N	98°20'W
Playa de Guanabo, Cuba (plä-yä-dĕ-gwä-nä´bȯ)	11a	23°10'N	82°07'W
Playa de Santa Fé, Cuba	11a	23°05'N	82°31'W
Playa Vicente, Mex. (vē-sĕn´tȧ)	7	17°49'N	95°49'W
Playa Vicente, r., Mex.	7	17°36'N	96°13'W
Pluma Hidalgo, Mex. (plōō´mä ē-däl´gō)	7	15°54'N	96°23'W
Plymouth, Monts.	9b	16°43'N	62°12'W
Pochotitán, Mex. (pô-chȯ-tē-tȧ´n)	6	21°37'N	104°33'W
Pochutla, Mex.	7	15°46'N	96°28'W

ng-sing; ŋ-baŋk; ɴ-nasalized n; nŏd; cŏmmit; ōld; ȯbey; ôrder; oi-boil; fōōd; ȯ-as oo in foot; ou-out; s-soft; sh-dish; th-thin; pūre; ůnite; ûrn; stŭd; circŭs; ü-as in French tu; ´-indeterminate vowel.

50 Poços de Caldas – Romano
PRONOUNCING INDEX

PLACE (Pronunciation)	PAGE	LAT.	LONG.
Poços de Caldas, Braz. (pō-sôs-dĕ-käl′dås)	19	21°48′S	46°34′W
Pointe-à-Pitre, Guad. (pwăNt′ à pē-tr′)	5	16°15′N	61°32′W
Polochic, r., Guat. (pō-lō-chēk′)	8	15°19′N	89°45′W
Polpaico, Chile (pōl-pà′y-kō)	17b	33°10′S	70°53′W
Pomba, r., Braz. (pô′m-bå)	17a	21°28′S	42°28′W
Pomuch, Mex. (pô-mo͞o′ch)	8a	20°12′N	90°10′W
Ponce, P.R. (pōn′sā)	5	18°01′N	66°43′W
Ponta Grossa, Braz. (grō′så)	19	25°09′S	50°05′W
Ponte Nova, Braz. (pô′n-tĕ-nô′vå)	19	20°26′S	42°52′W
Poopó, Lago de, l., Bol.	18	18°45′S	67°07′W
Popayán, Col. (pō-pä-yän′)	18	2°21′N	76°43′W
Popocatépetl Volcán, Mex. (pō-pō-kä-tā′pĕ′t′l)	4	19°01′N	98°38′W
Porce, r., Col. (pôr-sĕ)	18a	7°11′N	74°55′W
Poriúncula, Braz.	17a	20°58′S	42°02′W
Porlamar, Ven. (pōr-lä-mär′)	18	11°00′N	63°55′W
Portachuelo, Bol. (pōrt-ä-chwā′lō)	18	17°20′S	63°12′W
Port Antonio, Jam.	5	18°10′N	76°25′W
Port-au-Prince, Haiti (prăNs′)	5	18°35′N	72°20′W
Port de Paix, Haiti (pĕ)	11	19°55′N	72°50′W
Portillo, Chile (pôr-tē′l-yō)	17b	32°51′S	70°09′W
Portland Bight, b., Jam.	10	17°45′N	77°05′W
Portland Point, c., Jam.	10	17°40′N	77°20′W
Port Maria, Jam. (må-rī′å)	10	18°20′N	76°55′W
Porto Acre, Braz. (ä′krĕ)	18	9°38′S	67°34′W
Porto Alegre, Braz. (ä-lā′grĕ)	20	29°58′S	51°11′W
Portobelo, Pan. (pôr′tō-bā′lō)	5	9°32′N	79°40′W
Pôrto de Pedras, Braz. (pā′dräzh)	19	9°09′S	35°20′W
Pôrto Feliz, Braz. (fē-lē′s)	17a	23°12′S	47°30′W
Port of Spain, Trin. (spān)	19	10°44′N	61°24′W
Porto Mendes, Braz. (mĕ′n-dĕs)	19	24°41′S	54°13′W
Porto Murtinho, Braz. (mòr-tēn′yò)	19	21°43′S	57°43′W
Porto Nacional, Braz. (nå-syō-näl′)	19	10°43′S	48°14′W
Porto Seguro, Braz. (sā-go͞o′rò)	19	16°26′S	38°59′W
Porto Velho, Braz. (vāl′yò)	18	8°45′S	63°43′W
Portoviejo, Ec. (pôr-tō-vyā′hō)	18	1°11′S	80°28′W
Port Royal, b., Jam. (roi′ăl)	10	17°50′N	76°45′W
Portsmouth, Dom.	9b	15°33′N	61°28′W
Posadas, Arg. (pō-sä′dhäs)	20	27°32′S	55°56′W
Potosí, Bol.	18	19°35′S	65°45′W
Potosi, r., Mex. (pō-tō-sē′)	2	25°04′N	99°36′W
Potrerillos, Hond. (pō-trå-rēl′yōs)	8	15°13′N	87°58′W
Pouso Alegre, Braz. (pō′zò ä-lā′grĕ)	19	22°13′S	45°56′W
Powell Point, c., Bah.	10	24°50′N	76°20′W
Poza Rica, Mex. (pô-zō-rē′kä)	7	20°32′N	97°25′W
Pozos, Mex. (pō′zōs)	6	22°05′N	100°50′W
Pradera, Col. (prä-dĕ′rä)	18a	3°24′N	76°13′W
Prado, Col. (prädō)	18a	3°44′N	74°55′W
Prados, Braz. (prà′dōs)	17a	21°05′S	44°04′W
Praia Funda, Ponta da, c., Braz. (pôn′tä-då-prä′yå-fo͞o′n-dä)	20b	23°04′S	43°34′W
Presidencia Rogue Sáenz Peña, Arg.	20	26°52′S	60°15′W
Presidente Epitácio, Braz. (prā-sē-dĕn′tĕ å-pē-tä′syò)	19	21°56′S	52°01′W
Presidio, Río del, r., Mex. (rē′ō-dĕl-prĕ-sē′dyô)	6	23°54′N	105°44′W
Prespuntal, r., Ven.	19b	9°55′N	64°32′W
Prinzapolca, Nic. (prēn-zä-pōl′kä)	9	13°18′N	83°35′W
Prinzapolca, r., Nic.	9	13°23′N	84°23′W
Progreso, Hond. (prō-grĕ′sō)	8	15°28′N	87°49′W
Progreso, Mex. (prō-grā′sō)	4	21°14′N	89°39′W
Progreso, Mex.	2	27°29′N	101°05′W
Providencia, Isla de, i., Col.	9	13°21′N	80°55′W
Providenciales, i., T./C. Is.	11	21°50′N	72°15′W
Puebla, Mex. (pwā′blä)	4	19°02′N	98°11′W
Puebla, state, Mex.	7	19°00′N	97°45′W
Pueblo Nuevo, Mex. (nwā′vō)	6	23°23′N	105°21′W
Pueblo Viejo, Mex. (vyā′hō)	7	17°23′N	93°46′W
Puente Alto, Chile (pwĕ′n-tĕ àl′tō)	17b	33°36′S	70°34′W
Puerto Aisén, Chile (pwĕ′r-tō ä′y-sĕ′n)	20	45°28′S	72°44′W
Puerto Ángel, Mex. (pwĕ′r-tō äŋ′hål)	7	15°42′N	96°32′W
Puerto Armuelles, Pan. (pwe′r-tō är-mo͞o-ā′lyās)	9	8°18′N	82°52′W
Puerto Barrios, Guat. (pwĕ′r-tō bär′rē-ōs)	4	15°43′N	88°36′W
Puerto Bermúdez, Peru (pwĕ′r-tō bĕr-mo͞o′dāz)	18	10°17′S	74°57′W
Puerto Berrío, Col. (pwĕ′r-tō bĕr-rē′ō)	18	6°29′N	74°27′W
Puerto Cabello, Ven. (pwĕ′r-tō kä-bĕl′yō)	18	10°28′N	68°01′W
Puerto Cabezas, Nic. (pwĕ′r-tō kä-bā′zäs)	9	14°01′N	83°26′W
Puerto Casado, Para. (pwĕ′r-tō kä-sä′dō)	20	22°16′S	57°57′W
Puerto Castilla, Hond. (pwĕ′r-tō käs-tēl′yō)	8	16°01′N	86°01′W
Puerto Chicama, Peru (pwĕ′r-tō chē-kä′mä)	18	7°46′S	79°18′W
Puerto Colombia, Col. (pwĕr′tō kō-lōm′bē-å)	18	11°08′N	75°09′W
Puerto Cortés, C.R. (pwĕ′r-tō kōr-tās′)	9	9°00′N	83°37′W
Puerto Cortés, Hond. (pwĕ′r-tō kōr-tās′)	4	15°48′N	87°57′W
Puerto Cumarebo, Ven. (pwĕ′r-tō ko͞o-mä-rĕ′bō)	18	11°25′N	69°17′W
Puerto de Nutrias, Ven. (pwĕ′r-tō dĕ no͞o-trē-äs′)	18	8°02′N	69°19′W
Puerto Deseado, Arg. (pwĕ′r-tō dā-så-ä′dhō)	20	47°38′S	66°00′W
Puerto Eten, Peru (pwĕ′r-tō ĕ-tĕ′n)	18	6°59′S	79°51′W
Puerto Jiménez, C.R. (pwĕ′r-tō κē-mĕ′nĕz)	9	8°35′N	83°23′W
Puerto La Cruz, Ven. (pwĕ′r-tō lä kro͞o′z)	18	10°14′N	64°38′W
Puerto Madryn, Arg. (pwĕ′r-tō mä-drēn′)	20	42°45′S	65°01′W
Puerto Maldonado, Peru (pwĕ′r-tō mäl-dō-nä′dō)	18	12°43′S	69°01′W
Puerto Miniso, Mex. (pwĕ′r-tō mē-nē′sō)	6	16°06′N	98°02′W
Puerto Montt, Chile (pwĕ′r-tō mō′nt)	20	41°29′S	73°00′W
Puerto Natales, Chile (pwĕ′r-tō nä-tà′lĕs)	20	51°48′S	72°01′W
Puerto Niño, Col. (pwĕ′r-tō nē′n-yô)	18a	5°57′N	74°36′W
Puerto Padre, Cuba (pwĕ′r-tō pä′drā)	10	21°10′N	76°40′W
Puerto Peñasco, Mex. (pwĕ′r-tō pĕn-yä′s-kō)	4	31°39′N	113°15′W
Puerto Pinasco, Para. (pwĕ′r-tō pē-nà′s-kō)	20	22°31′S	57°50′W
Puerto Píritu, Ven. (pwĕ′r-tō pē′rē-to͞o)	19b	10°05′N	65°04′W
Puerto Plata, Dom. Rep. (pwĕ′r-tō plä′tä)	5	19°50′N	70°40′W
Puerto Rico, dep., N.A. (pwĕr′tō rē′kō)	5	18°16′N	66°50′W
Puerto Rico Trench, deep	5	19°45′N	66°30′W
Puerto Salgar, Col. (pwĕ′r-tō säl-gär′)	18a	5°30′N	74°39′W
Puerto Santa Cruz, Arg. (pwĕ′r-tō sän′tä kro͞oz′)	20	50°04′S	68°32′W
Puerto Suárez, Bol. (pwĕ′r-tō swä′råz)	19	18°55′S	57°39′W
Puerto Tejada, Col. (pwĕ′r-tō tĕ-κä′dä)	18	3°13′N	76°23′W
Puerto Vallarta, Mex. (pwĕ′r-tō väl-yär′tä)	6	20°36′N	105°13′W
Puerto Varas, Chile (pwĕ′r-tō vä′räs)	20	41°16′S	73°03′W
Puerto Wilches, Col. (pwĕ′r-tō vēl′c-hĕs)	18	7°19′N	73°54′W
Pulacayo, Bol. (po͞o-lä-kä′yō)	18	20°12′N	66°33′W
Punata, Bol. (po͞o-nä′tä)	18	17°43′S	65°43′W
Puno, Peru (po͞o′nō)	18	15°58′S	70°02′W
Punta Arenas, Chile (po͞o′n-tä-rĕ′näs)	20	53°09′S	70°48′W
Punta de Piedras, Ven. (po͞o′n-tä dĕ pyĕ′dräs)	19b	10°54′N	64°06′W
Punta Gorda, Belize (pòn′tä gôr′dä)	8	16°07′N	88°50′W
Punta Gorda, Río, r., Nic. (po͞o′n-tä gō′r-dä)	9	11°34′N	84°13′W
Punta Indio, Canal, strt., Arg. (po͞o′n-tä- ē′n-dyō)	17c	34°56′S	57°20′W
Puntarenas, C.R. (pònt-ä-rā′näs)	5	9°59′N	84°49′W
Punto Fijo, Ven. (po͞o′n-tō fē′κô)	18	11°48′N	70°14′W
Puquio, Peru (po͞o′kyō)	18	14°43′S	74°02′W
Purépero, Mex. (po͞o-rā′på-rō)	6	19°56′N	102°02′W
Purial, Sierra de, mts., Cuba (sē-ĕ′r-rå-dĕ-po͞o-rē-äl′)	11	20°15′N	74°40′W
Purificación, Col. (po͞o-rē-fē-kä-syōn′)	18	3°52′N	74°54′W
Purificación, Mex. (po͞o-rē-fē-kä-syō′n)	6	19°44′N	104°38′W
Purificación, r., Mex.	6	19°30′N	104°54′W
Puruandiro, Mex. (pò-ro͞o-än′dē-rō)	6	20°04′N	101°33′W
Purús, r., S.A. (po͞o-ro͞o′s)	18	6°45′S	64°34′W
Pustunich, Mex. (po͞os-to͞o′nēch)	7	19°10′N	90°29′W
Putaendo, Chile (po͞o-tä-ĕn-dō)	17b	32°37′S	70°42′W
Putla de Guerrero, Mex. (po͞o′tlä-dĕ-gĕr-rĕ′rō)	7	17°03′N	97°55′W
Putumayo, r., S.A. (pò-to͞o-mä′yō)	18	1°02′S	73°50′W

Q

PLACE (Pronunciation)	PAGE	LAT.	LONG.
Queimados, Braz. (kā-må′dōs)	20b	22°42′S	43°34′W
Quemado de Güines, Cuba (kā-mä′dhä-dĕ-gwē′nĕs)	10	22°45′N	80°20′W
Quepos, C.R. (kā′pōs)	9	9°26′N	84°10′W
Quepos, Punta, c., C.R. (po͞o′n-tä)	9	9°23′N	84°20′W
Querétaro, Mex. (kå-rā′tä-rō)	4	20°37′N	100°25′W
Querétaro, state, Mex.	6	21°00′N	100°00′W
Quetame, Col. (kĕ-tä′mĕ)	18a	4°20′N	73°50′W
Quezaltenango, Guat. (kå-zäl′tå-näŋ′gō)	4	14°50′N	91°30′W
Quezaltepeque, El Sal. (kĕ-zäl′tĕ′pĕ-kĕ)	8	13°50′N	89°17′W
Quezaltepeque, Guat. (kå-zäl′tå-pā′kå)	8	14°39′N	89°26′W
Quibdo, Col. (kēb′dō)	18	5°42′N	76°41′W
Quilimari, Chile (kē-lē-mä′rē)	17b	32°06′S	71°28′W
Quillota, Chile (kēl-yō′tä)	20	32°52′S	71°14′W
Quilmes, Arg. (kēl′mās)	17c	34°43′S	58°16′W
Quimbaya, Col. (kēm-bä′yä)	18a	4°38′N	75°46′W
Quintana Roo, state, Mex. (rô′ō)	4	19°30′N	88°30′W
Quintero, Chile (kēn-tĕ′rō)	17b	32°48′S	71°30′W
Quiroga, Mex. (kē-rō′gä)	6	19°39′N	101°30′W
Quito, Ec. (kē′tō)	18	0°17′S	78°32′W

R

PLACE (Pronunciation)	PAGE	LAT.	LONG.
Raccoon Cay, i., Bah.	11	22°25′N	75°50′W
Rafaela, Arg. (rä-fä-å′lä)	20	31°15′S	61°21′W
Rainbow City, Pan.	4a	9°20′N	79°53′W
Rama, Nic. (rä′mä)	9	12°11′N	84°14′W
Ramallo, Arg. (rä-mä′l-yō)	17c	33°28′S	60°02′W
Ramos, Mex. (rä′mōs)	6	22°46′N	101°52′W
Ramos Arizpe, Mex. (ä-rēz′på)	2	25°33′N	100°57′W
Rancagua, Chile (rän-kä′gwä)	20	34°10′S	70°43′W
Rancho Boyeros, Cuba (rä′n-chō-bō-yĕ′rōs)	11a	23°00′N	82°23′W
Rapel, r., Chile (rä-pål′)	17b	34°05′S	71°30′W
Rauch, Arg. (rà′o͞och)	20	36°47′S	59°05′W
Raúl Soares, Braz. (rä-o͞o′l-sōä′rĕs)	17a	20°05′S	42°28′W
Rawson, Arg. (rô′sŭn)	20	43°16′S	65°09′W
Rawson, Arg.	17c	34°36′S	60°03′W
Rayón, Mex. (rä-yōn′)	6	21°49′N	99°39′W
Realengo, Braz. (rĕ-ä-län-gô)	17a	23°50′S	43°25′W
Recife (Pernambuco), Braz. (rå-sē′fĕ)	19	8°09′S	34°59′W
Reconquista, Arg. (rā-kōn-kēs′tä)	20	29°01′S	59°41′W
Redenção da Serra, Braz. (rĕ-dĕn-souN-dä-sĕ′r-rä)	17a	23°17′S	45°31′W
Redonda, Isla, i., Braz. (ē′s-lä-rĕ-dō′n-dä)	20b	23°05′S	43°11′W
Redonda Island, i., Antig. (rĕ-dŏn′då)	9b	16°55′N	62°28′W
Regla, Cuba (rāg′lä)	10	23°08′N	82°20′W
Reina Adelaida, Archipiélago, is., Chile	20	52°00′S	74°15′W
Remedios, Col. (rĕ-mĕ′dyōs)	18a	7°03′N	74°42′W
Remedios, Cuba (rā-mā′dhē-ōs)	10	22°30′N	79°35′W
Remedios, Pan. (rĕ-mĕ′dyōs)	9	8°14′N	81°46′W
Rengo, Chile (rĕn′gō)	17b	34°22′S	70°50′W
Resende, Braz. (rĕ-sĕ′n-dĕ)	17a	22°30′S	44°26′W
Resende Costa, Braz. (kôs-tä)	17a	20°55′S	44°12′W
Resistencia, Arg. (rā-sēs-tĕn′syä)	20	27°24′S	58°54′W
Restrepo, Col. (rĕs-trĕ′pō)	18a	3°49′N	76°31′W
Restrepo, Col.	18a	4°16′N	73°32′W
Retalhuleu, Guat. (rā-täl-o͞o-lān′)	8	14°31′N	91°41′W
Reventazón, Río, r., C.R. (rå-vĕn-tä-zōn′)	9	10°10′N	83°30′W
Revillagigedo, Islas, is., Mex. (ē′s-läs-rĕ-vēl-yä-hē′gĕ-dō)	4	18°45′N	111°00′W
Rey, l., Mex. (rā)	2	27°00′N	103°33′W
Rey, Isla del, i., Pan. (ē′s-lä-dĕl-rā′ē)	9	8°20′N	78°40′W
Reyes, Bol. (rā′yĕs)	18	14°19′S	67°16′W
Reynosa, Mex. (rā-ē-nō′sä)	2	26°05′N	98°21′W
Riachão, Braz. (rē-ä-chouN′)	19	7°15′S	46°30′W
Ribeirão Prêto, Braz. (rē-bā-rouN-prĕ′tô)	19	21°11′S	47°47′W
Riberalta, Bol. (rē-bå-räl′tä)	18	11°06′S	66°02′W
Riding Rocks, is., Bah.	10	25°20′N	79°10′W
Rincón de Romos, Mex. (rēn-kōn dā rō-mōs′)	6	22°13′N	102°21′W
Río Abajo, Pan. (rē′ō-ä-bä′κō)	4a	9°01′N	78°30′W
Río Balsas, Mex. (rē′ō-bäl-säs)	6	17°59′N	99°45′W
Riobamba, Ec. (rē′ō-bäm-bä)	18	1°45′S	78°37′W
Rio Bonito, Braz. (rē′ò bō-nē′tò)	17a	22°44′S	42°38′W
Rio Branco, Braz. (rē′ò bräŋ′kò)	18	9°57′S	67°50′W
Río Branco, Ur. (rĭō brăncō)	20	32°33′S	53°29′W
Rio Casca, Braz. (rē′ō-kà′s-kä)	17a	20°15′S	42°39′W
Río Chico, Ven. (rē′ō chē′kŏ)	19b	10°20′N	65°58′W
Rio Claro, Braz. (rē′ò klä′rò)	19	22°25′S	47°33′W
Río Cuarto, Arg. (rē′ō kwär′tō)	20	33°05′S	64°15′W
Rio das Flores, Braz. (rē′ō-däs-flô-rĕs)	17a	22°10′S	43°35′W
Rio de Janeiro, Braz. (rē′ò dā zhä-nå′ē-rò)	20b	22°50′S	43°20′W
Rio de Janeiro, state, Braz.	19	22°27′S	42°43′W
Río de Jesús, Pan.	9	7°54′N	80°59′W
Río Frío, Mex. (rē′ō-frē′ō)	7a	19°21′N	98°40′W
Río Gallegos, Arg. (rē′ō gä-lā′gōs)	20	51°43′S	69°15′W
Rio Grande, Braz. (rē′ò grän′dĕ)	20	31°04′S	52°14′W
Río Grande, Mex. (rē′ō grän′dā)	6	23°51′N	102°59′W
Rio Grande do Norte, state, Braz.	19	5°26′S	37°20′W
Rio Grande do Sul, state, Braz. (rē′ò grän′dĕ-dō-so͞o′l)	20	29°00′S	54°00′W
Ríohacha, Col. (rē′ō-ä′chä)	18	11°30′N	72°54′W
Río Hato, Pan. (rē′ō-ä′tō)	9	8°19′N	80°11′W
Ríonegro, Col. (rē′ō-nĕ′grō)	18a	6°09′N	75°22′W
Río Negro, prov., Arg. (rē′ō nā′grō)	20	40°15′S	68°15′W
Río Negro, dept., Ur. (rē′ō-nĕ′grō)	17c	32°48′S	57°45′W
Río Negro, Embalse del, res., Ur.	20	32°45′S	55°50′W
Rio Novo, Braz. (rē′ō-nô′vō)	17a	21°30′S	43°08′W
Rio Pardo de Minas, Braz. (rē′ō pär′dō-dĕ-mē′näs)	19	15°43′S	42°24′W
Rio Pombo, Braz. (rē′ō pôm′bä)	17a	21°17′S	43°09′W
Rio Sorocaba, Represa do, res., Braz.	17a	23°37′S	47°19′W
Ríosucio, Col. (rē′ō-so͞o′syō)	18a	5°25′N	75°41′W
Río Tercero, Arg. (rē′ō dĕr-sĕ′rō)	20	32°12′S	63°59′W
Rio Verde, Braz. (vĕr′dĕ)	19	17°47′S	50°49′W
Ríoverde, Mex. (rē′ō-vĕr′då)	4	21°54′N	99°59′W
Risaralda, dept., Col.	18a	5°15′N	76°00′W
Ritacuva, Alto, mtn., Col. (ä′l-tō-rē-tä-ko͞o′vä)	18	6°22′N	72°13′W
Riva, Dom. Rep. (rē′vä)	11	19°10′N	69°55′W
Rivas, Nic. (rē′väs)	8	11°25′N	85°51′W
Rivera, Ur. (rē-vā′rä)	20	30°52′S	55°32′W
Roatan, Hond. (rō-ä-tän′)	8	16°18′N	86°33′W
Roatán, i., Hond.	8	16°19′N	86°46′W
Rocas, Atol das, atoll, Braz. (ä-tōl-däs-rō′kås)	19	3°50′S	33°46′W
Rocha, Ur. (rō′chås)	20	34°26′S	54°14′W
Roche à Bateau, Haiti (rōsh à bà-tō′)	11	18°10′N	74°00′W
Rock Sound, strt., Bah.	10	24°50′N	76°05′W
Rockstone, Guy. (rŏk′stōn)	19	5°55′N	57°27′W
Rodas, Cuba (rō′dhås)	10	22°20′N	80°35′W
Rodeo, Mex. (rō-dā′ō)	2	25°12′N	104°34′W
Rogoaguado, l., Bol. (rō′gō-ä-gwä-dō)	18	12°42′S	66°46′W
Rojas, Arg. (rō′häs)	17c	34°11′S	60°42′W
Rojo, Cabo, c., Mex. (rō′hō)	7	21°35′N	97°16′W
Rojo, Cabo, c., P.R. (rō′hō)	5b	17°55′N	67°14′W
Roldanillo, Col. (rôl-dä-nē′l-yō)	18a	4°24′N	76°09′W
Rolleville, Bah.	10	23°40′N	76°00′W
Romano, Cayo, i., Cuba (kä′yō-rô-må′nō)	10	22°15′N	78°00′W

ăt; finăl; rāte; senåte; ärm; åsk; sofå; fâre; ch-choose; dh-as th in other; bē; ēvent; bĕt; recĕnt; cratẽr; g-gō; gh-guttural g; bĭt; ĭ-short neutral; rīde; κ-guttural k as ch in German ich;

PRONOUNCING INDEX

Romita – San Martín 51

PLACE (Pronunciation)	PAGE	LAT.	LONG.
Romita, Mex. (rō-mē'tä)	6	20°53'N	101°32'W
Roncador, Serra do, mts., Braz. (sĕr'rá dò rôn-kä-dōr')	19	12°44'S	52°19'W
Rondônia, state, Braz.	18	10°15'S	63°07'W
Roosevelt, r., Braz. (rô'sĕ-vĕlt)	19	9°22'S	60°28'W
Roque Pérez, Arg. (rô'kĕ-pĕ'rĕz)	17c	35°23'S	59°22'W
Roques, Islas los, is., Ven.	18	12°25'N	67°40'W
Roraima, state, Braz. (rô'rīy-mä)	18	2°00'N	62°15'W
Roraima, Mount, mtn., S.A. (rô-rä-ē'mä)	19	5°12'N	60°52'W
Rosales, Mex. (rō-zä'läs)	2	28°15'N	100°43'W
Rosamorada, Mex. (rō'zä-mō-rä'dhä)	6	22°06'N	105°16'W
Rosaria, Laguna, l., Mex. (lä-gō'nä-rō-sá'ryä)	7	17°50'N	93°51'W
Rosario, Arg. (rō-zä'rē-ō)	20	32°58'S	60°42'W
Rosario, Braz. (rô zä'rê ó)	19	2°49'S	44°15'W
Rosario, Mex.	2	26°31'N	105°40'W
Rosario, Mex.	6	22°58'N	105°54'W
Rosario, Ur.	17c	34°19'S	57°24'E
Rosario, Cayo, i., Cuba (kä'yō-rō-sä'ryō)	10	21°40'N	81°55'W
Rosário do Sul, Braz. (rô-zä'rē-ò-dō-sōō'l)	20	30°17'S	54°52'W
Rosário Oeste, Braz. (ō'ĕst'ĕ)	19	14°47'S	56°20'W
Roseau, Dom.	9b	15°17'N	61°23'W
Rosignol, Guy. (rōs-ĭg-nél)	19	6°16'N	57°37'W
Rovira, Col. (rô-vē'rä)	18a	4°14'N	75°13'W
Royal, i., Bah.	10	25°30'N	76°50'W
Ruiz, Mex. (rōē'z)	6	21°55'N	105°09'W
Ruiz, Nevado del, vol., Col. (nĕ-vá'dô-dĕl-rōōē'z)	18a	4°52'N	75°20'W
Rum Cay, i., Bah.	11	23°40'N	74°50'W
Russas, Braz. (rōō's-säs)	19	4°48'S	37°50'W

S

PLACE (Pronunciation)	PAGE	LAT.	LONG.
Saavedra, Arg. (sä-ä-vä'drä)	20	37°45'S	62°23'W
Saba, i., Neth. Ant. (sä'bä)	9b	17°39'N	63°20'W
Sabana, Archipiélago de, is., Cuba	10	23°05'N	80°00'W
Sabana, Río, r., Pan. (sä-bä'nä)	9	8°40'N	78°02'W
Sabana de la Mar, Dom. Rep. (sä-bä'nä dā lä mär')	11	19°05'N	69°30'W
Sabana de Uchire, Ven. (sä-bà'nä dĕ ōō-chē'rĕ)	19b	10°02'N	65°32'W
Sabanagrande, Hond. (sä-bä'nä-grä'n-dĕ)	8	13°47'N	87°16'W
Sabanalarga, Col. (sä-bá'nä-lär'gä)	18	10°38'N	75°02'W
Sabanas Páramo, mtn., Col. (sä-bá'näs pá'rä-mô)	18a	6°28'N	76°08'W
Sabancuy, Mex. (sä-bäŋ-kwē')	7	18°58'N	91°09'W
Sabinal, Cayo, i., Cuba (kä'yō sä-bē-näl')	10	21°40'N	77°20'W
Sabinas, Mex.	4	28°05'N	101°30'W
Sabinas, r., Mex. (sä-bē'näs)	2	26°37'N	99°52'W
Sabinas, Río, r., Mex. (rē'ō sä-bē'näs)	2	27°25'N	100°33'W
Sabinas Hidalgo, Mex. (ē-däl'gō)	2	26°30'N	100°10'W
Saco, r., Braz. (sä'kō)	20b	22°20'S	43°26'W
Sacramento, Mex.	2	25°45'N	103°22'W
Sacramento, Mex.	2	27°05'N	101°45'W
Sagua de Tánamo, Cuba (sä-gwä dĕ tá'nä-mō)	11	20°40'N	75°15'W
Sagua la Grande, Cuba (sä-gwä lä grä'n-dĕ)	10	22°45'N	80°05'W
Sahuayo de Dias, Mex.	6	20°03'N	102°43'W
Sain Alto, Mex. (sä-ēn' äl'tō)	6	23°35'N	103°13'W
Sainte Anne, Guad.	9b	16°15'N	61°23'W
Saint Ann's Bay, Jam.	10	18°25'N	77°15'W
Saint Barthélemy, i., Guad.	9b	17°55'N	62°32'W
Saint Catherine, Mount, mtn., Gren.	9b	12°10'N	61°42'W
Saint Christopher-Nevis *see* Saint Kitts and Nevis, nation, N.A.	5	17°24'N	63°30'W
Saint Croix, i., V.I.U.S. (sånt kroi')	5	17°40'N	64°43'W
Saint Georges, Fr. Gu.	19	3°48'N	51°47'W
Saint George's, Gren.	9b	12°02'N	61°57'W
Saint John, i., V.I.U.S.	5b	18°16'N	64°48'W
Saint Johns, Antig.	9b	17°07'N	61°50'W
Saint Joseph, Dom.	9b	15°25'N	61°26'W
Saint Kitts, i., St. K./N. (sånt kĭtts)	5	17°24'N	63°30'W
Saint Kitts and Nevis, nation, N.A.	5	17°24'N	63°30'W
Saint Laurent, Fr. Gu.	19	5°27'N	53°56'W
Saint Lucia, nation, N.A.	5	13°54'N	60°40'W
Saint Lucia Channel, strt., N.A. (lū'shĭ-á)	9b	14°15'N	61°00'W
Saint Marc, Haiti (săN' märk')	11	19°10'N	72°40'W
Saint-Marc, Canal de, strt., Haiti	11	19°05'N	73°15'W
Saint Martin, i., N.A. (mär'tĭn)	9b	18°06'N	62°54'W
Saint Michel-de-l'Atalaye, Haiti	11	19°25'N	72°20'W
Saint Nicolas, Cap, c., Haiti	11	19°45'N	73°35'W
Saint Pierre, Mart. (săn'pyâr')	9b	14°45'N	61°12'W
Sainte Rose, Guad.	9b	16°19'N	61°45'W
Saint Thomas, i., V.I.U.S.	5	18°22'N	64°57'W
Saint Thomas Harbor, b., V.I.U.S. (tŏm'ás)	5c	18°19'N	64°56'W
Saint Vincent and the Grenadines, nation, N.A.	5	13°20'N	60°50'W
Saint Vincent Passage, strt., N.A.	9b	13°35'N	61°10'W
Sajama, Nevada, mtn., Bol. (nĕ-vá'dä-sä-há'mä)	18	18°13'S	68°53'W
Sal, Cay, i., Bah. (kē säl)	10	23°45'N	80°25'W
Saladillo, Arg. (sä-lä-dēl'yō)	20	35°38'S	59°48'W
Salado, Hond. (sä-lä'dhō)	8	15°44'N	87°03'W
Salado, r., Arg.	17c	35°53'S	58°12'W
Salado, r., Arg.	20	37°00'S	67°00'W

PLACE (Pronunciation)	PAGE	LAT.	LONG.
Salado, r., Arg. (sä-lä'dō)	20	26°05'S	63°35'W
Salado, r., Mex.	4	28°00'N	102°00'W
Salado, r., Mex. (sä-lä'dō)	7	18°30'N	97°29'W
Salado de los Nadadores, Río, r., Mex. (dĕ-lôs-nä-dä-dō'rĕs)	2	27°26'N	101°35'W
Salamanca, Chile (sä-lä-mä'n-kä)	17b	31°48'S	70°57'W
Salamanca, Mex.	4	20°36'N	101°10'W
Salamina, Col. (sä-lä-mē'-nä)	18a	5°25'N	75°29'W
Salaverry, Peru (sä-lä-vā'rē)	18	8°16'S	78°54'W
Salcedo, Dom. Rep. (säl-sā'dō)	11	19°25'N	70°30'W
Saldaña, r., Col. (säl-dà'n-yä)	18a	3°42'N	75°16'W
Salina Cruz, Mex. (sä-lē'nä krōōz')	4	16°10'N	95°12'W
Salina Point, c., Bah.	11	22°10'N	74°20'W
Salinas, Mex.	4	22°38'N	101°42'W
Salinas, P.R.	5b	17°58'N	66°16'W
Salinas, r., Mex. (sa-lē'nas)	7	16°15'N	90°31'W
Salinas, Bahía de, b., N.A. (bä-ē'ä-dĕ-sä-lē'nás)	8	11°05'N	85°55'W
Salinas Victoria, Mex. (sä-lē'näs vēk-tō'rē-ä)	2	25°59'N	100°19'W
Salta, Arg. (säl'tä)	20	24°50'S	65°16'W
Salta, prov., Arg.	20	25°15'S	65°00'W
Salt Cay, i., T./C. Is.	11	21°20'N	71°15'W
Saltillo, Mex. (säl-tēl'yō)	4	25°24'N	100°59'W
Salto, Arg. (säl'tō)	17c	34°17'S	60°15'W
Salto, Ur.	20	31°18'S	57°45'W
Salto, r., Mex.	6	22°16'N	99°18'W
Salto, Serra do, mtn., Braz. (sĕ'r-rä-dō)	17a	20°26'S	43°28'W
Salto Grande, Braz. (grän'dä)	19	22°57'S	49°58'W
Salud, Mount, mtn., Pan. (sä-lōō'th)	4a	9°14'N	79°42'W
Salvador (Bahia), Braz. (säl-vä-dōr') (bă-ē'á)	19	12°59'S	38°27'W
Salvador Point, c., Bah.	10	24°30'N	77°45'W
Salvatierra, Mex. (säl-vä-tyĕr'rä)	6	20°13'N	100°52'W
Samana, Cabo, c., Dom. Rep.	5	19°20'N	69°00'W
Samana or Atwood Cay, i., Bah.	11	23°05'N	73°45'W
Samborombón, r., Arg.	17c	35°20'S	57°52'W
Samborombón, Bahía, b., Arg. (bä-ē'ä-säm-bô-rôm-bô'n)	17c	35°57'S	57°05'W
San Ambrosio, Isla, i., Chile (ē's-lä-dĕ-sän äm-brō'zē-ō)	15	26°40'S	80°00'W
San Andrés, Col. (sän-än-drĕ's)	18a	6°57'N	75°41'W
San Andrés, Mex. (sän än-drās')	7a	19°15'N	99°10'W
San Andrés, i., Col.	9	12°32'N	81°34'W
San Andrés, Laguna de, l., Mex.	7	22°40'N	97°50'W
San Andrés Tuxtla, Mex. (sän-än-drā's-tōōs'tlä)	4	18°27'N	95°12'W
San Antonio, Chile (sän-än-tō'nyō)	20	33°34'S	71°36'W
San Antonio, Col.	18a	2°57'N	75°06'W
San Antonio, Col.	18a	3°55'N	75°28'W
San Antonio, Cabo, c., Cuba (kä'bō-sän-än-tō'nyō)	5	21°55'N	84°55'W
San Antonio de Areco, Arg. (dā ä-rä'kō)	17c	34°16'S	59°30'W
San Antonio de las Vegas, Cuba	11a	22°51'N	82°23'W
San Antonio de los Baños, Cuba (dā lōs bän'yōs)	10	22°54'N	82°30'W
San Antonio de los Cobres, Arg. (dā lōs kō'brās)	20	24°15'S	66°29'W
San Antônio de Pádua, Braz. (dĕ-pá'dwä)	17a	21°32'S	42°09'W
San Antonio de Tamanaco, Ven.	19b	9°42'N	66°03'W
San Antonio Oeste, Arg. (sän-nä-tō'nyō ō-ĕs'tå)	20	40°49'S	64°56'W
Sanarate, Guat. (sä-nä-rä'tĕ)	8	14°47'N	90°12'W
San Bartolo, Mex. (sän bär-tō'lō)	7a	19°36'N	99°43'W
San Bartolo, Mex.	2	24°43'N	103°12'W
San Bernardo, Chile (sän bĕr-när'dō)	17b	33°35'S	70°42'W
San Blas, Mex. (sän bläs')	4	21°33'N	105°19'W
San Blas, Cordillera de, mts., Pan.	9	9°17'N	78°20'W
San Blas, Golfo de, b., Pan.	9	9°33'N	78°42'W
San Blas, Punta, c., Pan.	9	9°35'N	78°55'W
San Buenaventura, Mex. (bwā'ná-vĕn-tōō'rá)	2	27°07'N	101°30'W
San Carlos, Chile (sän-kä'r-lōs)	20	36°23'S	71°58'W
San Carlos, Col.	18a	6°11'N	74°58'W
San Carlos, Mex. (sän kär'lōs)	7	17°49'N	92°33'W
San Carlos, Mex.	2	24°36'N	98°52'W
San Carlos, Nic. (sän-kä'r-lôs)	9	11°08'N	84°48'W
San Carlos, Ven.	18	9°36'N	68°35'W
San Carlos, r., C.R.	9	10°36'N	84°18'W
San Carlos de Bariloche, Arg.	20	41°15'S	71°26'W
San Casimiro, Ven. (kä-sē-mē'rō)	19b	10°01'N	67°02'W
Sánchez, Dom. Rep. (sän'chĕz)	5	19°15'N	69°40'W
Sanchez, Río de los, r., Mex. (rē'ō-dĕ-lôs)	6	20°31'N	102°29'W
Sánchez Román, Mex. (rô-má'n)	6	21°48'N	103°20'W
San Cristóbal, Dom. Rep. (krēs-tō'bäl)	11	18°25'N	70°05'W
San Cristóbal, Guat.	8	15°22'N	90°26'W
San Cristóbal, Ven.	18	7°43'N	72°15'W
San Cristóbal de las Casas, Mex.	4	16°44'N	92°39'W
Sancti Spíritus, Cuba (säŋk'tê spē'rē-tōōs)	5	21°55'N	79°25'W
Sancti Spíritus, prov., Cuba	10	22°05'N	79°20'W
San Diego de la Unión, Mex. (sän dē-å-gō dā lä ōō-nyōn')	6	21°27'N	100°52'W
San Dimas, Mex. (dē-mäs')	6	24°08'N	105°57'W
San Estanislao, Para. (ĕs-tä-nēs-lá'ō)	20	24°38'S	56°20'W
San Esteban, Hond. (ĕs-tĕ'bän)	8	15°13'N	85°53'W
San Felipe, Chile (fā-lē'på)	20	32°45'S	70°43'W
San Felipe, Mex. (fē-lē'pĕ)	6	21°29'N	101°13'W
San Felipe, Mex.	6	22°21'N	105°26'W
San Felipe, Ven. (fĕ-lē'pĕ)	18	10°13'N	68°45'W
San Felipe, Cayos de, is., Cuba (kä'yōs-dĕ-sän fĕ lē'pĕ)	10	22°00'N	83°30'W

PLACE (Pronunciation)	PAGE	LAT.	LONG.
San Félix, Isla, i., Chile (ē's-lä-dĕ-sän fĕ-lēks')	15	26°20'S	80°10'W
San Fernando, Arg. (fĕr-ná'n-dō)	20a	34°26'S	58°34'W
San Fernando, Chile	17b	35°36'S	70°58'W
San Fernando, Mex. (fĕr-nän'dō)	2	24°52'N	98°10'W
San Fernando, r., Mex. (sän fĕr-nän'dō)	2	25°07'N	98°25'W
San Fernando de Apure, Ven. (sän-fĕr-nä'n-dō-dĕ-ä-pōō'rä)	18	7°46'N	67°29'W
San Fernando de Atabapo, Ven. (dĕ-ä-tä-bä'pō)	18	3°58'N	67°41'W
San Francisco, Arg. (săn från'sĭs'kō)	20	31°23'S	62°09'W
San Francisco, El Sal.	8	13°48'N	88°11'W
San Francisco del Oro, Mex. (dĕl ō'rō)	4	27°00'N	106°37'W
San Francisco del Rincón, Mex. (del rēn-kōn')	6	21°01'N	101°51'W
San Francisco de Macaira, Ven. (dĕ-mä-kī'rä)	19b	9°58'N	66°17'W
San Francisco de Macoris, Dom. Rep. (dä-mä-kō'rēs)	11	19°20'N	70°15'W
San Francisco de Paula, Cuba (dā pou'lä)	11a	23°04'N	82°18'W
San Gabriel Chilac, Mex. (sän-gä-brē-ĕl-chē-läk')	7	18°19'N	97°22'W
San Gil, Col. (sän-кē'l)	18	6°32'N	73°13'W
San Ignacio, Belize	8a	17°11'N	89°04'W
San Isidro, Arg. (ē-sē'drō)	17c	34°28'S	58°31'W
San Isidro, C.R.	9	9°24'N	83°43'W
San Javier, Chile (sän-hà-vē'ĕr)	17b	35°35'S	71°43'W
San Jerónimo, Mex.	7a	19°31'N	98°46'W
San Jerónimo de Juárez, Mex. (hä-rō'nē-mō dā hwä'råz)	6	17°08'N	100°30'W
San Joaquín, Ven.	19b	10°16'N	67°47'W
San Jorge, Golfo, b., Arg. (gôl-fō-sän-кô'r-кĕ)	20	46°15'S	66°45'W
San José, C.R. (sän hô-sā')	5	9°57'N	84°05'W
San José, i., Mex. (кō-sĕ')	4	25°00'N	110°35'W
San José, Isla de, i., Pan. (ē's-lä-dĕ-sän hô-sā')	9	8°17'N	79°20'W
San José de Feliciano, Arg. (då lä ĕs-kē'ná)	20	30°26'S	58°44'W
San José de Gauribe, Ven. (sän-hô-sĕ'dĕ-gáōō-rē'bĕ)	19b	9°51'N	65°49'W
San José de las Lajas, Cuba (sän-кô-sĕ'dĕ-läs-lá'käs)	11a	22°58'N	82°10'W
San José Iturbide, Mex. (ē-tōōr-bē'dĕ)	6	21°00'N	100°24'W
San Juan, Arg. (hwän')	20	31°36'S	68°29'W
San Juan, Col. (hòá'n)	18a	3°23'N	73°48'W
San Juan, Dom. Rep. (sän hwän')	11	18°50'N	71°15'W
San Juan, P.R. (sän hwän')	5	18°30'N	66°10'W
San Juan, prov., Arg.	20	31°00'S	69°30'W
San Juan, r., Mex. (sän-hōō-än')	7	18°10'N	95°23'W
San Juan, r., N.A.	5	10°58'N	84°18'W
San Juan, Cabezas de, c., P.R.	5b	18°29'N	65°30'W
San Juan, Pico, mtn., Cuba (pē'kô-sän-kòá'n)	10	21°55'N	80°00'W
San Juan, Río, r., Mex. (rē'ō-sän-hwän)	2	25°35'N	99°15'W
San Juan Bautista, Para. (sän hwän' bou-tēs'tä)	20	26°48'S	57°09'W
San Juan Capistrano, Mex. (sän-hōō-än' kä-pēs-trä'nō)	6	22°41'N	104°07'W
San Juan de Guadalupe, Mex. (sän hwan då gwä-dhä-lōō'pā)	2	24°37'N	102°43'W
San Juan del Norte, Nic.	9	10°55'N	83°44'W
San Juan del Norte, Bahía de, b., Nic.	9	11°12'N	83°40'W
San Juan de los Lagos, Mex. (sän-hōō-än'då lōs lä'gôs)	6	21°15'N	102°18'W
San Juan de los Lagos, r., Mex. (då lōs lä'gôs)	6	21°13'N	102°12'W
San Juan de los Morros, Ven. (dĕ-lôs-mô'r-rôs)	19b	9°54'N	67°22'W
San Juan del Río, Mex.	6	20°21'N	99°59'W
San Juan del Río, Mex. (sän hwän del rē'ō)	2	24°47'N	104°29'W
San Juan del Sur, Nic. (dĕl sōōr)	4	11°15'N	85°53'W
San Juan Evangelista, Mex. (sän-hōō-ä'n-å-väŋ-kå-lēs'ta')	7	17°57'N	95°08'W
San Juan Ixtenco, Mex. (ēx-tĕ'n-kō)	7	19°14'N	97°52'W
San Juan Martínez, Cuba	10	22°15'N	83°50'W
San Julián, Arg. (sän hōō-lyá'n)	20	49°17'S	68°02'W
San Justo, Arg. (hōōs'tō)	20a	34°40'S	58°33'W
San Lázaro, Cabo, c., Mex. (sän-lá'zä-rō)	4	24°58'N	113°30'W
San Lorenzo, Arg. (sän lō-rĕn'zō)	20	32°46'S	60°44'W
San Lorenzo, Hond. (sän lō-rĕn'zō)	8	13°24'N	87°24'W
San Lucas, Bol. (lōō'kás)	18	20°12'S	65°06'W
San Lucas, Cabo, c., Mex.	4	22°45'N	109°45'W
San Luis, Arg. (lō-ēs')	20	33°16'S	66°15'W
San Luis, Col. (lòē's)	18a	6°03'N	74°57'W
San Luis, Cuba	11	20°15'N	75°50'W
San Luis, Guat.	8	14°38'N	89°42'W
San Luis, prov., Arg.	20	32°45'S	66°00'W
San Luis de la Paz, Mex. (då lä päz')	6	21°17'N	100°32'W
San Luis del Cordero, Mex. (dĕl kôr-dä'rō)	2	25°25'N	104°20'W
San Luis Potosí, Mex.	4	22°08'N	100°58'W
San Luis Potosí, state, Mex.	4	22°45'N	101°45'W
San Marcos, Guat. (mär'kōs)	8	14°57'N	91°49'W
San Marcos, Mex.	6	16°46'N	99°23'W
San Marcos de Colón, Hond. (sän-má'r-kōs-dĕ-kō-lōn')	8	13°17'N	86°50'W
San Martín, Col. (sän mär-tē'n)	18a	3°42'N	73°44'W

52 San Martín – Silver Bank Passage PRONOUNCING INDEX

PLACE (Pronunciation)	PAGE	LAT.	LONG.
San Martín, vol., Mex. (mär-tē´n)	7	18°36'N	95°11'W
San Martín, l., S.A.	20	48°15'S	72°30'W
San Martín Chalchicuautla, Mex.	6	21°22'N	98°39'W
San Martín Hidalgo, Mex. (sän mär-tē´n-ē-däl´go̊)	6	20°27'N	103°55'W
San Mateo, Mex.	7	16°59'N	97°04'W
San Mateo, Ven. (sän mȧ-tĕ´ō)	19b	9°45'N	64°34'W
San Matías, Golfo, b., Arg. (sän mä-tē´äs)	20	41°30'S	63°45'W
San Miguel, El Sal. (sän me̊-gål´)	4	13°28'N	88°11'W
San Miguel, Mex. (sän me̊-gål´)	7	18°18'N	97°09'W
San Miguel, Pan.	9	8°26'N	78°55'W
San Miguel, Ven. (sän mē-gĕ´l)	19b	9°56'N	64°58'W
San Miguel, vol., El Sal.	8	13°27'N	88°17'W
San Miguel, r., Bol. (sän-mē-gĕl´)	18	13°34'S	63°58'W
San Miguel, r., N.A. (sän me̊-gål´)	7	15°27'N	92°00'W
San Miguel, Bahía, b., Pan. (bä-ē´ä-sän me̊-gål´)	9	8°17'N	78°26'W
San Miguel de Allende, Mex. (då ä-lyĕn´då)	6	20°54'N	100°44'W
San Miguel el Alto, Mex. (ĕl äl´to̊)	6	21°03'N	102°26'W
San Nicolás, Arg. (sän nē-kô-lȧ´s)	20	33°20'S	60°14'W
San Nicolás, r., Mex.	6	19°40'N	105°08'W
San Pablo, Ven. (sän-pȧ´blô)	19b	9°46'N	65°04'W
San Pablo, r., Pan. (sän päb´lo̊)	9	8°12'N	81°12'W
San Pedro, Arg. (sän pā´dro̊)	20	24°15'S	64°15'W
San Pedro, Arg.	17c	33°41'S	59°42'W
San Pedro, Chile (sän pĕ´dro̊)	17b	33°54'S	71°27'W
San Pedro, El Sal. (sän pā´dro̊)	8	13°49'N	88°58'W
San Pedro, Mex. (sän pā´dro̊)	7	18°38'N	92°25'W
San Pedro, Para. (sän-pĕ´dro̊)	20	24°13'S	57°00'W
San Pedro, r., Cuba (sän-pĕ´dro̊)	10	21°05'N	78°15'W
San Pedro, r., Mex. (sän pā´dro̊)	6	22°08'N	104°59'W
San Pedro, r., Mex.	2	27°56'N	105°50'W
San Pedro, Río de, r., Mex.	6	21°51'N	102°24'W
San Pedro, Río de, r., N.A.	7	18°23'N	92°13'W
San Pedro de las Colonias, Mex. (dĕ-läs-ko̊-lô´nyäs)	2	25°47'N	102°58'W
San Pedro de Macorís, Dom. Rep. (sän-pĕ´drō-då mä-ko̊-rēs´)	11	18°30'N	69°30'W
San Pedro Lagunillas, Mex. (sän pā´dro̊ lä-go͞o-nēl´yäs)	6	21°12'N	104°47'W
San Pedro Sula, Hond. (sän pā´dro̊ so͞o´lä)	8	15°29'N	88°01'W
San Rafael, Arg. (sän rä-fä-āl´)	20	34°30'S	68°13'W
San Rafael, Col. (sän-rä-fä-ĕ´l)	18a	6°18'N	75°02'W
San Rafael, Cabo, c., Dom. Rep. (kȧ´bo̊)	11	19°00'N	68°50'W
San Ramón, C.R.	9	10°07'N	84°30'W
San Roque, Col. (sän-rô´kĕ)	18a	6°29'N	75°00'W
San Salvador, El Sal. (sän säl-vä-dōr´)	4	13°45'N	89°11'W
San Salvador (Watling), i., Bah. (săn săl´vȧ-dôr)	11	24°05'N	74°30'W
San Salvador, i., Ec.	18	0°14'S	90°50'W
San Salvador, r., Ur. (sän-säl-vä-dô´r)	17c	33°42'S	58°04'W
San Sebastián, Ven. (sän-sĕ-bäs-tyȧ´n)	19b	9°58'N	67°11'W
Santa Ana, El Sal.	4	14°02'N	89°35'W
Santa Ana, Mex. (sän´tä ä´nä)	6	19°18'N	98°10'W
Santa Bárbara, Braz. (sän-tä-bȧ´r-bä-rä)	19	19°57'S	43°25'W
Santa Bárbara, Hond.	8	14°52'N	88°20'W
Santa Bárbara, Mex.	2	26°48'N	105°50'W
Santa Branca, Braz. (sän-tä-brä´ɴ-kä)	17a	23°25'S	45°52'W
Santa Catalina, Cerro de, mtn., Pan.	9	8°39'N	81°36'W
Santa Catarina, Mex. (sän´tȧ kä-tä-rē´nä)	2	25°41'N	100°27'W
Santa Catarina, state, Braz. (sän-tä-kä-tä-re̊´nȧ)	20	27°15'S	50°30'W
Santa Catarina, r., Mex.	6	16°31'N	98°39'W
Santa Clara, Cuba (sän´tȧ klä´rȧ)	5	22°25'N	80°00'W
Santa Clara, Mex.	2	24°29'N	103°22'W
Santa Clara, Ur.	20	32°46'S	54°51'W
Santa Clara, vol., Nic.	8	12°44'N	87°00'W
Santa Clara, Bahía de, b., Cuba (bä-ē´ä-dĕ-sän-tä-klä-rä)	10	23°05'N	80°50'W
Santa Clara, Sierra, mts., Mex. (sē-ĕ´r-rä-sän´tȧ klä´rȧ)	4	27°30'N	113°50'W
Santa Cruz, Bol. (sän´tȧ kro͞oz´)	18	17°45'S	63°03'W
Santa Cruz, Braz. (sän-tä-kro͞o´s)	20	29°43'S	52°15'W
Santa Cruz, Braz.	20b	22°55'S	43°41'W
Santa Cruz, Chile	17b	34°38'S	71°21'W
Santa Cruz, C.R.	8	10°16'N	85°37'W
Santa Cruz, Mex.	2	25°50'N	105°25'W
Santa Cruz, prov., Arg.	20	48°00'S	70°00'W
Santa Cruz, i., Ec. (sän-tä-kro͞o´z)	18	0°38'S	90°20'W
Santa Cruz, r., Arg. (sän´tȧ kro͞oz´)	20	50°05'S	71°00'W
Santa Cruz Barillas, Guat. (sän-tä-kro͞o´z-bä-re̊´l-yäs)	8	15°47'N	91°22'W
Santa Cruz del Sur, Cuba (sän-tä-kro͞o´s-dĕl-sȯ´r)	10	20°45'N	78°00'W
Santa Domingo, Cay, i., Bah.	11	21°50'N	75°45'W
Santa Fe, Arg. (sän´tä fā´)	20	31°33'S	60°45'W
Santa Fé, Cuba (sän-tä-fĕ´)	10	21°45'N	82°40'W
Santa Fe, prov., Arg. (sän´tä fā´)	20	32°00'S	61°15'W
Santa Fe de Bogotá *see* Bogotá, Col.	18	4°36'N	74°05'W
Santa Filomena, Braz. (sän-tä-fē-lo̊-mĕ´nä)	19	9°09'S	44°45'W
Santa Genoveva, mtn., Mex. (sän-tä-hĕ-nō-vĕ´vä)	4	23°30'N	110°00'W
Santa Inés, Ven. (sän´tä ē-nĕ´s)	19b	9°54'N	64°21'W
Santa Inés, i., Chile (sän´tä e̊-nās´)	20	53°45'S	74°15'W
Santa Lucía, Cuba (sän´tȧ lo͞o-sē´ȧ)	10	21°15'N	77°30'W
Santa Lucía, Ur. (sän-tä-lo͞o-se̊´ä)	20	34°27'S	56°23'W
Santa Lucía, Ven.	19b	10°18'N	66°40'W
Santa Lucía, r., Ur.	17c	34°19'S	56°13'W
Santa Lucia Bay, b., Cuba (sän´tȧ lo͞o-sē´ȧ)	10	22°55'N	84°20'W
Santa Margarita, i., Mex. (sän´tȧ mär-gȧ-rē´tȧ)	4	24°15'N	112°00'W
Santa Maria, Braz. (sän´tȧ mä-rē´ȧ)	20	29°40'S	54°00'W
Santa María, vol., Guat.	8	14°45'N	91°33'W
Santa María, r., Mex. (sän´tȧ mä-rē´ȧ)	6	21°33'N	100°17'W
Santa Maria, Cape, c., Bah.	11	23°45'N	75°30'W
Santa María, Cayo, i., Cuba	10	22°40'N	79°00'W
Santa María del Oro, Mex. (sän´tä-mä-re̊´ä-dĕl-ô-ro̊)	6	21°21'N	104°35'W
Santa María de los Ángeles, Mex. (dĕ-lo̊s-ȧ´n-hĕ-lĕs)	6	22°10'N	103°34'W
Santa María del Río, Mex.	6	21°46'N	100°43'W
Santa María de Ocotán, Mex.	6	22°56'N	104°30'W
Santa Maria Madalena, Braz.	17a	22°00'S	42°00'W
Santa Marta, Col. (sän´tȧ mär´tȧ)	18	11°15'N	74°13'W
Santana, r., Braz. (sän-tä´nä)	20b	22°33'S	43°37'W
Santander, Col. (sän-tän-dĕr´)	18a	3°00'N	76°25'W
Santarém, Braz. (sän-tä-rĕɴ´)	19	2°28'S	54°37'W
Santaren Channel, strt., Bah. (săn-t*a*-rĕn´)	10	24°15'N	79°30'W
Santa Rita do Sapucai, Braz. (sä-pȯ-kȧ´ē)	17a	22°15'S	45°41'W
Santa Rosa, Arg. (sän-tä-ro̊-sä)	20	36°45'S	64°10'W
Santa Rosa, Col. (sän-tä-ro̊-sä)	18a	6°38'N	75°26'W
Santa Rosa, Ec.	18	3°29'S	79°55'W
Santa Rosa, Guat. (sän´tȧ rō´s*ȧ*)	8	14°21'N	90°16'W
Santa Rosa, Hond.	8	14°45'N	88°51'W
Santa Rosa, Ven. (sän-tä-ro̊-sä)	19b	9°37'N	64°10'W
Santa Rosa de Cabal, Col. (sän-tä-ro̊-sä-dĕ-kä-bä´l)	18a	4°53'N	75°38'W
Santa Rosa de Viterbo, Braz. (sän-tä-rô-sȧ-dĕ-vē-tĕr´-bo̊)	17a	21°30'S	47°21'W
Santa Rosalía, Mex. (sän´tȧ ro̊-zä´le̊-ȧ)	4	27°13'N	112°15'W
Santa Teresa, Arg. (sän-tä-tĕ-rĕ´sä)	17c	33°27'S	60°47'W
Santa Teresa, Ven.	19b	10°14'N	66°40'W
Santa Vitória do Palmar, Braz. (sän-tä-vē-tŏ´ryä-do̊-päl-mȧr)	20	33°30'S	53°16'W
Santiago, Braz. (sän-tyȧ´gô)	20	29°05'S	54°46'W
Santiago, Chile (sän-te̊-ä´go̊)	20	33°26'S	70°40'W
Santiago, Pan.	5	8°07'N	80°58'W
Santiago, prov., Chile (sän-tyȧ´gō)	17b	33°28'S	70°55'W
Santiago de Cuba, Cuba (sän-tyȧ´go̊-dā ko͞o´bȧ)	5	20°00'N	75°50'W
Santiago de Cuba, prov., Cuba	11	20°20'N	76°05'W
Santiago de las Vegas, Cuba (sän-tyȧ´go̊-dĕ-läs-vĕ´gäs)	11a	22°58'N	82°23'W
Santiago del Estero, Arg.	20	27°50'S	64°14'W
Santiago del Estero, prov., Arg. (sän-tē-ȧ´gō-dĕl ĕs-tā-ro̊)	20	27°15'S	63°30'W
Santiago de los Caballeros, Dom. Rep.	5	19°30'N	70°45'W
Santiago Rodríguez, Dom. Rep. (sän-tyȧ´go̊-rô-dre̊´gĕz)	11	19°30'N	71°25'W
Santiago Tuxtla, Mex. (sän-tyȧ´gô-to͞o´x-tlä)	7	18°28'N	95°18'W
Santiaguillo, Laguna de, l., Mex. (lä-oô´nä-dĕ-sän-te̊-a-gēl´yo̊)	2	24°51'N	104°43'W
Santo Amaro, Braz. (sän´tȯ ä-mä´rȯ)	19	12°32'S	38°33'W
Santo Amaro de Campos, Braz.	17a	22°01'S	41°05'W
Santo André, Braz.	17a	23°40'S	46°31'W
Santo Angelo, Braz. (sän-tô-ȧ´n-zhĕ-lo̊)	20	28°16'S	53°59'W
Santo Antônio do Monte, Braz. (sän-tô-än-tô´nyo̊-do̊-môn´tĕ)	17a	20°06'S	45°18'W
Santo Domingo, Cuba (sän´to̊-dōmĭŋ´go̊)	10	22°35'N	80°20'W
Santo Domingo, Dom. Rep. (sän´to̊ do̊-mĭn´go̊)	5	18°30'N	69°55'W
Santo Domingo, Nic. (sän-to̊-do̊-me̊´n-gō)	9	12°15'N	84°56'W
Santos, Braz. (sän´tozh)	19	23°58'S	46°20'W
Santos Dumont, Braz. (sän´to̊s-dȯ-mô´nt)	19	21°28'S	43°33'W
San Urbano, Arg. (sän-ȯr-bä´no̊)	17c	33°39'S	61°28'W
San Valentín, Monte, mtn., Chile (sän-vä-lĕn-tē´n)	20	46°41'S	73°30'W
San Vicente, Arg. (sän-vē-sĕn´tĕ)	17c	35°00'S	58°26'W
San Vicente, Chile	17b	34°25'S	71°06'W
San Vicente, El Sal. (sän ve̊-sĕn´tå)	8	13°41'N	88°43'W
São Bernardo do Campo, Braz. (souɴ-bĕr-nȧr´do̊-do̊-kȧ´m-pô)	17a	23°44'S	46°33'W
São Borja, Braz. (souɴ-bôr-zhä)	20	28°44'S	55°59'W
São Carlos, Braz. (souɴ kär´lōzh)	19	22°02'S	47°54'W
São Cristovão, Braz. (souɴ-kre̊s-tō-vouɴ)	19	11°04'S	37°11'W
São Fidélis, Braz. (souɴ-fē-dĕ´lēs)	17a	21°41'S	41°45'W
São Francisco, Braz. (souɴ fräɴ-sēsh´kȯ)	19	15°59'S	44°42'W
São Francisco, r., Braz. (souɴ fräɴ-sēsh´kȯ)	19	8°56'S	40°20'W
São Francisco do Sul, Braz. (souɴ fräɴ-sēsh´kȯ-do̊-so͞o´l)	20	26°15'S	48°42'W
São Gabriel, Braz. (souɴ´gä-bre̊-ĕl´)	20	30°28'S	54°11'W
São Geraldo, Braz. (souɴ-zhĕ-rä´l-do̊)	17a	21°01'S	42°49'W
São Gonçalo, Braz. (souɴ´gōŋ-sä´lȯ)	17a	22°55'S	43°04'W
São João da Barra, Braz. (souɴ-zho̊uɴ-dä-bȧ´rä)	17a	21°40'S	41°03'W
São João da Boa Vista, Braz. (souɴ-zho̊uɴ-dä-bôä-ve̊´s-tä)	17a	21°58'S	46°45'W
São João del Rei, Braz. (souɴ zho̊-ouɴ´dĕl-rā)	20	21°08'S	44°14'W
São João de Meriti, Braz. (souɴ-zho̊uɴ-dĕ-mĕ-rē-te̊)	20b	22°47'S	43°22'W
São João do Araguaia, Braz. (souɴ zho̊-ouɴ´do̊-ä-rä-gwä´yä)	19	5°29'S	48°44'W
São João Nepomuceno, Braz. (souɴ-zho̊uɴ-nĕ-po̊-mo͞o-sĕ-nō)	17a	21°33'S	43°00'W
São José do Rio Pardo, Braz. (souɴ-zhô-sĕ´do̊-rē´ō-pȧ´r-dō)	17a	21°36'S	46°50'W
São José do Rio Prêto, Braz. (souɴ zho̊-zĕ´do̊-re´ō-prĕ-tō)	19	20°57'S	49°12'W
São José dos Campos, Braz. (souɴ zho̊-zā´dōzh käɴ pōzh´)	17a	23°12'S	45°53'W
São Leopoldo, Braz. (souɴ-lĕ-ō-pôl´do̊)	20	29°46'S	51°09'W
São Luis, Braz.	19	2°31'S	43°14'W
São Luis do Paraitinga, Braz. (souɴ-lo͞oe̊´s-do̊-pä-rä-ē-tē´n-gä)	17a	23°15'S	45°18'W
São Manuel, r., Braz.	19	8°28'S	57°07'E
São Mateus, Braz. (souɴ mä-tā´ȯzh)	19	18°44'S	39°45'W
São Mateus, Braz.	20b	22°49'S	43°23'W
São Miguel Arcanjo, Braz. (souɴ-mē-gĕ´l-ȧr-kän-zhō)	17a	23°54'S	47°59'W
Saona, i., Dom. Rep. (sä-ō´nä)	11	18°10'N	68°55'W
São Paulo, Braz. (souɴ´ pou´lȯ)	19	23°34'S	46°38'W
São Paulo, state, Braz. (souɴ pou´lȯ)	19	21°45'S	50°47'W
São Paulo de Olivença, Braz. (souɴ´pou´lȯdå ō-le̊-vĕn´s*ȧ*)	18	3°32'S	68°46'W
São Pedro, Braz. (souɴ-pĕ´dro̊)	17a	22°34'S	47°54'W
São Pedro de Aldeia, Braz. (souɴ-pĕ´dro̊-dĕ-äl-dĕ´yä)	17a	22°50'S	42°04'W
São Pedro e São Paulo, Rocedos, rocks, Braz.	15	1°50'N	30°00'W
São Raimundo Nonato, Braz. (souɴ´ rī-mȯ´n-do no̊-nä´tȯ)	19	9°09'S	42°32'W
São Roque, Braz. (souɴ´ rō´kĕ)	17a	23°32'S	47°08'W
São Roque, Cabo de, c., Braz. (kä´bo-dĕ-souɴ´ rō´kĕ)	19	5°06'S	35°11'W
São Sebastião, Braz. (souɴ så-bäs-te̊-ouɴ´)	17a	23°48'S	45°25'W
São Sebastião, Ilha de, i., Braz.	17a	23°52'S	45°22'W
São Sebastião do Paraíso, Braz.	17a	20°54'S	46°58'W
São Simão, Braz. (souɴ-sē-mouɴ)	17a	21°30'S	47°33'W
São Vicente, Braz. (souɴ ve-se´n-tĕ)	19	23°57'S	46°25'W
Sapucaí, r., Braz. (sä-po͞o-kȧ-ē´)	17a	22°20'S	45°53'W
Sapucaia, Braz. (sä-po͞o-kä´yȧ)	17a	22°01'S	42°54'W
Sapucaí Mirim, r., Braz. (sä-po͞o-kȧ-ē´mē-re̊ɴ)	17a	21°06'S	47°03'W
Saquarema, Braz. (sä-kwä-rĕ-mä)	17a	22°56'S	42°32'W
Sarandi, Arg. (sä-rän´de̊)	20a	34°41'S	58°21'W
Sarandi Grande, Ur. (sä-rän´dē-grän´dĕ)	17c	33°42'S	56°21'W
Sarmiento, Monte, mtn., Chile (mo̊´n-tĕ-sär-myĕn´tō)	20	54°28'S	70°40'W
Sarstun, r., N.A. (särs-to͞o´n)	8	15°50'N	89°26'W
Saumatre, Étang, l., Haiti	11	18°40'N	72°10'W
Savanna la Mar, Jam. (sȧ-vän´ȧ lȧ mär´)	10	18°10'N	78°10'W
Sayula, Mex. (sä-yo͞o´lä)	7	17°51'N	94°56'W
Sayula, Mex.	6	19°50'N	103°33'W
Sayula, Laguna de, l., Mex. (lä-gȯ´nä-dĕ)	6	20°00'N	103°33'W
Seal Cays, is., Bah.	11	22°40'N	75°55'W
Seal Cays, is., T./C. Is.	11	21°10'N	71°45'W
Sebaco, Nic. (sĕ-bä´ko̊)	8	12°50'N	86°03'W
Sebastián Vizcaíno, Bahía, b., Mex.	4	28°45'N	115°15'W
Seco, r., Mex. (sĕ´ko̊)	7	18°11'N	93°18'W
Segovia, Col. (sĕ-go̊´vēä)	18a	7°08'N	74°42'W
Seibo, Dom. Rep. (sĕ´y-bō)	11	18°45'N	69°05'W
Seio do Venus, mtn., Braz. (sĕ-yo̊-do̊-vĕ´no͞os)	20b	22°28'S	43°12'W
Senador Pompeu, Braz. (sĕ-nä-dōr-pôm-pĕ´ȯ)	19	5°34'S	39°18'W
Senhor do Bonfim, Braz. (sĕn-yo̊r do̊ bôɴ-fē´ɴ)	19	10°21'S	40°09'W
Sensuntepeque, El Sal. (sĕn-so͞on-tå-pā´kå)	8	13°53'N	88°34'W
Sepetiba, Baía de, b., Braz. (bäē´ä dĕ så-på-tē´b*ȧ*)	20b	23°01'S	43°42'W
Septentrional, Cordillera, mts., Dom. Rep.	11	19°50'N	71°15'W
Sergipe, state, Braz. (sĕr-zhē´pĕ)	19	10°27'S	37°04'W
Serodino, Arg. (sĕ-ro̊-de̊´nō)	17c	32°36'S	60°56'W
Seropédica, Braz. (sĕ-ro̊-pĕ´dē-kä)	20b	22°44'S	43°43'W
Serrinha, Braz. (sĕr-rēn´yȧ)	19	11°43'S	38°49'W
Sertânia, Braz. (sĕr-tȧ´nyä)	19	8°28'S	37°13'W
Sertãozinho, Braz. (sĕr-touɴ-ze̊´n-yo̊)	17a	21°10'S	47°58'W
Sete Lagoas, Braz. (sĕ-tĕ lä-gô´äs)	19	19°23'S	43°58'W
Sete Pontes, Braz.	20b	22°51'S	43°05'W
Settlement Point, c., Bah. (sĕt´l-m*ĕ*nt)	10	26°40'N	79°00'W
Sevilla, Col. (sĕ-ve̊´l-yä)	18a	4°16'N	75°56'W
Sewell, Chile (sĕ´ȯ-ĕl)	20	34°01'S	70°18'W
Seybaplaya, Mex. (sā-e̊-bä-plä´yä)	7	19°38'N	90°40'W
Ship Channel Cay, i., Bah. (shĭp chă-n*ĕ*l kē)	10	24°50'N	76°50'W
Shroud Cay, i., Bah.	10	24°20'N	76°40'W
Sico, r., Hond. (se̊-ko̊)	8	15°32'N	85°42'W
Sierra Mojada, Mex. (sē-ĕ´r-rä-mo̊-ᴋä´dä)	2	27°22'N	103°42'W
Sigsig, Ec. (se̊g-sēg´)	18	3°04'S	78°44'W
Siguanea, Ensenada de la, b., Cuba	10	21°45'N	83°15'W
Siguatepeque, Hond. (sē-gwȧ´tĕ-pĕ-kĕ)	8	14°33'N	87°51'W
Silao, Mex. (se̊-lä´ō)	6	20°56'N	101°25'W
Silocayoápan, Mex. (sē-lo̊-kä-yo̊-ȧ´pän)	6	17°29'N	98°09'W
Silva Jardim, Braz. (se̊´l-vä-zhär-de̊ɴ)	17a	22°40'N	42°24'W
Silvânia, Braz. (sēl-vȧ´nyä)	19	16°43'S	48°33'W
Silver Bank, bk.	11	20°40'N	69°40'W
Silver Bank Passage, strt., N.A.	11	20°40'N	70°20'W

ăt; fin*ă*l; rāte; senåte; ärm; ȧsk; sof*ȧ*; fâre; ch-choose; dh-as th in other; bē; e̊vent; bĕt; rec*ĕ*nt; cratẽr; g-gō; gh-guttural g; bĭt; *ĭ*-short neutral; rīde; ᴋ-guttural k as ch in German ich;

PLACE (Pronunciation)	PAGE	LAT.	LONG.
Silver City, Pan.	9	9°20'N	79°54'W
Simms Point, c., Bah.	10	25°00'N	77°40'W
Simojovel, Mex. (sē-mō-hō-vĕl')	7	17°12'N	92°43'W
Simonésia, Braz. (sē-mō-nĕ'syä)	17a	20°04'S	41°53'W
Sinaloa, state, Mex. (sē-nä-lô-ä)	4	25°15'N	107°45'W
Sincelejo, Col. (sēn-så-lā'hō)	18	9°12'N	75°30'W
Sinnamary, Fr. Gu.	19	5°15'N	52°52'W
Sino, Pedra de, mtn., Braz. (pĕ'drä-dō-sē'nō)	20b	22°27'S	43°02'W
Sint Eustatius, i., Neth. Ant.	9b	17°32'N	62°45'W
Siqueros, Mex. (sē-kā'rōs)	6	23°19'N	106°14'W
Siquía, Río, r., Nic. (sē-kē'ä)	9	12°23'N	84°36'W
Sirama, El Sal. (Sē-rä-mä)	8	13°23'N	87°55'W
Sisal, Mex. (sē-säl')	4	21°09'N	90°03'W
Sixaola, r., C.R.	9	9°31'N	83°07'W
Skeldon, Guy. (skĕl'dŭn)	19	5°49'N	57°15'W
Skyring, Seno de, b., Chile (sĕ'nô-s-krē'ng)	20	52°35'S	72°30'W
Snap Point, c., Bah.	10	23°45'N	77°30'W
Sobral, Braz. (sō-brä'l)	19	3°39'S	40°16'W
Socoltenango, Mex. (sō-kōl-tĕ-näŋ'gō)	7	16°17'N	92°20'W
Socorro, Braz. (sŏ-kō'r-rō)	17a	22°35'S	46°32'W
Socorro, Col. (sō-kôr'rō)	18	6°23'N	73°19'W
Sogamoso, Col. (sô-gä-mō'sō)	18	5°42'N	72°51'W
Sola de Vega, Mex.	7	16°31'N	96°58'W
Soledad, Col. (sŏ-lĕ-dä'd)	18	10°47'N	75°00'W
Soledad Díez Gutiérrez, Mex.	6	22°19'N	100°54'W
Solentiname, Islas de, is., Nic. (ē's-läs-dĕ-sō-lĕn-tē-nä'må)	8	11°15'N	85°16'W
Solimões *see* Amazon, r., Braz.	18	2°45'S	67°44'W
Sololá, Guat. (sō-lō'lä)	8	14°45'N	91°12'W
Sombrerete, Mex. (sōm-brå-rā'tå)	6	23°38'N	103°37'W
Sombrero, Cayo, i., Ven. (kä-yō-sôm-brĕ'rō)	19b	10°52'N	68°12'W
Somoto, Nic. (sō-mō'tō)	8	13°28'N	86°37'W
Sonora, state, Mex.	4	29°45'N	111°15'W
Sonora, r., Mex.	4	28°45'N	111°35'W
Sonsón, Col. (sōn-sōn')	18	5°42'N	75°28'W
Sonsonate, El Sal. (sōn-sō-nä'tå)	8	13°46'N	89°43'W
Sopetrán, Col. (sō-pĕ-trä'n)	18a	6°30'N	75°44'W
Sordo, r., Mex. (sō'r-dō)	7	16°39'N	97°33'W
Soriano, dept., Ur. (sŏ-rēä'nō)	17c	33°25'S	58°00'W
Sorocaba, Braz. (sō-rō-kä'bȧ)	19	23°29'S	47°27'W
Sota la Marina, Mex. (sō-tä-lä-mä-rē'nä)	6	23°45'N	98°11'W
Soteapan, Mex. (sō-tå-ä'pän)	7	18°14'N	94°51'W
Soto la Marina, Río, r., Mex. (rē'ō-so'tō lä mä-rē'nä)	6	23°55'N	98°30'W
Sotuta, Mex. (sō-tōō'tä)	8a	20°35'N	89°00'W
Soublette, Ven. (sō-ōō-blĕ'tē)	19b	9°55'N	66°06'W
Soufrière, St. Luc. (sōō-frē-âr')	9b	13°50'N	61°03'W
Soufrière, mtn., St. Vin.	9b	13°19'N	61°12'W
Soufrière, vol., Guad. (sōō-frē-ăr')	9b	16°06'N	61°42'W
South America, cont.	15	15°00'S	60°00'W
South Bay, b., Bah.	11	20°55'N	73°35'W
South Bight, b., Bah.	10	24°20'N	77°35'W
South Bimini, i., Bah. (bē'mē-nē)	10	25°40'N	79°20'W
South Caicos, i., T./C. Is. (kī'kōs)	11	21°30'N	71°35'W
South Georgia, i., S. Geor. (jôr'jȧ)	15	54°00'S	37°00'W
South Negril Point, c., Jam. (nå-grēl')	10	18°15'N	78°25'W
South Orkney Islands, is., Ant.	15	57°00'S	45°00'W
South Point, c., Barb.	9b	13°00'N	59°43'W
South Sandwich Islands, is., S. Geor. (sȧnd'wĭch)	15	58°00'S	27°00'W
South Sandwich Trench, deep	15	55°00'S	27°00'W
South Shetland Islands, is., Ant.	15	62°00'S	70°00'W
Southwest Point, c., Bah.	10	25°50'N	77°10'W
Southwest Point, c., Bah.	11	23°55'N	74°30'W
Spanish Town, Jam.	5	18°00'N	76°55'W
Staniard Creek, Bah.	10	24°50'N	77°55'W
Stanley, Falk. Is.	20	51°46'S	57°59'W
Stann Creek, Belize (stăn krēk)	8a	17°01'N	88°14'W
Stormy Point, c., V.I.U.S. (stôr'mē)	5c	18°22'N	65°01'W
Suchiapa, Mex. (sōō-chē-ä'pä)	7	16°38'N	93°08'W
Suchiapa, r., Mex.	7	16°27'N	93°26'W
Suchitoto, El Sal. (sōō-chē-tō'tō)	8	13°58'N	89°03'W
Sucio, r., Col. (sōō'syō)	18a	6°55'N	76°15'W
Sucre, Bol. (sōō'krā)	18	19°06'S	65°16'W
Sucre, dept., Ven. (sōō'krĕ)	19b	10°18'N	64°12'W
Sud, Canal du, strt., Haiti	11	18°40'N	73°15'W
Suipacha, Arg. (swē-pä'chä)	17c	34°45'S	59°43'W
Sulaco, r., Hond. (sōō-lä'kō)	8	14°55'N	87°31'W
Sullana, Peru (sōō-lyä'nä)	18	4°57'S	80°47'W
Sultepec, Mex. (sōōl-tå-pĕk')	6	18°50'N	99°51'W
Sumidouro, Braz. (sōō-mē-dō'rō)	17a	22°04'S	42°41'W
Superior, Laguna, l., Mex. (lä-gōō'nȧ sōō-pā-rē-ōr')	7	16°20'N	94°55'W
Suriname (Netherlands Guiana), nation, S.A. (sōō-rē-näm')	19	4°00'N	56°00'W

T

PLACE (Pronunciation)	PAGE	LAT.	LONG.
Tabasará, Serranía de, mts., Pan.	9	8°29'N	81°22'W
Tabasco, Mex. (tä-bäs'kō)	6	21°47'N	103°04'W
Tabasco, state, Mex.	4	18°10'N	93°00'W
Taboga, i., Pan. (tä-bō'gä)	4a	8°48'N	79°35'W
Taboguilla, i., Pan. (tä-bō-gē'l-yä)	4a	8°48'N	79°31'W
Tacámbaro, r., Mex. (tä-käm'bä-rō)	6	18°55'N	101°25'W
Tacámbaro de Codallos, Mex.	6	19°12'N	101°28'W
Tacarigua, Laguna de la, l., Ven.	19b	10°18'N	65°43'W
Tacna, Peru (täk'nä)	18	18°34'S	70°16'W
Tacotalpa, Mex. (tä-kō-täl'pä)	7	17°37'N	92°51'W
Tacotalpa, r., Mex.	7	17°24'N	92°38'W
Taitao, Península de, pen., Chile	20	46°20'S	77°15'W
Tajano de Morais, Braz. (tĕ-zhä'nō-dĕ-mô-rä'ēs)	17a	22°05'S	42°04'W
Tajumulco, vol., Guat. (tä-hōō-mōōl'kō)	8	15°03'N	91°53'W
Tala, Mex. (tä'lä)	6	20°39'N	103°42'W
Talagante, Chile (tä-lä-gȧ'n-tĕ)	17b	33°39'S	70°54'W
Talamanca, Cordillera de, mts., C.R.	9	9°37'N	83°55'W
Talanga, Hond. (tä-lä'n-gä)	8	14°21'N	87°09'W
Talara, Peru (tä-lä'rä)	18	4°32'S	81°17'W
Talca, Chile (täl'kä)	20	35°25'S	71°39'W
Talca, prov., Chile	17b	35°23'S	71°15'W
Talca, Punta, c., Chile (pōō'n-tä-täl'kä)	17b	33°25'S	71°42'W
Talcahuano, Chile (täl-kä-wä'nō)	20	36°41'S	73°05'W
Talea de Castro, Mex. (tä'lå-ä dā käs'trō)	7	17°22'N	96°14'W
Talpa de Allende, Mex. (täl'pä dā äl-yĕn'då)	6	20°25'N	104°48'W
Taltal, Chile (täl-täl')	20	25°26'S	70°32'W
Tamanaco, r., Ven. (tä-mä-nä'kō)	19b	9°32'N	66°00'W
Tamaulipas, state, Mex. (tä-mä-ōō-lē'päs')	4	23°45'N	98°30'W
Tamazula de Gordiano, Mex.	6	19°44'N	103°09'W
Tamazulapan del Progreso, Mex.	7	17°41'N	97°34'W
Tamazunchale, Mex. (tä-mä-zōn-chä'lå)	6	21°16'N	98°46'W
Tambador, Serra do, mts., Braz. (sĕ'r-rä-dō-täm'bä-dōr)	19	10°33'S	41°16'W
Tamiahua, Mex. (tä-myä-wä)	7	21°17'N	97°26'W
Tamiahua, Laguna, l., Mex. (lä-gō'nä-tä-myä-wä)	7	21°38'N	97°33'W
Tampico, Mex. (täm-pē'kō)	4	22°14'N	97°51'W
Tampico Alto, Mex. (täm-pē'kō äl'tō)	7	22°07'N	97°48'W
Tamuín, Mex. (tä-mōō-ē'n)	6	22°04'N	98°47'W
Tancítaro, Mex. (tän-sē'tä-rō)	6	19°16'N	102°24'W
Tancítaro, Cerro de, mtn., Mex. (sĕ'r-rō-dĕ)	6	19°24'N	102°19'W
Tancoco, Mex. (tän-kō'kō)	7	21°16'N	97°45'W
Tandil, Arg. (tän-dēl')	20	36°16'S	59°01'W
Tandil, Sierra del, mts., Arg.	20	38°40'S	59°40'W
Tangancícuaro, Mex. (täŋ-gän-sē'kwa-rō)	6	19°52'N	102°13'W
Tanquijo, Arrecife, i., Mex. (är-rĕ-sē'fē-tän-kē'kô)	7	21°07'N	97°16'W
Tantoyuca, Mex. (tän-tō-yōō'kä)	6	21°22'N	98°13'W
Tapachula, Mex.	8	14°55'N	92°20'W
Tapajós, r., Braz. (tä-pä-zhô's)	19	3°27'S	55°33'W
Tapalque, Arg. (tä-päl-kĕ')	17c	36°22'S	60°05'W
Tapanatepec, Mex. (tä-pä-nä-tĕ-pĕk)	7	16°22'N	94°19'W
Taquara, Serra de, mts., Braz. (sĕ'r-rä-dĕ-tä-kwä'rä)	19	15°28'S	54°33'W
Taquari, r., Braz. (tä-kwä'rĭ)	19	18°35'S	56°50'W
Tarapoto, Peru (tä-rä-pō'tō)	18	6°29'S	76°26'W
Tarata, Bol. (tä-rä'tä)	18	17°43'S	66°00'W
Tarija, Bol. (tär-rē'hä)	18	21°42'S	64°52'W
Tarma, Peru (tär'mä)	18	11°20'S	75°40'W
Tarpum Bay, b., Bah. (tär'pŭm)	10	25°05'N	76°20'W
Tartagal, Arg. (tär-tä-gȧ'l)	20	23°31'S	63°47'W
Tasquillo, Mex. (täs-kē'lyō)	6	20°34'N	99°21'W
Taviche, Mex. (tä-vĕ'chĕ)	7	16°43'N	96°35'W
Taxco de Alarcón, Mex. (täs'ko dĕ ä-lär-kō'n)	6	18°34'N	99°37'W
Teabo, Mex. (tĕ-ä'bō)	8a	20°25'N	89°14'W
Teapa, Mex. (tā-ä'pä)	7	17°35'N	92°56'W
Tecalitlán, Mex. (tā-kä-lē-tlän')	6	19°28'N	103°17'W
Tecoanapa, Mex. (tāk-wä-nä-pä')	6	16°33'N	98°46'W
Tecoh, Mex. (tĕ-kō)	8a	20°46'N	89°27'W
Tecolotlán, Mex. (tā-kō-lō-tlän')	6	20°13'N	103°57'W
Tecolutla, Mex. (tā-kō-lōō'tlä)	7	20°33'N	97°00'W
Tecolutla, r., Mex.	7	20°16'N	97°14'W
Tecomán, Mex. (tā-kō-män')	6	18°53'N	103°53'W
Tecómitl, Mex. (tĕ-kŏ'mētl)	7a	19°13'N	98°59'W
Tecozautla, Mex. (tā'kŏ-zä-ōō'tlä)	6	20°33'N	99°38'W
Tecpán de Galeana, Mex. (tĕk-pän' dā gä-lā-ä'nä)	6	17°13'N	100°41'W
Tecpatán, Mex. (tĕk-pä-tȧ'n)	7	17°08'N	93°18'W
Tecuala, Mex. (tĕ-kwä-lä)	6	22°24'N	105°29'W
Tegucigalpa, Hond. (tå-gōō-sē-gäl'pä)	4	14°08'N	87°15'W
Tehuacán, Mex. (tā-wä-kän')	4	18°27'N	97°23'W
Tehuantepec, Mex.	4	16°20'N	95°14'W
Tehuantepec, r., Mex.	7	16°30'N	95°23'W
Tehuantepec, Golfo de, b., Mex. (gōl-fô dĕ)	4	15°45'N	95°00'W
Tehuantepec, Istmo de, isth., Mex. (ē'st-mô dĕ)	7	17°55'N	94°35'W
Tehuehuetla, Arroyo, r., Mex. (tĕ-wĕ-wĕ'tlä ȧr-rō-yō)	6	17°54'N	100°26'W
Tehuitzingo, Mex. (tā-wē-tzĭŋ'gō)	6	18°21'N	98°16'W
Tejupan, (Santiago) Mex. (tĕ-koo'-pän) (sän-tyȧ'gô)	7	17°39'N	97°34'W
Tejupan, Punta, c., Mex.	6	18°19'N	103°30'W
Tejupilco de Hidalgo, Mex. (tå-hōō-pēl'kō dā ē-dhäl'gō)	6	18°52'N	100°07'W
Tekax de Alvaro Obregon, Mex.	8a	20°12'N	89°11'W
Tekit, Mex. (tĕ-kē't)	8a	20°35'N	89°18'W
Tela, Hond. (tā'lä)	4	15°45'N	87°25'W
Tela, Bahía de, b., Hond.	8	15°53'N	87°29'W
Telica, vol., Nic. (tå-lē'kä)	8	12°38'N	86°52'W
Tello, Col. (tĕ'l-yō)	18a	3°05'N	75°08'W
Teloloapan, Mex. (tā'lō-lō-ä'pän)	6	18°19'N	99°54'W
Temascalcingo, Mex. (tā'mäs-käl-sĭŋ'gō)	6	19°55'N	100°00'W
Temascaltepec, Mex. (tā'mäs-käl-tå pĕk)	6	19°00'N	100°03'W
Temax, Mex. (tĕ'mäx)	4	21°10'N	88°51'W
Temoaya, Mex. (tĕ-mô-a-um-yä)	7a	19°28'N	99°36'W
Temperley, Arg. (tĕ'm-pĕr-lā)	20a	34°47'S	58°24'W
Tempoal, r., Mex. (tĕm-pô-ä'l)	6	21°38'N	98°23'W
Temuco, Chile (tå-mōō'kō)	20	38°46'S	72°38'W
Tenamaxtlán, Mex. (tā'nä-mäs-tlän')	6	20°13'N	104°06'W
Tenancingo, Mex. (tå-nän-sēŋ'gō)	6	18°54'N	99°36'W
Tenango, Mex. (tå-näŋ'gō)	7a	19°09'N	98°51'W
Teno, r., Chile (tĕ'nō)	17b	34°55'S	71°00'W
Tenosique, Mex. (tā-nō-sē'kå)	7	17°27'N	91°25'W
Teocaltiche, Mex. (tā'ō-käl-tē'chå)	6	21°27'N	102°38'W
Teocelo, Mex. (tā-ō-sā'lō)	7	19°22'N	96°57'W
Teocuitatlán de Corona, Mex.	6	20°06'N	103°22'W
Teófilo Otoni, Braz. (tĕ-ô'fē-lō-tô'nē)	19	17°49'S	41°18'W
Teoloyucan, Mex. (tā'ō-lō-yōō'kän)	6	19°43'N	99°12'W
Teopisca, Mex. (tā-ō-pēs'kä)	7	16°30'N	92°33'W
Teotihuacán, Mex. (tĕ-ô-tē-wȧ-kȧ'n)	7a	19°40'N	98°52'W
Teotitlán del Camino, Mex. (tā-ō-tē-tlän' dĕl kä-mē'nō)	7	18°07'N	97°04'W
Tepalcatepec, Mex. (tā'päl-kä-tă'pĕk)	6	19°11'N	102°51'W
Tepalcatepec, r., Mex.	6	18°54'N	102°25'W
Tepalcingo, Mex. (tā-päl-sēŋ'gō)	6	18°34'N	98°49'W
Tepatitlán de Morelos, Mex. (tā-pä-tē-tlän' dā mō-rā'los)	4	20°55'N	102°47'W
Tepeaca, Mex. (tā-på-ä'kä)	7	18°57'N	97°54'W
Tepecoacuilco de Trujano, Mex.	6	18°15'N	99°29'W
Tepeji del Río, Mex. (tā-på-ke' dĕl rē'ō)	6	19°55'N	99°22'W
Tepelmeme, Mex. (tā'pĕl-mā'må)	7	17°51'N	97°23'W
Tepetlaoxtoc, Mex. (tā'på-tlä'ōs-tōk')	6	19°34'N	98°49'W
Tepezala, Mex. (tā-på-zä-lä')	6	22°12'N	102°12'W
Tepic, Mex. (tā-pēk')	4	21°32'N	104°53'W
Teposcolula, Mex.	7	17°33'N	97°29'W
Tequendama, Salto de, wtfl., Col. (sä'l-tō dĕ tĕ-kĕn-dä'mä)	18	4°34'N	74°18'W
Tequila, Mex. (tå-kē'lä)	6	20°53'N	103°48'W
Tequisistlán, r., Mex. (tĕ-kē-sēs-tlȧ'n)	7	16°20'N	95°40'W
Tequisquiapan, Mex. (tå-kēs-kē-ä'pän)	6	20°33'N	99°57'W
Teresina, Braz. (tĕr-å-sē'nȧ)	19	5°04'S	42°42'W
Teresópolis, Braz. (tĕr-ā-sō'pō-lēzh)	17a	22°25'S	42°59'W
Términos, Laguna de, l., Mex. (lä-gō'nä dĕ ĕ'r-mē-nōs)	4	18°37'N	91°32'W
Tesecheacán, Mex. (tĕ-sĕ-chĕ-ä-kȧ'n)	7	18°10'N	95°41'W
Texcaltitlán, Mex. (tās-käl'tē-tlän')	6	18°54'N	99°51'W
Texcoco, Mex. (tās-kō'kō)	6	19°31'N	98°53'W
Texcoco, Lago de, l., Mex.	7a	19°30'N	99°00'W
Texistepec, Mex. (tĕk-sēs-tā-pĕk')	7	17°51'N	94°46'W
Texmelucan, Mex.	6	19°17'N	98°26'W
Texontepec, Mex. (tå-zōn-tå-pĕk')	6	19°52'N	98°48'W
Texontepec de Aldama, Mex. (dā äl-dä'mä)	6	20°19'N	99°19'W
Teziutlán, Mex. (tå-zē-ōō-tlän')	7	19°48'N	97°21'W
Thatch Cay, i., V.I.U.S. (thăch)	5c	18°22'N	64°53'W
Tibagi, Braz. (tē'bȧ-zhē)	19	24°40'S	50°35'W
Tiburon, Haiti	11	18°35'N	74°25'W
Tiburón, i., Mex.	4	28°45'N	113°10'W
Tiburón, Cabo, c., (kȧ'bō)	9	8°42'N	77°19'W
Ticul, Mex. (tē-kōō'l)	8a	20°22'N	89°32'W
Tierra Blanca, Mex. (tye r-ra-bla n-kä)	7	18°28'N	96°19'W
Tierra del Fuego, i., S.A. (tyĕr'rä dĕl fwā'gō)	20	53°50'S	68°45'W
Tigre, r., Peru	18	2°20'S	75°41'W
Tihuatlán, Mex. (tē-wä-tlän')	7	20°43'N	97°34'W
Tijuana, Mex. (tē-hwä'nä)	4	32°32'N	117°02'W
Tijuca, Pico da, mtn., Braz. (pē'kō-dä-tē-zhōō'kȧ)	20b	22°56'S	43°17'W
Tikal, hist., Guat. (tē-käl')	8a	17°16'N	89°49'W
Tina, Monte, mtn., Dom. Rep. (mō'n-tĕ-tē'nȧ)	11	18°50'N	70°40'W
Tinaguillo, Ven. (tē-nä-gē'l-yō)	19b	9°55'N	68°18'W
Tingo María, Peru (tē'ngō-mä-rē'ä)	18	9°15'S	76°04'W
Tinguindio, Mex.	6	19°38'N	102°02'W
Tinguiririca, r., Chile (tē'n-gē-rē-rē'kä)	17b	34°48'S	70°45'W
Tinogasta, Arg. (tē-nō-gäs'tä)	20	28°07'S	67°30'W
Tipitapa, Nic. (tē-pē-tä'pä)	8	12°14'N	86°05'W
Tipitapa, r., Nic.	8	12°13'N	85°57'W
Titicaca, Lago, l., S.A. (lä'gô-tē-tē-kä'kä)	18	16°12'S	70°33'W
Titiribi, Col. (tē-tē-rē-bē')	18a	6°05'N	75°47'W
Tixkokob, Mex. (tēx-kō-kō'b)	8a	21°01'N	89°23'W
Tixtla de Guerrero, Mex. (tē'x tlä-dĕ-gĕr-rĕ'rô)	6	17°36'N	99°24'W
Tizimín, Mex. (tē-zē-mē'n)	8a	21°08'N	88°10'W
Tiznados, r., Ven. (tēz-nä'dōs)	19b	9°53'N	67°49'W
Tlacolula de Matamoros, Mex.	7	16°56'N	96°29'W
Tlacotálpan, Mex. (tlä-kō-täl'pän)	7	18°39'N	95°40'W
Tlacotepec, Mex. (tlä-kō-tå-pĕ'k)	6	17°46'N	99°57'W
Tlacotepec, Mex.	7	18°41'N	97°40'W
Tlacotepec, Mex.	6	19°11'N	99°41'W
Tláhuac, Mex. (tlä-wäk')	7a	19°16'N	99°00'W
Tlajomulco de Zúñiga, Mex. (tlä-hô-mōō'l-ko-dĕ-zōō'n-yē-gä)	6	20°30'N	103°27'W
Tlalchapa, Mex. (tläl-chä'pä)	6	18°26'N	100°29'W
Tlalixcoyán, Mex. (tlä-lēs'kō-yän')	7	18°53'N	96°04'W
Tlalmanalco, Mex. (tläl-mä-nä'l-kô)	7a	19°12'N	98°48'W
Tlalnepantla, Mex.	7a	19°32'N	99°13'W
Tlalnepantla, Mex. (tläl-nå-pȧn'tlä)	7a	18°59'N	99°01'W
Tlalpán, Mex. (tläl-pä'n)	6	19°17'N	99°10'W
Tlalpujahua, Mex. (tläl-pōō-kä'wä)	6	19°50'N	100°10'W
Tlapa, Mex. (tlä'pä)	6	17°30'N	98°30'W
Tlapacoyán, Mex. (tlä-pä-kō-yȧ'n)	7	19°57'N	97°11'W
Tlapehuala, Mex. (tlä-på-wä'lä)	6	18°17'N	100°30'W
Tlaquepaque, Mex. (tlä-kĕ-pä'kĕ)	6	20°39'N	103°17'W
Tlatlaya, Mex. (tlä-tlä'yä)	6	18°36'N	100°14'W
Tlaxcala, Mex. (tläs-kä'lä)	4	19°16'N	98°14'W
Tlaxcala, state, Mex.	6	19°30'N	98°15'W
Tlaxco, Mex. (tläs'kō)	6	19°37'N	98°06'W
Tlaxiaco Santa María Asunción, Mex.	7	17°16'N	97°41'W
Tlayacapán, Mex. (tlä-yä-kä-pa'n)	7a	18°57'N	99°00'W

ng-sing; ŋ-baŋk; ɴ-nasalized n; nŏd; cŏmmit; ōld; ōbey; ôrder; oi-boil; fōōd; ȯ-as oo in foot; ou-out; s-soft; sh-dish; th-thin; pūre; ůnite; ûrn; stŭd; circŭs; ü-as in French tu; '-indeterminate vowel.

54 Toa – Virgin Islands PRONOUNCING INDEX

PLACE (Pronunciation)	PAGE	LAT.	LONG.
Toa, r., Cuba (tō'ä)	11	20°25'N	74°35'W
Toar, Cuchillas de, mts., Cuba (kōō-chē'l-lyäs-dĕ-tô-ä'r)	11	20°20'N	74°50'W
Tobago, i., Trin. (tô-bā'gō)	5	11°15'N	60°30'W
Tocaima, Col. (tô-kä'y-mä)	18a	4°28'N	74°38'W
Tocantinópolis, Braz. (tō-kän-tē-nō'pō-lēs)	19	6°27'S	47°18'W
Tocantins, state, Braz.	19	10°00'S	48°00'W
Tocantins, r., Braz. (tō-kän-tēns')	19	3°28'S	49°22'W
Tocoa, Hond. (tō-kō'ä)	8	15°37'N	86°01'W
Tocopilla, Chile (tō-kō-pēl'yä)	20	22°03'S	70°08'W
Tocuyo de la Costa, Ven. (tô-kōō'yō-dĕ-lä-kôs'tä)	19b	11°03'N	68°24'W
Tolcayuca, Mex. (tôl-kä-yōō'kä)	6	19°55'N	98°54'W
Tolima, dept., Col. (tô-lē'mä)	18a	4°07'N	75°20'W
Tolima, Nevado del, mtn., Col. (nĕ-vä-dô-dĕl-tô-lē'mä)	18a	4°40'N	75°20'W
Tolimán, Mex. (tô-lē-män')	6	20°54'N	99°54'W
Toluca, Mex. (tô-lōō'kȧ)	4	19°17'N	99°40'W
Toluca, Nevado de, mtn., Mex. (nĕ-vä-dô-dĕ-tô-lōō'kä)	4	19°09'N	99°42'W
Tomatlán, Mex. (tô-mä-tlä'n)	6	19°54'N	105°14'W
Tombador, Serra do, mts., Braz. (sĕr'rȧ dô tōm-bä-dôr')	19	11°31'S	57°33'W
Tombos, Braz. (tô'm-bōs)	17a	20°53'S	42°00'W
Tonala, Mex.	6	20°38'N	103°14'W
Tonalá, r., Mex.	7	18°05'N	94°08'W
Tongoy, Chile (tōn-goi')	20	30°16'S	71°29'W
Tonto, r., Mex. (tōn'tō)	7	18°15'N	96°13'W
Topilejo, Mex. (tô-pē-lĕ'hô)	7a	19°12'N	99°09'W
Topolobampo, Mex. (tō-pō-lô-bä'm-pô)	4	25°45'N	109°00'W
Toribío, Col. (tô-rē-bē'ô)	18a	2°58'N	76°14'W
Toronto, res., Mex.	2	27°35'N	105°37'W
Torra, Cerro, mtn., Col. (sĕ'r-rô-tô'r-rä)	18a	4°41'N	76°22'W
Torreón, Mex. (tôr-rå-ōn')	4	25°32'N	103°26'W
Tortola, i., Br. Vir. Is. (tôr-tō'lä)	5b	18°34'N	64°40'W
Tortue, Canal de la, strt., Haiti (tôr-tü')	11	20°05'N	73°20'W
Tortue, Île de la, i., Haiti	11	20°10'N	73°00'W
Tostado, Arg. (tôs-tä'dô)	20	29°10'S	61°43'W
Totness, Sur.	19	5°51'N	56°17'W
Totonicapán, Guat. (tôtō-nē-kä'pän)	4	14°55'N	91°20'W
Totoras, Arg. (tō-tô'räs)	17c	32°33'S	61°13'W
Treinta y Tres, Ur. (trå-ēn'tä ē trās')	20	33°14'S	54°17'W
Trelew, Arg. (trē'lū)	20	43°15'S	65°25'W
Trenque Lauquén, Arg. (trĕn'kĕ-lä'ô-kĕ'n)	20	35°50'S	62°44'W
Tres Arroyos, Arg. (träs'är-rō'yōs)	20	38°18'S	60°16'W
Três Corações, Braz. (trĕ's kō-rä-zô'ĕs)	17a	21°41'S	45°14'W
Tres Cumbres, Mex. (trĕ's kōō'm-brĕs)	7a	19°03'N	99°14'W
Três Lagoas, Braz. (trĕ's lä-gô'äs)	19	20°48'S	51°42'W
Três Marias, Reprêsa, res., Braz.	19	18°15'S	45°30'W
Tres Morros, Alto de, mtn., Col. (ä'l-tō dĕ trĕ's mô'r-rôs)	18a	7°08'N	76°10'W
Três Pontas, Braz. (trĕ'pô'n-täs)	17a	21°22'S	45°30'W
Três Rios, Braz. (trĕ's rē'ōs)	17a	22°07'S	43°13'W
Trinidad, Bol. (trē-nē-dhädh')	18	14°48'S	64°43'W
Trinidad, Cuba (trē-nē-dhädh')	5	21°50'N	80°00'W
Trinidad, Ur.	20	33°29'S	56°55'W
Trinidad, i., Trin. (trĭn'ĭ-dăd)	19	10°00'N	61°00'W
Trinidad, r., Pan.	4a	8°55'N	80°01'W
Trinidad, Sierra de, mts., Cuba (sē-ĕ'r-rä dĕ trē-nē-dä'd)	10	21°50'N	79°55'W
Trinidad and Tobago, nation, N.A. (trĭn'ĭ-dăd) (tô-bā'gō)	5	11°00'N	61°00'W
Trinitaria, Mex. (trē-nē-tä'ryä)	7	16°09'N	92°04'W
Triste, Golfo, b., Ven. (gôl-fô trē's-tĕ)	19b	10°40'N	68°05'W
Tronador, Cerro, mtn., S.A. (sĕ'r-rô trō-nä'dôr)	20	41°17'S	71°56'W
Troncoso, Mex. (trôn-kô'sō)	6	22°43'N	102°22'W
Trujillo, Col. (trō-kē'l-yō)	18a	4°10'N	76°20'W
Trujillo, Peru	18	8°08'S	79°00'W
Trujillo, Ven.	18	9°15'N	70°28'W
Trujillo, r., Mex.	6	23°12'N	103°10'W
Trujín, Lago, l., Dom. Rep. (trōō-kĕn')	11	17°45'N	71°25'W
Tubarão, Braz. (tōō-bä-rouɴ')	20	28°23'N	48°56'W
Tucacas, Ven. (tōō-kä'käs)	18	10°48'N	68°20'W
Tucumán, Arg. (tōō-kōō-män')	20	26°52'S	65°08'W
Tucumán, prov., Arg.	20	26°30'S	65°30'W
Tucupita, Ven. (tōō-kōō-pē'tä)	18	9°00'N	62°09'W
Tula, Mex. (tōō'lä)	6	20°04'N	99°22'W
Tula, r., Mex. (tōō'lä)	6	20°40'N	99°27'W
Tulancingo, Mex. (tōō-län-sĭŋ'gō)	4	20°04'N	98°24'W
Tulcán, Ec. (tōōl-kän')	18	0°44'N	77°52'W
Tulcingo, Mex. (tōōl-sĭŋ'gō)	6	18°03'N	98°27'W
Tulpetlac, Mex. (tōōl-på-tlák')	7a	19°33'N	99°04'W
Tulum, Mex. (tōō-lō'm)	8a	20°17'N	87°26'W
Tuma, r., Nic. (tōō'mä)	8	13°07'N	85°32'W
Tumbes, Peru (tōō'm-bĕs)	18	3°39'S	80°27'W
Tumbiscatío, Mex. (tōōm-bē-skä-tē'ō)	6	18°32'N	102°23'W
Tumeremo, Ven. (tōō-må-rā'mō)	19	7°15'N	61°28'W
Tumuc-Humac Mountains, mts., S.A. (tōō-mŏk'ōō-mäk')	19	2°15'N	54°50'W
Tunas de Zaza, Cuba (tōō'näs dā zä'zä)	10	21°40'N	79°35'W
Tunja, Col. (tōō'n-hä)	18	5°32'N	73°19'W
Tupinambaranas, Ilha, i., Braz.	19	3°04'S	58°09'W
Tupiza, Bol. (tōō-pē'zä)	18	21°26'S	65°43'W
Tupungato, Cerro, vol., S.A.	20	33°30'S	69°52'W
Tuquerres, Col. (tōō-kĕ'r-rĕs)	18	1°12'N	77°44'W
Turbio, r., Mex. (tōōr-byô)	6	20°28'N	101°40'W
Turbo, Col. (tōō'bō)	18	8°02'N	76°43'W
Turicato, Mex. (tōō-rē-kä'tō)	6	19°03'N	101°24'W
Turiguano, i., Cuba (tōō-rē-gwä'nō)	10	22°20'N	78°35'W
Turks, is., T./C. Is. (tûrks)	5	21°40'N	71°45'W
Turks Island Passage, strt., T./C. Is.	11	21°15'N	71°25'W
Turneffe, i., Belize	4	17°25'N	87°43'W
Turner Sound, strt., Bah.	10	24°20'N	78°05'W
Turquino, Pico, mtn., Cuba (pē'kō dā tōōr-kē'nō)	10	20°00'N	76°50'W
Turrialba, C.R. (tōōr-ryä'l-bä)	9	9°54'N	83°41'W
Tutitlán, Mex. (tōō-tē-tlä'n)	7a	19°38'N	99°10'W
Tutóia, Braz. (tōō-tō'yȧ)	19	2°42'S	42°21'W
Túxpan, Mex. (tōōs'pän)	6	19°34'N	103°22'W
Túxpan, Mex.	4	20°57'N	97°26'W
Túxpan, r., Mex. (tōōs'pän)	7	20°55'N	97°52'W
Túxpan, Arrecife, i., Mex. (är-rē-sē'fē-tōō'x-pä'n)	7	21°01'N	97°12'W
Tuxtepec, Mex. (tōs-tå-pĕk')	7	18°06'N	96°09'W
Tuxtla Gutiérrez, Mex. (tōs'tlä gōō-tyår'rĕs)	4	16°44'N	93°08'W
Tuy, r., Ven. (tōō'ē)	19b	10°15'N	66°03'W
Tuyra, r., Pan. (tōō-ē'rȧ)	9	7°55'N	77°37'W
Tzucacab, Mex. (tzōō-kä-kä'b)	8a	20°06'N	89°03'W

U

PLACE (Pronunciation)	PAGE	LAT.	LONG.
Uaupés, Braz. (wä-ōō'pās)	18	0°02'S	67°03'W
Ubatuba, Braz. (ōō-bȧ-tōō'bȧ)	17a	23°25'S	45°06'W
Uberaba, Braz. (ōō-bå-rä'bȧ)	19	19°47'S	47°47'W
Uberlândia, Braz. (ōō-bĕr-lá'n-dyä)	19	18°54'S	48°11'W
Ucayali, r., Peru (ōō'kä-yä'lē)	18	8°58'S	74°13'W
Ulua, r., Hond. (ōō-lōō'ȧ)	8	15°49'N	87°45'W
Umán, Mex. (ōō-män')	8a	20°52'N	89°44'W
Unare, r., Ven.	19b	9°45'N	65°12'W
Unare, Laguna de, l., Ven. (lä-gô'nä-de-ōō-nä'rĕ)	19b	10°07'N	65°23'W
Uncia, Bol. (ōōn'sē-ä)	18	18°28'S	66°32'W
União da Vitória, Braz. (ōō-nē-ouɴ' dä vē-tō'ryä)	20	26°17'S	51°13'W
Unión de Reyes, Cuba	10	22°45'N	81°30'W
Unión de San Antonio, Mex.	6	21°07'N	101°56'W
Unión de Tula, Mex.	6	19°57'N	104°14'W
Unión Hidalgo, Mex. (ē-dä'lgô)	7	16°29'N	94°51'W
Uno, Canal Número, can., Arg.	17c	36°43'S	58°14'W
Upata, Ven. (ōō-pä'tä)	18	7°58'N	62°27'W
Urdinarrain, Arg. (ōōr-dē-när-räē'n)	17c	32°43'S	58°53'W
Urrao, Col. (ōōr-rà'ô)	18	6°19'N	76°11'W
Urubamba, r., Peru (ōō-rōō-bäm'bä)	18	11°48'S	72°34'W
Uruguaiana, Braz.	20	29°45'S	57°00'W
Uruguay, nation, S.A. (ōō-rōō-gwī') (ū'rōō-gwā)	20	32°45'S	56°00'W
Uruguay, r., S.A. (ōō-rōō-gwī')	20	27°05'S	55°15'W
Ushuaia, Arg. (ōō-shōō-ī'ä)	20	54°46'S	68°24'W
Uspallata Pass, p., S.A. (ōōs-pä-lyä'tä)	20	32°47'S	70°08'W
Uspanapa, r., Mex. (ōōs-pä-nä'pä)	7	17°43'N	94°14'W
Usulután, El Sal. (ōō-sōō-lä-tän')	8	13°22'N	88°25'W
Usumacinta, r., N.A. (ōō'sōō-mä-sēn'tō)	7	18°24'N	92°30'W
Utila, i., Hond. (ōō-tē'lä)	8	16°07'N	87°05'W
Utuado, P.R. (ōō-tōō-ä'dhō)	5b	18°16'N	66°40'W
Uxmal, hist., Mex. (ōō'x-mä'l)	8a	20°22'N	89°44'W
Uyuni, Bol. (ōō-yōō'nē)	18	20°28'S	66°45'W
Uyuni, Salar de, pl., Bol. (sä-lär-dē)	18	20°58'S	67°09'W

V

PLACE (Pronunciation)	PAGE	LAT.	LONG.
Vache, Île à, i., Haiti	11	18°05'N	73°40'W
Valdés, Península, pen., Arg. (vȧl-dĕ's)	20	42°15'S	63°15'W
Valdivia, Chile (väl-dē'vē-ä)	20	39°47'S	73°13'W
Valdivia, Col. (väl-dē'vēä)	18a	7°10'N	75°26'W
Valença, Braz. (vä-lĕn'sȧ)	19	13°43'S	38°58'W
Valencia, Ven. (vä-lĕn'syä)	18	10°11'N	68°00'W
Valencia, Lago de, l., Ven.	19b	10°11'N	67°45'W
Valera, Ven. (vä-lĕ'rä)	18	9°12'N	70°45'W
Valladolid, Mex. (väl-yä-dhō-lēdh')	4	20°39'N	88°13'W
Valle de Allende, Mex. (väl'yä dā äl-yĕn'dā)	2	26°55'N	105°25'W
Valle de Bravo, Mex. (brä'vô)	6	19°12'N	100°07'W
Valle de Guanape, Ven. (vä'l-yĕ-dĕ-gwä-nä'pĕ)	19b	9°54'N	65°41'W
Valle de la Pascua, Ven. (lä-pä's-kōōä)	18	9°12'N	65°08'W
Valle del Cauca, dept., Col. (vä'l-yĕ del kou'ĸä)	18a	4°03'N	76°13'W
Valle de Santiago, Mex. (sän-tē-ä'gô)	6	20°23'N	101°11'W
Valledupar, Col. (dōō-pär')	18	10°13'N	73°39'W
Valle Grande, Bol. (grän'dä)	18	18°27'S	64°03'W
Vallejo, Sierra de, mts., Mex. (sē-ĕ'r-rä-dĕ-väl-yĕ'ĸô)	6	21°00'N	105°10'W
Vallenar, Chile (väl-yå-när')	20	28°39'S	70°52'W
Valles, Ciudad de, Mex.	4	21°59'N	99°02'W
Vallière, Haiti (väl-yâr')	11	19°30'N	71°55'W
Vallimanca, r., Arg. (väl-yē-mä'n-kä)	17c	36°21'S	60°55'W
Valparaíso, Chile (väl'pä-rä-ē'sô)	20	33°02'S	71°32'W
Valparaíso, Mex.	6	22°49'N	103°33'W
Valparaíso, prov., Chile	17b	32°58'S	71°23'W
Vanegas, Mex. (vä-nĕ'gäs)	4	23°54'N	100°54'W
Varginha, Braz. (vär-zhē'n-yä)	19	21°33'S	45°25'W
Vassouras, Braz. (väs-sō'räzh)	17a	22°25'S	43°40'W
Vaupés, r., S.A. (vá'ōō-pĕ's)	18	1°18'N	71°14'W
Veadeiros, Chapadas dos, hills, Braz. (shä-pä'däs-dôs-vĕ-á-dā'rōs)	19	14°00'S	47°00'W
Vedia, Arg. (vĕ'dyä)	17c	34°29'S	61°30'W
Vega de Alatorre, Mex. (vā'gä dā ä-lä-tōr'rā)	7	20°02'N	96°39'W
Vega Real, reg., Dom. Rep. (vĕ'gä-rē-ä'l)	11	19°30'N	71°05'W
Veinticinco de Mayo, Arg.	17c	35°26'S	60°09'W
Venadillo, Col. (vĕ-nä-dē'l-yō)	18a	4°43'N	74°55'W
Venado, Mex. (vå-mä'dō)	6	22°54'N	101°07'W
Venado Tuerto, Arg. (vĕ-nä'dô-tōōĕ'r-tô)	20	33°28'S	61°47'W
Venezuela, nation, S.A. (vĕn-ē-zwē'lȧ)	18	8°00'N	65°00'W
Venezuela, Golfo de, b., S.A. (gôl-fô-dĕ)	18	11°34'N	71°02'W
Ventana, Sierra de la, mts., Arg. (sē-ĕ-rä-dĕ-lä-vĕn-tá'nä)	20	38°00'S	63°00'W
Ventuari, r., Ven. (vĕn-tōōá'rē)	18	4°47'N	65°56'W
Venustiano Carranza, Mex. (vĕ-nōōs-tyä'nō-kär-rä'n-zä)	6	19°44'N	103°48'W
Venustiano Carranzo, Mex. (kär-rä'n-zô)	7	16°21'N	92°36'W
Vera, Arg. (vĕ-rä)	20	29°22'S	60°09'W
Veracruz, Mex.	4	19°13'N	96°07'W
Veracruz, state, Mex. (vā-rä-krōōz')	4	20°30'N	97°15'W
Verde, r., Mex.	6	21°48'N	99°50'W
Verde, r., Mex.	6	20°50'N	103°00'W
Verde, r., Mex.	7	16°05'N	97°44'W
Verde, Cap, c., Bah.	11	22°50'N	75°00'W
Verde, Cay, i., Bah.	11	22°00'N	75°05'W
Viacha, Bol. (vēä'chä)	18	16°43'S	68°16'W
Viana, Braz. (vē-ä'nä)	19	3°09'S	44°44'W
Vicente López, Arg. (vē-sĕ'n-tĕ-lō'pĕz)	20a	34°31'S	58°29'W
Viçosa, Braz. (vē-sô'sä)	17a	20°46'S	42°51'W
Victoria, Arg. (vēk-tō'rēä)	20	32°36'S	60°09'W
Victoria, Chile (vēk-tô-rēä)	20	38°15'S	72°16'W
Victoria, Col. (vēk-tô'rēä)	18a	5°19'N	74°54'W
Victoria de las Tunas, Cuba (vēk-tō'rē-ä dā läs tōō'näs)	10	20°55'N	77°05'W
Victoria Peak, mtn., Belize (vēk-tōrĭ'ȧ)	8a	16°47'N	88°40'W
Viedma, Arg. (vyād'mä)	20	40°55'S	63°03'W
Viedma, l., Arg.	20	49°40'S	72°35'W
Viejo, r., Nic. (vyā'hō)	8	12°45'N	86°19'W
Vieques, P.R. (vyā'kås)	5b	18°09'N	65°27'W
Vieques, i., P.R. (vyā'kås)	5b	18°05'N	65°28'W
Viesca, Mex. (vē-ās'kä)	2	25°21'N	102°47'W
Viesca, Laguna de, l., Mex. (lä-ō'nä-dĕ)	2	25°30'N	102°40'W
Villa Acuña, Mex. (vēl'yä-kōō'n-yä)	2	29°20'N	100°56'W
Villa Ahumada, Mex. (ä-ōō-mä'dä)	2	30°43'N	106°30'W
Villa Alta, Mex. (äl'tä)(sän ēl-dä-fōn'sō)	7	17°20'N	96°08'W
Villa Ángela, Arg. (vē'l-yä á'n-ĸĕ-lä)	20	27°31'S	60°42'W
Villa Ballester, Arg. (vē'l-yä-bál-yĕs-tĕr)	20a	34°33'S	58°33'W
Villa Bella, Bol. (bĕ'l-yä)	18	10°25'S	65°22'W
Villa Clara, prov., Cuba	10	22°40'N	80°10'W
Villa Constitución, Arg. (kōn-stē-tōō-syōn')	17c	33°15'S	60°19'W
Villa Coronado, Mex. (kō-rō-nä'dhō)	2	26°45'N	105°10'W
Villa Cuauhtémoc, Mex. (vēl'yä-kōō-äō-tē'mōk)	7	22°11'N	97°50'W
Villa de Allende, Mex. (vēl'yä'dā äl-yĕn'dā)	2	25°18'N	100°01'W
Villa de Álvarez, Mex. (vēl'yä-dĕ-ä'l-vä-rĕz)	6	19°17'N	103°44'W
Villa de Cura, Ven. (dĕ-kōō'rä)	19b	10°03'N	67°29'W
Villa de Guadalupe, Mex. (dĕ-gwä-dhä-lōō'pā)	6	23°22'N	100°44'W
Villa de Mayo, Arg.	20a	34°31'S	58°41'W
Villa Dolores, Arg. (vēl'yä dô-lō'räs)	20	31°50'S	65°05'W
Villa Escalante, Mex. (vēl'yä-ĕs-kä-lán'tĕ)	6	19°24'N	101°36'W
Villa Flores, Mex. (vēl'yä-flō'rās)	7	16°13'N	93°17'W
Villa García, Mex. (gär-sē'ä)	6	22°07'N	101°55'W
Villagrán, Mex.	2	24°28'N	99°30'W
Villaguay, Arg. (vē'l-yä-gwī)	20	31°47'S	58°53'W
Villa Hayes, Para. (vēl'yä äyås)(hāz)	20	25°07'S	57°31'W
Villahermosa, Mex. (vēl'yä-ĕr-mō'sä)	4	17°59'N	92°56'W
Villa Hidalgo, Mex. (vēl'yäē-däl'gô)	6	21°39'N	102°41'W
Villaldama, Mex. (vēl-yäl-dä'mä)	4	26°30'N	100°26'W
Villa López, Mex. (vēl'yä lō'pĕz)	2	27°00'N	105°02'W
Villa María, Arg. (vē'l-yä-mä-rē'ä)	20	32°17'S	63°08'W
Villa Mercedes, Arg. (mĕr-sā'dås)	20	33°38'S	65°16'W
Villa Montes, Bol. (vē'l-yä-mô'n-tĕs)	18	21°13'S	63°26'W
Villa Morelos, Mex. (mō-rĕ'lomcs)	6	20°01'N	101°24'W
Villanueva, Col. (vē'l-yä-nòĕ'vä)	18	10°44'N	73°08'W
Villanueva, Hond. (vēl'yä-nwä'vä)	8	15°19'N	88°02'W
Villanueva, Mex. (vēl'yä-nòĕ'vä)	6	22°25'N	102°53'W
Villa Obregón, Mex. (vē'l-yä-ô-brĕ-gô'n)	7a	19°21'N	99°11'W
Villa Ocampo, Mex. (ô-käm'pō)	2	26°26'N	105°30'W
Villa Pedro Montoya, Mex. (vēl'yä-pĕ'drô-mōn-tô'yä)	6	21°38'N	99°51'W
Villarrica, Para. (vēl-yä-rē'kä)	20	25°55'S	56°23'W
Villa Unión, Mex. (vēl'yä-ōō-nyōn')	6	23°10'N	106°14'W
Villavicencio, Col. (vē'l-yä-vē-sĕ'n-syō)	18	4°09'N	73°38'W
Villavieja, Col. (vē'l-yä-vē-ĕ'ĸä)	18a	3°13'N	75°13'W
Villazón, Bol. (vē'l-yä-zô'n)	18	22°02'S	65°42'W
Villeta, Col. (vē'l-yĕ'tä)	18a	5°02'N	74°29'W
Viña del Mar, Chile (vē'nyä dĕl mär')	20	33°00'S	71°33'W
Virgin Islands, is., N.A. (vûr'jĭn)	5	18°15'N	64°00'W

ăt; finăl; rāte; senåte; ärm; åsk; sofȧ; fâre; ch-choose; dh-as th in other; bē; ēvent; bĕt; recĕnt; cratẽr; g-gō; gh-guttural g; bĭt; ĭ-short neutral; rīde; ĸ-guttural k as ch in German ich;

PLACE (Pronunciation)	PAGE	LAT.	LONG.
Vitória, Braz. (vě-tō′rě-ä)	19	20°09′S	40°17′W
Vitória de Conquista, Braz.			
(vě-tō′rě-ä-dä-kôn-kwě′s-tä)	19	14°51′S	40°44′W
Volta Redonda, Braz.			
(vōl′tä-rā-dôn′dä)	19	22°32′S	44°05′W
Vuelta Abajo, reg., Cuba			
(vwěl′tä ä-bä′hō)	10	22°20′N	83°45′W

W

PLACE (Pronunciation)	PAGE	LAT.	LONG.
Water, i., V.I.U.S. (wô′tĕr)	5c	18°20′N	64°57′W
Water Cay, i., Bah.	11	22°55′N	75°50′W
Wellington, i., Chile (ŏĕ′lĕng-tŏn)	20	49°30′S	76°30′W
West, Mount, mtn., Pan.	4a	9°10′N	79°52′W
West Caicos, i., T./C. Is.			
(kāē′kō) (kī′kŏs)	11	21°40′N	72°30′W
West End, Bah.	10	26°40′N	78°55′W
West Indies, is., (ĭn′dēz)	5	19°00′N	78°30′W
West Sand Spit, i., T./C. Is.	11	21°25′N	72°10′W
Whale Cay, i., Bah.	10	25°20′N	77°45′W
Whale Cay Channels, strt., Bah.	10	26°45′N	77°10′W
Wheelwright, Arg. (ŏē′l-rē′gt)	17c	33°46′S	61°14′W
Wilhelmina Gebergte, mts., Sur.	19	4°30′N	57°00′W
Willemstad, Neth. Ant.	18	12°12′N	68°58′W
Williams, i., Bah.	10	24°30′N	78°30′W
Windward Islands, is., N.A.			
(wind′wĕrd)	5	12°45′N	61°40′W
Windward Passage, strt., N.A.	5	19°30′N	74°20′W
Wismar, Guy. (wĭs′mär)	19	5°58′N	58°15′W

X

PLACE (Pronunciation)	PAGE	LAT.	LONG.
Xagua, Banco, bk., Cuba			
(bä′n-kō-sä′gwä)	10	21°35′N	80°50′W
Xcalak, Mex. (sä-lä′k)	8a	18°15′N	87°50′W
Xicotencatl, Mex. (sē-kô-tĕn-kät′′l)	6	23°00′N	98°58′W
Xilitla, Mex. (sĕ-lē′tlä)	6	21°24′N	98°59′W
Xingu, r., Braz. (zhĕŋ-gò′)	19	6°20′S	52°34′W

PLACE (Pronunciation)	PAGE	LAT.	LONG.
Xochihuehuetlán, Mex.			
(sô-chē-wĕ-wĕ-tlä′n)	7	17°53′N	98°29′E
Xochimilco, Mex. (sō-chě-mĕl′kô)	7a	19°15′N	99°06′W

Y

PLACE (Pronunciation)	PAGE	LAT.	LONG.
Yacuiba, Bol. (yà-kōō-ē′bá)	18	22°02′S	63°44′W
Yaguajay, Cuba (yä-guä-hä′ě)	10	22°20′N	79°20′W
Yahualica, Mex. (yä-wä-lē′kä)	6	21°08′N	102°53′W
Yajalón, Mex. (yä-hä-lōn′)	7	17°16′N	92°20′W
Yambi, Mesa de, mtn., Col.			
(mě′sä-dĕ-yá′m-bē)	18	1°55′N	71°45′W
Yaque del Norte, r., Dom. Rep.			
(yä′kå dĕl nôr′tà)	5	19°40′N	71°25′W
Yaque del Sur, r., Dom. Rep.			
(yä-kĕ′-dĕl-sōō′r)	11	18°35′N	71°05′W
Yaqui, r., Mex. (yä′kē)	4	28°15′N	109°40′W
Yaracuy, dept., Ven. (yä-rä-kōō′ē)	19b	10°10′N	68°31′W
Yarumal, Col. (yä-rōō-mäl′)	18	6°57′N	75°24′W
Yateras, Cuba (yä-tä′räs)	11	20°00′N	75°00′W
Yautepec, Mex. (yä-ōō-tä-pĕk′)	6	18°53′N	99°04′W
Yojoa, Lago de, l., Hond.			
(lä′gô dĕ yô-hō′ä)	8	14°49′N	87°53′W
Yolaina, Cordillera de, mts., Nic.	9	11°34′N	84°34′W
Yoro, Hond. (yō′rô)	8	15°09′N	87°05′W
Young, Ur. (yô-ōō′ng)	17c	32°42′S	57°38′W
Yucatán, state, Mex. (yōō-kä-tän′)	4	20°45′N	89°00′W
Yucatán Channel, strt., N.A.	4	22°30′N	87°00′W
Yucatán Peninsula, pen., N.A.	8	19°30′N	89°00′W
Yuma, r., Dom. Rep.	11	19°05′N	70°05′W
Yurécuaro, Mex. (yōō-rā′kwä-rô)	6	20°21′N	102°16′W
Yurimaguas, Peru (yōō-rĕ-mä′gwäs)	18	5°59′S	76°12′W
Yuriria, Mex. (yōō′rĕ rē′ä)	6	20°11′N	101°00′W
Yuscarán, Hond. (yōōs-kä-rän′)	8	13°57′N	86°48′W
Yuty, Para. (yōō-tē′)	20	26°45′S	56°13′W

Z

PLACE (Pronunciation)	PAGE	LAT.	LONG.
Zaachila, Mex. (sä-ä-chē′lá)	7	16°56′N	96°45′W
Zacapa, Guat. (sä-kä′pä)	8	14°56′N	89°30′W

PLACE (Pronunciation)	PAGE	LAT.	LONG.
Zacapoaxtla, Mex. (sä-kä-pō-äs′tlä)	7	19°51′N	97°34′W
Zacatecas, Mex. (sä-kä-tä′käs)	4	22°44′N	102°32′W
Zacatecas, state, Mex.	4	24°00′N	102°45′W
Zacatecoluca, El Sal.			
(sä-kä-tä-kô-lōō′kä)	8	13°31′N	88°50′W
Zacatelco, Mex.	6	19°12′N	98°12′W
Zacatepec, Mex.			
(sä-kä-tä-pĕk′) (sän-tē-ä′gô)	7	17°10′N	95°53′W
Zacatlán, Mex. (sä-kä-tlän′)	7	19°55′N	97°57′W
Zacoalco de Torres, Mex.			
(sä-kô-äl′kô dä tōr′rěs)	6	20°12′N	103°33′W
Zacualpán, Mex. (sä-kô-äl-pän′)	6	18°43′N	99°46′W
Zacualtipán, Mex. (sä-kô-äl-tē-pän′)	6	20°38′N	98°39′W
Zamora, Mex. (sä-mō′rä)	4	19°59′N	102°16′W
Zanatepec, Mex.	7	16°30′N	94°22′W
Zapala, Arg. (zä-pä′lä)	20	38°53′S	70°02′W
Zapata, Ciénaga de, sw., Cuba			
(syě′nä-gä-dĕ-zä-pá′tä)	10	22°30′N	81°20′W
Zapata, Península de, pen., Cuba			
(pě-ně′n-sōō-lä-dĕ-zä-pá′tä)	10	22°20′N	81°30′W
Zapatera, Isla, i., Nic.			
(ě′s-lä-sä-pä-tä′rō)	8	11°45′N	85°45′W
Zapopan, Mex. (sä-pō′pän)	6	20°42′N	103°23′W
Zapotiltic, Mex. (sä-pō-tĕl-tēk′)	6	19°37′N	103°25′W
Zapotitlán, Mex. (sä-pô-tē-tlän′)	6	19°13′N	98°58′W
Zapotitlán, Punta, c., Mex.	7	18°34′N	94°48′W
Zapotlanejo, Mex. (sä-pô-tlä-nä′hô)	6	20°38′N	103°05′W
Zaragoza, Mex. (sä-rä-gō′sä)	6	23°59′N	99°45′W
Zaragoza, Mex.	6	22°02′N	100°45′W
Zárate, Arg. (zä-rä′tä)	20	34°05′S	59°05′W
Zarzal, Col. (zär-zá′l)	18a	4°23′N	76°04′W
Zaza, r., Cuba (zá′zá)	10	21°40′N	79°25′W
Zempoala, Punta, c., Mex.			
(pōō′n-tä-sěm-pô-ä′lä)	7	19°30′N	96°18′W
Zempoatlépetl, mtn., Mex.			
(sěm-pô-ä-tlä′pět′l)	7	17°13′N	95°59′W
Zimapon, Mex. (sē-mä′pän)	6	20°43′N	99°23′W
Zimatlán de Álvarez, Mex.	7	16°52′N	96°47′W
Zinacatepec, Mex. (zē-nä-kä-tě′pěk)	7	18°19′N	97°15′W
Zinapécuaro, Mex. (sē-nä-pä′kwä-rô)	6	19°50′N	100°49′W
Zirandaro, Mex. (sē-rän-dä′rō)	6	18°28′N	101°02′W
Zitacuaro, Mex. (sē-tä-kwä′rō)	6	19°25′N	100°22′W
Zitlala, Mex. (sě-tlä′lä)	6	17°38′N	99°09′W
Zoquitlán, Mex. (sō-kēt-län′)	7	18°09′N	97°02′W
Zulueta, Cuba (zōō-lò-ē′tä)	10	22°20′N	79°35′W
Zumpango, Mex. (sòm-pän-gō)	6	19°48′N	99°06′W

ng-sing; ŋ-baŋk; N-nasalized n; nŏd; cŏmmit; ōld; ôbey; ôrder; oi-boil; fōōd; ò-as oo in foot; ou-out; s-soft; sh-dish; th-thin; pūre; ûnite; ûrn; stŭd; circŭs; ü-as in French tu; ′-indeterminate vowel.

SUBJECT INDEX

Listed below are major topics covered by the thematic maps, graphs and/or statistics.
Page citations are for world, continent and country maps and for world tables.

SOURCES

The following sources have been consulted during the process of creating and updating the thematic maps and statistics for the 21st Edition.

Air Carrier Traffic at Canadian Airports, Statistics Canada

Annual Coal Report, U.S. Dept. of Energy, Energy Information Administration

Armed Conflicts Report, Project Ploughshares

Atlas of Canada, Natural Resources Canada

Canadian Minerals Yearbook, Statistics Canada

Census of Canada, Statistics Canada

Census of Population, U.S. Census Bureau

Chromium Industry Directory, International Chromium Development Association

Coal Fields of the Conterminous United States, U.S. Geological Survey

Coal Quality and Resources of the Former Soviet Union, U.S. Geological Survey

Coal-Bearing Regions and Structural Sedimentary Basins of China and Adjacent Seas, U.S. Geological Survey

Commercial Service Airports in the United States with Percent Boardings Change, Federal Aviation Administration (FAA)

Completed Peacekeeping Operations, Center for Defense Information

Conventional Arms Transfers to Developing Nations, Library of Congress, Congressional Research Service

Current Status of the World's Major Episodes of Political Violence: Hot Wars and Hot Spots, Center for Systemic Peace

Dependencies and Areas of Special Sovereignty, U.S. Dept. of State, Bureau of Intelligence and Research

Earth's Seasons—Equinoxes, Solstices, Perihelion, and Aphelion, U.S. Naval Observatory

EarthTrends: The Environmental Information Portal, World Resources Institute and World Conservation Monitoring Centre 2003. Available at http://earthtrends.wri.org/ Washington, D.C.: World Resources Institute

Economic Census, U.S. Census Bureau

Employment, Hours, and Earnings from the Current Employment Statistics Survey, U.S. Dept. of Labor, Bureau of Labor Statistics

Energy Statistics Yearbook, United Nations Dept. of Economic and Social Affairs

Epidemiological Fact Sheets by Country, Joint United Nations Program on HIV/AIDS (UNAIDS), World Health Organization, United Nations Children's Fund (UNICEF)

Estimated Water Use in the United States, U.S. Geological Survey

Estimates of Health Personnel, World Health Organization

FAO Food Balance Sheet, Food and Agriculture Organization of the United Nations (FAO)

FAO Statistical Databases (FAOSTAT), Food and Agriculture Organization of the United Nations (FAO)

Fishstat Plus, Food and Agriculture Organization of the United Nations (FAO)

Geothermal Resources Council Bulletin, Geothermal Resources Bulletin

Geothermal Resources in China, Bob Lawrence and Associates, Inc.

Global Alcohol Database, World Health Organization

Global Forest Resources Assessment, Food and Agriculture Organization of the United Nations (FAO), Forest Resources Assessment Programme

Great Lakes Factsheet Number 1, U.S. Environmental Protection Agency

The Hop Atlas, Joh. Barth & Sohn GmbH & Co. KG

Human Development Report 2003, United Nations Development Programme, © 2003 by United Nations Development Programme. Used by permission of Oxford University Press, Inc.

Installed Generating Capacity, International Geothermal Association

International Database, U.S. Census Bureau

International Energy Annual, U.S. Dept. of Energy, Energy Information Administration

International Journal on Hydropower and Dams, International Commission on Large Dams

International Petroleum Encyclopedia, PennWell Publishing Co.

International Sugar and Sweetener Report, F.O. Licht, Licht Interactive Data

International Trade Statistics, World Trade Organization

International Water Power and Dam Construction Yearbook, Wilmington Publishing

Iron and Steel Statistics, U.S. Geological Survey, Thomas D. Kelly and Michael D. Fenton

Lakes at a Glance, LakeNet

Land Scan Global Population Database, U.S. Dept. of Energy, Oak Ridge National Laboratory (© 2003 UT-Battelle, LLC. All rights reserved. Notice: These data were produced by UT-Battelle, LLC under Contract No. DE-AC05-00OR22725 with the Department of Energy. The Government has certain rights in this data. Neither UT-Battelle, LLC nor the United States Department of Energy, nor any of their employees, makes any warranty, express or implied, or assumes any legal liability or responsibility for the accuracy, completeness, or usefulness of any data, apparatus, product, or process disclosed, or represents that its use would not infringe privately owned rights.)

Largest Rivers in the United States, U.S. Geological Survey

Lengths of the Major Rivers, U.S. Geological Survey

Likely Nuclear Arsenals Under the Strategic Offensive Reductions Treaty, Center for Defense Information

Major Episodes of Political Violence, Center for Systemic Peace

Maps of Nuclear Power Reactors, International Nuclear Safety Center

Mineral Commodity Summaries, U.S. Geological Survey, Bureau of Mines

Mineral Industry Surveys, U.S. Geological Survey, Bureau of Mines

Minerals Yearbook, U.S. Geological Survey, Bureau of Mines

National Priorities List, U.S. Environmental Protection Agency

National Tobacco Information Online System (NATIONS), U.S. Dept. of Health and Human Services, Centers for Disease Control and Prevention (CDC)

Natural Gas Annual, U.S. Dept. of Energy, Energy Information Administration

New and Recent Conflicts of the World, The History Guy

Nuclear Power Reactors in the World, International Atomic Energy Agency

Oil and Gas Journal DataBook, PennWell Publishing Co.

Oil and Gas Resources of the World, Oilfield Publications, Ltd.

Petroleum Supply Annual, U.S. Dept. of Energy, Energy Information Administration

Population of Capital Cities and Cities of 100,000 and More Inhabitants, United Nations Dept. of Economic and Social Affairs

Preliminary Estimate of the Mineral Production of Canada, Natural Resources Canada

Red List of Threatened Species, International Union for Conservation and Natural Resources

Significant Earthquakes of the World, U.S. Geological Survey

State of Food Insecurity in the World, Food and Agriculture Organization of the United Nations (FAO)

State of the World's Children, United Nations Children's Fund (UNICEF)

Statistical Abstract of the United States, U.S. Census Bureau

Statistics on Asylum-Seekers, Refugees and Others of Concern to UNHCR, United Nations High Commissioner for Refugees (UNHCR)

Survey of Energy Resources, World Energy Council

Tables of Nuclear Weapons Stockpiles, Natural Resources Defense Council

TeleGeography Research, PriMetrica, Inc. (www.primetrica.com)

Tobacco Atlas, World Health Organization

Tobacco Control Country Profiles, World Health Organization

Transportation in Canada, Minister of Public Works and Government Services, Transport Canada

UNESCO Statistical Tables, United Nations Educational, Scientific and Cultural Organization (UNESCO)

United Nations Commodity Trade Statistics (COMTRADE), United Nations Dept. of Economic and Social Affairs

United Nations Peacekeeping in the Service of Peace, United Nations Dept. of Peacekeeping Operations

United Nations Peacekeeping Operations, United Nations Dept. of Peacekeeping Operations

Uranium: Resources, Production and Demand, United Nations Organization for Economic Co-operation and Development (OECD)

Volcanoes of the World, Smithsonian National Museum of Natural History

Water Account for Australia, Australian Bureau of Statistics

Women in National Parliaments, Inter-Parliamentary Union

Women's Suffrage, Inter-Parliamentary Union

The World at War, Center for Defense Information, The Defense Monitor

The World at War, Federation of American Scientists, Military Analysis Network

World Conflict List, National Defense Council Foundation

World Contraceptive Use, United Nations Dept. of Economic and Social Affairs

The World Factbook, U.S. Dept. of State, Central Intelligence Agency (CIA)

World Facts and Maps, Rand McNally

World Lakes Database, International Lake Environment Committee

World Population Prospects, United Nations Dept. of Economic and Social Affairs

World Urbanization Prospects, United Nations Dept. of Economic and Social Affairs

World Water Resources and Their Use, State Hydrological Institute of Russia/UNESCO

The World's Nuclear Arsenal, Center for Defense Information

Special Acknowledgements

The American Geographical Society, for permission to use the Miller cylindrical projection.

The Association of American Geographers, for permission to use R. Murphy's landforms map.

The McGraw-Hill Book Company, for permission to use G. Trewartha's climatic regions map.

The University of Chicago Press, for permission to use Goode's Homolosine equal-area projection.